Data-Driven Modelling with Fuzzy Sets

Fuzzy sets have long been employed to handle imprecise and uncertain information in the real world, but their limitations in dealing with incomplete and inconsistent data led to the emergence of neutrosophic sets. In this thought-provoking book, titled *Data-Driven Modelling with Fuzzy Sets: A Neutrosophic Perspective*, the authors delve into the theories and extensive applications of neutrosophic sets, ranging from neutrosophic graphs to single-valued trapezoidal neutrosophic sets and their practical implications in knowledge management, including student learning assessment, academic performance evaluation, and technical article screening. This comprehensive resource is intended to benefit mathematicians, physicists, computer experts, engineers, scholars, practitioners, and students seeking to deepen their understanding of neutrosophic sets and their practical applications in diverse fields.

This book comprises 11 chapters that provide a thorough examination of neutrosophic set theory and its extensions. Each chapter presents valuable insights into various aspects of data-driven modeling with neutrosophic sets and explores their applications in different domains. The book covers a wide range of topics. The specific topics covered in the book include neutrosophic submodules, applications of neutrosophic sets, solutions to differential equations with neutrosophic uncertainty, cardinalities of neutrosophic sets, neutrosophic cylindrical coordinates, applications to graphs and climatic analysis, neutrosophic differential equation approaches to growth models, neutrosophic aggregation operators for decision making, and similarity measures for Fermatean neutrosophic sets. The diverse contributions from experts in the field, coupled with the constructive feedback from reviewers, ensure the book's high quality and relevance.

This book:

- presents a qualitative assessment of big data in the education sector using linguistic quadripartitioned single-valued neutrosophic soft sets.
- showcases application of n-cylindrical fuzzy neutrosophic sets in education using neutrosophic affinity degree and neutrosophic similarity index.
- covers scientific evaluation of student academic performance using single-valued neutrosophic Markov chain.
- illustrates multi-granulation single-valued neutrosophic probabilistic rough sets for teamwork assessment.
- examines estimation of distribution algorithms based on multiple-attribute group decision-making to evaluate teaching quality.

With its wealth of knowledge, this book aims to inspire further research and innovation in the field of neutrosophic sets and their extensions, providing a valuable resource for scholars, practitioners, and students alike.

Intelligent Data-Driven Systems and Artificial Intelligence

Series Editor: Harish Garg

Data Driven Technologies and Artificial Intelligence in Supply Chain
Tools and Techniques
Mahesh Chand, Vineet Jain and Puneeta Ajmera

Data-Driven Modelling with Fuzzy Sets

A Neutrosophic Perspective

Edited by
Said Broumi, D. Nagarajan,
Michael Gr. Voskoglou, and S. A. Edalatpanah

CRC Press
Taylor & Francis Group
Boca Raton London New York

CRC Press is an imprint of the
Taylor & Francis Group, an **informa** business

Designed cover image: Sorapop/PIXTA

First edition published 2025
by CRC Press
2385 NW Executive Center Drive, Suite 320, Boca Raton FL 33431

and by CRC Press
4 Park Square, Milton Park, Abingdon, Oxon, OX14 4RN

CRC Press is an imprint of Taylor & Francis Group, LLC

Library of Congress Cataloging-in-Publication Data
Names: Broumi, Said, 1978– editor. | Nagarajan, D., editor. | Voskoglou, Michael, 1949– editor. | Edalatpanah, S. A. (Seyyed Ahmad), editor.
Title: Data-driven modelling with fuzzy sets : a neutrosophic perspective / edited by Said Broumi, D. Nagarajan, Michael Gr. Voskoglou, and S.A. Edalatpanah.
Description: First edition. | Boca Raton, FL : CRC Press, 2024. | Series: Intelligent data-driven systems and artificial intelligence | Includes bibliographical references and index.
Identifiers: LCCN 2024014925 (print) | LCCN 2024014926 (ebook) | ISBN 9781032782638 (hardback) | ISBN 9781032782713 (paperback) | ISBN 9781003487104 (ebook)
Subjects: LCSH: Fuzzy sets. | Neutrosophic logic. | Education—Mathematical models.
Classification: LCC QA248.5 .D28 2024 (print) | LCC QA248.5 (ebook) | DDC 511.3/223—dc23/eng/20240403
LC record available at https://lccn.loc.gov/2024014925
LC ebook record available at https://lccn.loc.gov/2024014926

ISBN: 9781032782638 (hbk)
ISBN: 9781032782713 (pbk)
ISBN: 9781003487104 (ebk)

DOI: 10.1201/9781003487104

Typeset in Times New Roman
by Apex CoVantage, LLC

Contents

Preface

Fuzzy sets have long been utilized to describe imprecise and fuzzy information in the real world. While they are effective in handling such data, their ability to handle incomplete and inconsistent information is limited. In recent years, an extension of fuzzy sets known as neutrosophic sets has gained prominence due to their enhanced versatility. Neutrosophic sets, building upon the foundation of fuzzy sets and intuitionistic fuzzy sets, introduce an indeterminacy function that allows for more precise handling of uncertain information. As a result, neutrosophic sets offer improved scalability and flexibility, making them well suited for various fields.

Data-Driven Modelling with Fuzzy Sets: A Neutrosophic Perspective delves into the theories and advancements in this field. It explores the wide-ranging applications of neutrosophic sets, including neutrosophic graphs, single-valued neutrosophic sets, single-valued trapezoidal neutrosophic sets, and n-cylindrical fuzzy neutrosophic sets. Additionally, the book delves into the practical applications of neutrosophic sets in the realm of knowledge management, such as evaluating student learning abilities, assessing academic performance, and screening technical articles.

The contents of this comprehensive resource are intended to benefit mathematicians, physicists, computer experts, engineers, scholars, practitioners, and students seeking to deepen their understanding of neutrosophic sets and their practical applications in diverse fields. The book comprises 11 chapters that cover a broad range of topics, all aligned with the overall vision of the book. Each chapter contributes valuable insights into the study of neutrosophic set theory and its extensions, while also exploring their applications in different fields. Here is a brief summary of the findings presented in the chapters:

Chapter 1 focuses on the generalization of fuzzy set theory to neutrosophic set theory and explores its applications in science, humanity, and education, presented by V. M. Gobinath.

Chapter 2 investigates ordinary differential equations in a pentagonal neutrosophic environment, proposed by Baisakhi Banik and Avishek Chakraborty.

Chapter 3 presents the neutrosophic M/M/c queueing model, which extends the conventional M/M/c model, studied by B. Vennila and C. Antony Crispin Sweety.

Chapter 4 studies the cardinality of neutrosophic sets and neutrosophic crisp sets in finite sets, along with counting formulae for neutrosophic chains of different lengths, proposed by Bhimraj Basumatary and J. Basumatary.

Chapter 5 explores the systems of neutrosophic cylindrical coordinates, studied by Prasen Boro et al.

Chapter 6 introduces the concept of interval quadripartitioned single-valued neutrosophic sets and their application to graphs and climatic analysis, presented by Satham S. Hussain et al.

Chapter 7 focuses on neutrosophic algebraic structures and operations, specifically their applicability to the study of classical algebraic structures, presented by Binu.

Chapter 8 formulates a logistic model with quota harvesting in a neutrosophic fuzzy environment, investigated by Ashish Acharya et al.

Chapter 9 explores the pentapartitioned neutrosophic geometric Bonferroni mean and pentapartitioned neutrosophic weighted geometric Bonferroni mean under pentapartitioned neutrosophic environments, studied by Radha.

Chapter 10 develops a neutrosophic Mayor–Torrens t-norms and t-conorms for multicriteria decision-making, presented by Augus Kurian and I. R. Sumathi.

Chapter 11 presents similarity measures between Fermatean neutrosophic sets based on the Euclidean distance, studied by R. Princy et al.

The successful publication of this book is indebted to the skills and efforts of numerous individuals. First and foremost, the advisory committee provided invaluable guidance to the editor throughout the editing process. The contributors, with their diverse expertise, shared their valuable perspectives on data-driven modeling with fuzzy sets and their applications to knowledge management. Additionally, the book reviewers played a crucial role by providing constructive feedback on the book chapters, ensuring their quality and relevance.

It is our sincere hope that this book will serve as a valuable resource for readers, allowing them to deepen their understanding of the theories, advancements, and practical applications of Neutrosophic sets and their extensions, particularly in the context of knowledge management. May this book inspire further research and innovation in the field, benefiting scholars, practitioners, and students alike.

Editors' Biographies

Said Broumi is serving as an associate assistant professor at the Regional Center for the Professions of Education and Training, Casablanca-Settat, Morocco. Dr. Broumi completed his MSc and PhD in computer science from University of Hassan II in Casablanca. He is a permanent member of the Laboratory of Information Processing, Faculty of Science Ben M'Sik, University Hassan II. His research interests are graph theory, extended fuzzy graphs, decision analysis, and neutrosophic theory. He has published diverse works on neutrosophic graph theory, soft set theory, multiattribute decision analysis, and some alternative theories of neutrosophic mathematics. He has published more than 200 research articles in international peer-reviewed ISI indexed /impact factor journals. Some of his papers have been published in high-impact journals including *Complex & Intelligent Systems*, *Computational and Applied Mathematics*, *Symmetry*, and *Journal of Neutrosophic Set and Systems*. His work has been cited more than 5,992 times (Scholar Google, H-index 41). He has presented his research work at some international as well as national conferences. He is a reviewer for a number of international journals and He is also an editor-in-chief of *Neutrosophic Set and Systems* and *International Journal of Neutrosophic Science.*

D. Nagarajan is a professor in the Department of Mathematics, Rajalakshmi Institute of Technology, Chennai, India. He received his Ph.D. from Manonmaniam Sundaranar University in 2007. He has a total of 23 years of experience in both academia and research international as well as national institutions. He has published 110 research articles in international reputed journals as well as serving as a reviewer for many peer-reviewed journals. He has two patents to his credit. His research interests include stochstic processes, fuzzy sets and system, neutrosophic control systems and image processing. He is also associated with various educational and research societies, like IMS, ISTE, IAENG, AMTI and ISRD. He is the guest editor of three journals and coauthor of three books published by Lambert and Notion Press. He is also associate editor of *Franklin Open.*

Michael Gr. Voskoglou (B.Sc., M.Sc., M.Phil., Ph.D. in Mathematics) is an emeritus professor of mathematical sciences at the School of Engineering of the Graduate Technological Educational Institute of Western Greece, which has recently been joined with the University of Peloponnese. He was a full professor at the same Institute from 1987 to 2010. He was also an instructor at the Hellenic Open University, in the mathematics department of the University of Patras, and at the Schools of Primary and Secondary In-Service Teachers' Training in Patras and a teacher of mathematics at the Greek Public Secondary Education (1972–1987). He worked as a visiting researcher at the Institute of Mathematics and Informatics of the Bulgarian Academy of Sciences in Sofia for three years (1997–2000), under sabbatical. He has lectured as a visiting professor in postgraduate courses at the School of Management of the University of Warsaw (2009), in the Department of Operational Mathematics of the University of Applied Sciences in Berlin (2010) and in the Mathematics Department of the National Institute of Technology of Durgapur (2016) under a grant of the Indian government. He is the author/editor of 18 books in Greek and in English and of more than 600 papers published in international journals, book chapters, and conference proceedings in more than 30 countries on five continents, with more than 2000 citations from other researchers. He used to be the Editor in Chief of the *International Journal of Applications of Fuzzy Sets and Artificial Intelligence* (2011–2020), and currently he is a reviewer for the American Mathematical Society, Editor in Chief of *Construction, Design, Maintenance* (www.wseas.com/journals/dcm/index.php) and member of the editorial boards of many other international scientific journals. He has conducted five programs of technological research on applications of quantitative methods to management (1989–1997), he has supervised many student dissertations, and he was an external examiner of Ph.D. dissertations at Universities of Egypt, India and Saudi Arabia. He is the recipient of many scholarships, distinctions, and honorary awards and a member of many scientific associations (AMS, HMS, ICTMA, IETI, etc.). His research interests include algebra, fuzzy logic, Markov chains, artificial intelligence, and mathematics education.

S. A. Edalatpanah is an associate professor at the Ayandegan Institute of Higher Education, Tonekabon, Iran. Dr. Edalatpanah received his Ph.D. in applied mathematics from the University of Guilan, Rasht, Iran. He is currently working as the Chief of R&D at the Ayandegan Institute. He is also an academic member of Guilan University and the Islamic Azad University of Iran. Dr. Edalatpanah's fields of interest include numerical computations, operational research, uncertainty, fuzzy set and its extensions, numerical linear algebra, soft computing, and optimization. He has published over 150 journal and conference proceedings papers in these research areas. He serves on the editorial boards of several international journals. He is also the Director-in-Charge of *Journal of Fuzzy Extension & Applications*. Currently, he is president of the International Society of Fuzzy Set Extensions and Applications.

Contributors

Ashish Acharya, Department of Mathematics, Swami Vivekananda Institute Of Modern Science, Karbala More, Kolkata, West Bengal, India

Shariful Alam, Department of Mathematics, Indian Institute of Engineering Science and Technology, Shibpur, Howrah, India

Shaima Abdullah Amer Al Shanfari, Math and Computing Unit, University of Technology and Applied Science, Salalah, Sultanate of Oman

Muhammad Aslam, Department of Statistics, Faculty of Science, King Abdulaziz University, Jeddah, Saudi Arabia

Baisakhi Banik, Department of Mathematics, Indian Institute of Engineering Science and Technology, Shibpur, Howrah, India

Bhimraj Basumatary, Department of Mathematical Sciences, Bodoland University, Kokrajhar, Assam, India

J. Basumatary, Department of Mathematical Sciences, Bodoland University, Kokrajhar, Assam, India

R. Binu, Department of Mathematics, Rajagiri School of Engineering & Technology, Kerala, India

Manajat Ali Biswas, Department of Mathematics, Gobardanga Hindu College, 24 Parganas (North), West Bengal, India

Prasen Boro, Department of Mathematical Sciences, Bodoland University, Kokrajhar, Assam, India

Avishek Chakraborty, Department of Engineering Science, Academy of Technology, Adisaptagram, Hooghly, India

N. Durga, School of Advanced Sciences, Division of Mathematics, Vellore Institute of Technology, Chennai, Tamil Nadu, India

V. M. Gobinath, Department of Mechanical Engineering, Rajalakshmi Institute of Technology, Chennai, India

Satham S. Hussain, School of Advanced Sciences, Division of Mathematics, Vellore Institute of Technology, Chennai, Tamil Nadu, India

M. Kishorekumar, Department of Mathematics, Karpagam College of Engineering, Coimbatore, Tamilnadu, India

S. Krishnaprakash, Department of Mathematics, Sri Krishna College of Engineering and Technology, Coimbatore, Tamilnadu, India

Augus Kurian, Department of Mathematics, Amrita School of Physical Sciences, Coimbatore, Amrita Vishwa Vidyapeetham, India

Animesh Mahata, Mahadevnagar High School, Maheshtala, Kolkata, West Bengal, India

Supriya Mukherjee, Department of Mathematics, Gurudas College, Kolkata, West Bengal, India

D. Nagarajan, Department of Mathematics, Rajalakshmi Institute of Technology, Chennai, India

Subrata Paul, Department of Mathematics, Arambagh Govt. Polytechnic, Arambagh, West Bengal, India

R. Princy, Department of Science and Humanities, Karpagam College of Engineering, Coimbatore, Tamil Nadu, India

R. Radha, Department of Mathematics, Karpagam College of Engineering, Coimbatore, Tamil Nadu, India

Hossein Rahmonlou, School of Physics, Damghan University, Damghan 3671641167, Damghan City, Iran

Nikhilesh Sil, Department of Mathematics, Narula Institute of Technology, Kolkata, India

I. R. Sumathi, Department of Mathematics, Amrita School of Physical Sciences, Amrita Vishwa Vidyapeetham, Coimbatore, India

C. Antony Crispin Sweety, Avinashilingam Institute For Home Science And Higher Education For Women, Coimbatore, Tamil Nadu, India

B. Vennila, Sri Eshwar College of Engineering, Coimbatore, Tamil Nadu, India

Chapter 1

Applications of Neutroscopic Sets in Science, the Humanities, and Education

V. M. Gobinath, D. Nagarajan, Said Broumi, and Shaima Abdullah Amer Al Shanfari

1.1 INTRODUCTION

A neutrosophic set is a mathematical framework for handling uncertainty brought on by insufficient or inaccurate information, in which membership functions have three independent parameters: the degree of truth membership, the degree of indeterminacy membership, and the degree of falsity membership. In many real-world applications such as decision-making, pattern recognition, image processing, and expert systems, vagueness, ambiguity, and contradiction are inherent. Neutrosophic sets give users more flexibility in handling these issues.

The importance of neutrosophic sets in various fields lies in their ability to handle the different types of uncertainties that arise in real-world applications. Neutrosophic sets have been applied in a wide range of fields, including

Decision making: Neutrosophic sets have been applied to multicriteria decision-making where the levels of uncertainty for the various criteria vary. For instance, there can be ambiguity regarding resource availability, the local labor market, and the political environment when deciding where to locate a new plant.

Pattern recognition: Neutrosophic sets have been used in picture segmentation, where the boundaries between objects may not be clearly defined, to recognize patterns. Neutrosophic sets can be utilized to model the level of uncertainty in the borders between things.

Expert systems: Neutrosophic sets have been utilized in knowledge representation, where they can model the level of uncertainty in the knowledge base. This may enhance the expert system's accuracy.

Medical diagnosis: Neutrosophic sets have been employed in this field, where they can be used to model the degree of diagnostic ambiguity. This can lessen the chance of misdiagnosis and increase the accuracy of the diagnosis.

Natural language processing: Neutrosophic sets can be used to simulate the ambiguity and vagueness present in human language. This may allow machine learning algorithms that analyze natural language to process it more accurately.

Finance: Neutrosophic sets have been utilized in financial forecasting, where they can be used to model the level of uncertainty in economic data. This can

DOI: 10.1201/9781003487104-1

help to increase the reliability of financial forecasts and lower the likelihood of financial loss.

Engineering: Neutrosophic sets have been utilized in system design in engineering, where there may be ambiguity in the design parameters. It is possible to model the level of uncertainty in these parameters using neutrosophic sets.

Neutrosophic sets offer the potential to increase the reliability and accuracy of pattern recognition, decision-making, and other applications in a variety of industries. They offer a potent mathematical framework for addressing the uncertainties that develop in complex systems, and it has been demonstrated that they perform well in various applications.

1.1.1 Applications of Neutrosophic Sets in Science

1.1.1.1 Neutrosophic Sets in Physics: Quantum Mechanics

A relatively recent mathematical tool called neutrosophic set theory deals systematically and rigorously with uncertain and ambiguous information. It has applications in several disciplines, including physics. Particularly in quantum physics, fuzzy and uncertain quantum phenomena have been modeled and examined using neutrosophic sets.

The investigation of quantum superposition and entanglement is one of the main uses of neutrosophic sets in quantum physics. A quantum state can be described as a combination of the three values – true, false, and indeterminate – in a neutrosophic set structure. The indeterminate value represents the ambiguity and irrationality that characterize quantum states that cannot be classified as either true or untrue. Quantum teleportation, quantum cryptography, and quantum information processing have all been studied using the neutrosophic set approach to quantum physics, for instance a quantum key distribution protocol based on neutrosophic sets. The protocol has been demonstrated to be resistant to a variety of attacks and uses neutrosophic sets to describe the uncertainty and noise in quantum communication channels.

A key aspect of quantum mechanics is entanglement, which allows two or more particles to exhibit correlations that cannot be explained by traditional physics. Entanglement can be expressed as a combination of true, false, and indeterminate values in a neutrosophic set framework, which allows for a more nuanced understanding of entanglement phenomena (Smarandache, 2009). Overall, neutrosophic sets offer a potent mathematical tool for simulating and deciphering fuzzy and uncertain quantum processes. The usage of neutrosophic sets in quantum mechanics is predicted to increase as quantum technology develops and progresses.

1.1.1.2 Neutrosophic Sets in Physics: Astrophysics

Neutrosophic sets have been employed in astrophysics to represent and examine a variety of phenomena, for instance to investigate black holes and their characteristics. Researchers introduced the idea of a neutrosophic black hole, which is a black hole that depending on the viewpoint of the observer, might have varying degrees of truth, indeterminacy, and untruth. They also demonstrated that a neutrosophic black hole's entropy is a neutrosophic number, reflecting the ambiguity and uncertainty connected to black holes.

The two most significant and enigmatic elements of the universe, dark matter and dark energy, have also been studied using neutrosophic sets. Wei et al. (2021) recently mimicked the characteristics of dark matter and dark energy while accounting for their uncertainties and indeterminacies and demonstrated that neutrosophic rather than fuzzy or classical sets can describe these occurrences more precisely and realistically. Moreover, two of the most fundamental theories in physics, quantum mechanics and quantum field theory, have been modeled using neutrosophic sets.

According to Smarandache (2003), a neutrosophic quantum state is a quantum condition with varying degrees of truth, indeterminacy, and falsehood depending on the observer's point of view. He demonstrated how neutrosophic quantum states, rather than classical or fuzzy sets, can be utilized to reflect the uncertainties and indeterminacies of quantum mechanics and can offer a more thorough and accurate account of quantum processes. Finally, neutrosophic sets have crucial applications in astrophysics and other branches of physics, offering a more precise and realistic representation of events that are marked by uncertainty and indeterminacy. The use of neutrosophic sets in physics is a promising and fascinating area of study that may eventually result in new understandings and discoveries (Wei and Xia, 2021).

1.1.1.3 Neutrosophic Sets in Chemistry: Chemical Bonding

Fuzzy sets can be generalized to include neutrosophic sets, which support the representation of uncertainty, indeterminacy, and inconsistent data. Neutrophophic sets have been used in chemistry to study a variety of topics, including chemical bonding.

A fundamental idea in chemistry, chemical bonding is the mechanism through which atoms unite to form molecules. It is possible to express the uncertainty and ambiguity present in chemical bonding using neutrosophic sets. For instance, the degree of covalent and ionic character in a bond, which might change based on parameters like electronegativity differences and bond length, can be modeled using neutrosophic sets (Wang and Smarandache, 2019).

Bond orders are frequently thought to have integer values in conventional chemistry, but because of the delocalization of electrons, they sometimes have fractional values. The uncertainty surrounding the precise value of the bond order can be represented using neutrosophic sets (Zhang et al., 2018).

Neutrosophic sets have also been used to analyze molecular diversity and similarity. In the process of discovering new pharmaceuticals, scientists search for molecules with structures resembling those of existing medications. The ambiguity and uncertainty that come with molecular similarities, as well as the variety of molecular structures, can be modeled using neutrosophic sets. In general, neutrosophic sets have demonstrated promise in the application of chemical bonding and other chemistry domains, enabling a more realistic depiction of uncertainty and indeterminacy.

1.1.1.4 Neutrosophic Sets in Chemistry: Chemical Analysis

Many uses for neutrosophic sets have been found in the world of chemistry, particularly in chemical analysis; neutrosophic sets enable representing inaccurate, ambiguous, or indeterminate information, which can aid in presenting a more full picture of the system under study. For instance, modeling chemical processes is one application of neutrosophic sets in chemical analysis.

Chemical reactions can be intricate, with numerous factors influencing the velocity of the reaction and the creation of the products. To more accurately reflect the system, neutrosophic sets can be utilized to model the uncertainty related to these variables. For instance, the reaction rate of the oxidation of several organic compounds in an aqueous solution was modeled using neutrosophic sets, where the reaction rate can be altered by the presence of various metal ions (Smarandache and Ali, 2014).

The analysis of spectroscopic data has also made use of neutrosophic sets. Data from spectroscopy can be complicated, with noise and overlapping signals impacting the data's accuracy. Neutrosophic sets can be used to model the spectral data uncertainty, enabling a more precise interpretation of the data. To examine the infrared spectra of some organic compounds, for instance, where the spectra can be influenced by solvent effects, temperature, and other experimental variables, neutrosophic sets have been utilized (Alkhazali, 2021).

1.1.1.5 Neutrosophic Sets in Biology: Genetics

Neutrosophic sets have gained significant attention in recent years as a powerful mathematical tool and framework for handling imprecise and indeterminate data in various fields, including biology. In genetics, neutrosophic sets can be applied to model uncertainty in gene expression levels, gene regulatory networks, and genetic association studies. The method can assist in finding genes that are expressed differently under various circumstances, such as in healthy tissues versus tissues with the disease, and it can help in understanding the underlying molecular mechanisms of complicated disorders (Smarandache, 2018).

One of the primary applications of neutrosophic sets in genetics is modeling the levels of gene expression. Although these techniques produce noisy and inconsistent results, they are routinely employed. Neutrosophic sets handle the uncertainty in gene expression levels by providing each gene with a membership function that concurrently displays the degree of the gene's membership to various expression levels.

Modeling gene regulatory networks is another area in which neutrosophic sets can be used in genetics; it is crucial to comprehending the functional interactions between genes since they are intricate systems that regulate gene expression levels in cells. Modeling these networks is a difficult undertaking, nevertheless, because of the incomplete information and the existence of noise in experimental data. By assigning truth, falsity, or indeterminacy to each gene connection, and neutrosophic sets offer a paradigm for modeling the uncertainty. This method can assist in locating the crucial genes and regulatory relationships that manage significant biological processes (Zahra, 2021).

Genetic association studies, which seek to find genetic variants related to complicated disorders, can also make use of neutrosophic sets. Large genetic data sets are usually analyzed in these studies, and the presence of noise and missing data might produce erroneous results.

To handle uncertainty and incomplete data in genetics, neutrosophic sets offer a potent framework. This is essential for comprehending intricate biological processes and creating new disease treatments. Neutrosophic sets may also help genetics research predictions be more accurate and lower the risk of making erroneous discoveries. They can therefore transform our understanding of the genetic basis of disease and are a crucial tool in the area of genetics.

1.1.1.6 Neutrosophic Sets in Biology: Evolution

Modeling biological systems with various levels of uncertainty is one way that neutrosophic sets are used in biology. A complex biological system like the immune system, for instance, reacts to various diseases with variable levels of sensitivity and specificity. To account for the inherent uncertainty in immunological responses, neutrosophic sets can be used to describe the immune system.

The investigation of evolution is another biological use of neutrosophic sets. The indeterminacy of evolutionary processes, which are prone to varying degrees of stochasticity and unpredictability, can be modeled using neutrosophic set theory. For instance, the link between genotype and phenotype is represented by the fitness landscape, which may be modeled using neutrosophic sets to describe the various levels of fitness uncertainty (Khoshnevisan, 2014) and the diversity of life.

1.2 APPLICATIONS OF NEUTROSOPHIC SETS IN THE HUMANITIES

Neutrosophic set theory has been utilized in psychology to model the ambiguity and uncertainty that are frequently present in psychological events. For instance, utilizing classical set theory or even fuzzy set theory can be challenging since emotions are complex (Smarandache, 1998). By allowing for the depiction of contradictory and unclear emotions, neutrosophic set theory offers a more complex method of representing emotions.

The investigation of personality traits is another area of psychology that uses neutrosophic set theory. Binary terms like introverted or extroverted are frequently used to categorize personality traits (Yang, 2015). Individuals, however, may display features that lie in between these categories or even contradictory traits. These subtleties in personality traits can be represented using neutrosophic set theory.

The absence of agreement on the suitable parameters to use is one of the difficulties in utilizing neutrosophic set theory in psychology. For instance, there is no established method for figuring out just how true, false, and uncertain a neutrosophic set is. Several approaches have been put forth by researchers, but work is currently being done to identify the most effective strategy.

The potential of neutrosophic set theory as a tool for representing the ambiguity and uncertainty inherent in psychological processes has been demonstrated. The approach and criteria used to apply neutrosophic set theory in psychology require further study.

1.2.1 Neutrosophic Sets in Psychology: Clinical Diagnosis

Based on clinical observations and interviews, clinical diagnosis entails detecting and categorizing mental diseases. Neutrosophic sets have been employed in clinical diagnostics to deal with ambiguity and inconsistent results; they allow for working around some of the drawbacks of conventional diagnosis techniques, which may rely on inflexible and dichotomous classifications, particularly when there is a lack of medical community consensus regarding the diagnosis.

Neutrosophic sets, for instance, were utilized to create a diagnostic model for major depressive disorder (Xu et al., 2019). The model outperformed conventional diagnostic techniques in terms of accuracy and was able to handle the uncertainty and inconsistent nature of the diagnosis process. Similar to this, neutrosophic sets were utilized to create a diagnosis model for autism spectrum disorder and achieved greater accuracy than conventional diagnostic techniques (Kaur and Kumar, 2017).

1.2.2 Neutrosophic Sets in Psychology: Social Psychology

Social psychology is the study of how social context affects people's thoughts, feelings, and behaviors, and social research typically entails assessing categories that are intrinsically vague or difficult to define, such as attitudes or beliefs. Neutrosophic sets have been employed to deal with the ambiguity and consistency issues in social research, for instance offering a more precise way to gauge people's sentiments about corruption (Chakraborty and Chattopadhyay, 2018). The consistency of social research data has also been examined using neutrosophic sets, for instance relating to inconsistent data on environmental attitudes and behaviors (Deshpande and Kodag, 2020)

In conclusion, neutrosophic sets in psychology can manage the ambiguity and inconsistent nature of clinical diagnosis, resulting in more precise diagnoses. Neutrosophic sets can manage the ambiguity and consistency in social research data, resulting in more precise measures of attitudes and actions in social psychology. Overall, neutosophic sets can improve the precision and depth of psychological study and practice.

1.2.3 Neutrosophic Sets in Sociology: Demography

The study of human populations, including their size, makeup, and distribution, is known as demography. The uncertainty and ambiguity that frequently occur in demographic data can be represented using neutrosophic sets. For instance, statistics on elements like birth rates, death rates, and migration patterns may be lacking or contradictory in a study of population increase in a certain area. The degree of truth, falsity, and indeterminacy connected to each piece of data can be represented using neutrosophic sets, enabling a more sophisticated examination of population dynamics. A study on population aging in Bangladesh by Karim and Smarandache (2014) illustrates the usage of neutrosophic sets to depict the ambiguity and uncertainty in the data on age distribution, mortality rates, and migratory patterns. They were able to demonstrate that compared with conventional approaches, the neutrosophic technique offered a more accurate portrayal of the population aging process.

1.2.4 Neutrosophic Sets in Sociology: Criminology

The study of crime, criminal behavior, and the criminal justice system is known as criminology. In criminology, neutrosophic sets can be used to symbolize the ambiguity and uncertainty that frequently occur in criminal data. For instance, statistics on aspects like the number of crimes avoided and the program's influence on various sorts of crimes may be lacking or contradictory in a study of the efficacy of a certain crime prevention program. The degree of truth, falsity, and indeterminacy connected to each piece of data can be represented as neutrosophic sets, allowing for a more detailed study of the program's efficacy.

A study on the connection between poverty and crime in India illustrates the usage of neutrosophic sets in criminology (Majumder et al., 2018). To reflect the ambiguity and uncertainty in the data on crime and poverty rates, the researchers employed neutrosophic sets to demonstrate that in comparison with conventional approaches, the neutrosophic technique provided a more realistic picture of the association between poverty and crime. Neutrosophic sets enable a more nuanced portrayal of the uncertainty and ambiguity inherent in sociology data, especially in areas like criminology and demography where there is frequently conflicting, erroneous, or missing data.

1.2.5 Neutrosophic Sets in Philosophy: Ethics

Neutrosophic sets are recently being used in philosophy, mostly in the study of ethics, which is concerned with the notions of correct versus wrong behavior. They allow for expressing ethical ideas that are not well-defined, such as moral dilemmas, competing values, and ambiguous ethical judgments. In contrast, conventional ethical theories like utilitarianism and deontology rely on binary reasoning and presuppose that moral judgments may only be true or false. Neutrosophic sets offer a more adaptable framework for accommodating ambiguity.

For instance, Gao and Smarandache (2003) used neutrosophic sets to examine the moral implications of cloning. They stated that cloning entails ethical ideals that are at odds with one another, such as the values of respecting individual autonomy and protecting human life, and used the sets to reflect these opposing ideals and examine the nebulous or unclear character of ethical judgments on cloning.

Analyzing ethical decision-making is another area in which neutrosophic sets have been used in the study of ethics. For instance, Khan and Smarandache (2018) modeled the variables influencing ethical decision-making in corporate organizations using neutrosophic sets. The authors contended that making ethical decisions requires a complex interplay of several elements, including individual values, organizational culture, and environmental circumstances. Overall, neutrosophic sets offer a helpful framework for studying and modeling ethical concepts that contain doubt, ambiguity, and contradictory values. They have been used in a variety of ethical contexts, such as moral conundrums, ethical disputes, and ethical decision-making.

1.2.6 Neutrosophic Sets in Philosophy: Epistemology

Neutrosophic sets can be used in epistemology to symbolize the ambiguity and uncertainty present in all human knowledge. As human knowledge is limited and incomplete, no conclusion can ever be guaranteed. In contrast to conventional binary sets, neutrosophic sets enable us to capture this ambiguity and uncertainty more subtly.

Neutosophic sets, for instance, can be employed in the philosophy of science to symbolize the uncertainty of conclusions about the natural world because scientists frequently have to cope with contradictory or partial facts and base their decisions on incomplete knowledge. Moreover, the philosophy of language and semantics can make use of neutrosophic sets as language is intrinsically ambiguous and dependent on context. Understanding how language affects our thinking and interpersonal communication can be gained through this. Overall, neutrosophic sets can aid us in better understanding the constraints and flaws of our thinking and communication by offering a more sophisticated means of portraying uncertainty and ambiguity.

1.3 APPLICATIONS OF NEUTROSOPHIC SETS IN EDUCATION

1.3.1 Neutrosophic Sets in Teaching and Learning: Curriculum Design

Neutrosophic sets are also used to represent the indeterminate or ambiguous information in education. Making choices regarding what to teach, how to teach it, and how to evaluate student learning are all part of curriculum design, and neutrosophic sets can assist educators in depicting the uncertainty surrounding the most pertinent course topics and curriculum content. Neutrophic sets can also reflect the ambiguous efficacy of different teaching methodologies and assessment techniques.

Moreover, neutrosophic sets can be utilized to create diverse curricula that better accommodate students' requirements, for instance accounting for the cultural origins, learning preferences, and prior knowledge of their different pupils. With the aid of this knowledge, courses may be created that take diverse learning styles and readiness levels into account.

Furthermore, the efficacy of teaching methods and curricula can be assessed using neutrosophic sets to illustrate the ambiguity around the link between instructional strategies and student learning results. Teachers can decide how to improve their lessons and better fulfill the requirements of their students by using neutrosophic sets to assess the success of various teaching tactics. Teachers can also evaluate student learning and test performance using neutrosophic sets, for instance, allowing a teacher to gauge whether or not a student genuinely knows the course content.

In a university-level statistics course, Saaty et al. (2015) used neutrosophic sets to assess the efficacy of various teaching methodologies. In contrast to conventional decision-making techniques, they discovered that neutrosophic decision-making can offer a more nuanced knowledge of the efficacy of various instructional methods. In a different study, Wang and Wei (2020) developed a curriculum for an artificial intelligence course using neutrosophic sets and were able to construct a curriculum that better accounted for the ambiguity around the applicability and difficulty of various courses. Overall, neutrosophic sets have enormous potential to raise the caliber and diversity of education.

1.3.2 Neutrosophic Sets in Teaching and Learning: Assessment and Evaluation

Assessment and evaluation are integral components of education, and they play a crucial role in determining the success of the teaching and learning process. Neutrosophic sets can be used to represent the uncertainty associated with assessment and evaluation, particularly when dealing with subjective measures such as grading or scoring. For instance, in a traditional grading system, a student's performance is often represented using a crisp value or a range of values. However, this approach fails to account for the fact that a student's grade can reflect a variety of factors such as the grader's subjectivity, the difficulty of the assessment task, or the student's background.

Using neutrosophic sets, it is possible to represent the uncertainty associated with the grading process by assigning membership degrees to each possible grade. These membership degrees can be computed based on various criteria, such as the student's performance on different aspects of the assessment task, the grader's level of confidence, or the difficulty of the task. The resulting neutrosophic set provides a more nuanced representation of the student's performance, which can be used to inform instructional decisions, such as providing additional support or offering more challenging tasks. Another application of neutrosophic sets in education is in evaluating learning outcomes: statements that describe what students are expected to know, understand, or be able to do as a result of their learning experiences; evaluating learning outcomes can be challenging due to the complex and multidimensional nature of the learning process, and neutrosophic sets can represent the uncertainty by assigning membership degrees to each possible outcome. These membership degrees can be computed based on criteria such as the alignment between the assessment task and the learning outcome, the reliability of the assessment instrument, or the consistency of the evaluation process.

The use of neutrosophic sets in education is a relatively new area of research, and much work remains for exploring their full potential. However, the existing literature provides some evidence of their usefulness in dealing with indeterminacy and uncertainty in various educational contexts. For instance, Akbulut and Kocak (2020) recently demonstrated the effectiveness of using neutrosophic sets in assessing the effectiveness of teaching strategies in a computer programming course. In another study, Zhang and Smarandache (2015) explored the use of neutrosophic sets in evaluating the quality of online learning resources.

Neutrosophic sets provide a powerful tool for dealing with indeterminacy and uncertainty in education, particularly in the areas of assessment and evaluation. By representing uncertainty using membership degrees, neutrosophic sets enable a more nuanced and flexible representation of the teaching and learning process, which can be used to inform instructional decisions and improve student outcomes.

1.3.3 Neutrosophic Sets in Educational Psychology: Motivation

Educational psychology is a field that can benefit from the application of neutrosophic set theory in many ways, including in the area of motivation. In educational psychology, motivation refers to the reasons why students engage in learning activities, which can be ambiguous or uncertain. For example, a student's motivation to learn a particular subject might be influenced by a combination of internal factors such as personal interest, curiosity, and the desire for achievement and external factors such as the perceived value of the subject in the job market, peer pressure, and the teacher's attitude toward the subject.

Neutrosophic set theory has been used in several studies to analyze motivational factors in educational psychology. For example, Xu et al. (2020) used neutrosophic sets to model the factors that influence students' motivation to learn English as a foreign language. They found that the teacher's attitude, the quality of teaching materials, and the perceived usefulness of the language were important motivational factors, but their degree of importance varied depending on the individual student.

Li and Xu (2021) used neutrosophic sets to model the factors that influence college students' motivation to participate in online learning during the COVID-19 pandemic. They found that the quality of online teaching, the level of interaction with peers and teachers, and the degree of self-regulation were important motivational factors, but their degree of importance varied depending on the students' prior experience with online learning.

In conclusion, neutrosophic set theory provides a powerful tool for analyzing and modeling motivational factors in educational psychology. By representing these factors as neutrosophic sets, researchers can capture their ambiguity and uncertainty and analyze their complex relationships. The application of neutrosophic set theory can lead to a better understanding of student motivation and more effective interventions to promote learning.

1.3.4 Neutrosophic Sets in Educational Psychology: Learning Styles

Learning styles are defined as "characteristic cognitive, affective, and physiological behaviors that serve as relatively stable indicators of how learners perceive, interact with and respond to the learning environment" (Felder and Silverman, 1988). There are several models of learning styles, with the most widely recognized being the VARK model, which identifies four types of learners: visual, auditory, reading/writing, and kinesthetic (Fleming and Mills, 1992). Neutrosophic sets can be used to represent the uncertainty and ambiguity that exists in identifying and measuring learning styles. For example, a student may exhibit characteristics of both visual and kinesthetic learners, and a neutrosophic set could be used to represent the student's learning style as being partially visual and partially kinesthetic.

Neutrosophic sets can also be used to represent the uncertainty and ambiguity that exists in the relationship between learning styles and academic performance. While some research has shown a positive relationship between matching a student's learning style with instructional methods and improved academic performance (Felder and Silverman, 1988), other studies have shown no relationship or even a negative relationship (Coffield et al., 2004). In this case, a neutrosophic set could be used to represent the uncertain relationship.

Overall, the application of neutrosophic sets in educational psychology provides a more nuanced way to represent the ambiguity and uncertainty that exists in various areas, including learning styles. This can lead to a more comprehensive understanding of these areas and potentially improve the effectiveness of educational interventions.

1.4 CONCLUSION

Neutrosophic sets have been gaining increasing attention in various fields, including science, humanity, and education. Neutrosophic set theory is an extension of fuzzy set theory, which can handle more complex and uncertain information. Here are some of the main applications of neutrosophic sets in these fields:

In science, neutrosophic sets have been applied to various fields such as medicine, engineering, and computer science. For instance, in medicine, neutrosophic sets have been used to handle complex and uncertain medical data such as medical images, diagnostic reports, and patient records. In engineering, neutrosophic sets have been used in the design of complex systems, especially in the case of conflicting and uncertain requirements. In computer science, neutrosophic sets have been used in decision-making systems, intelligent systems, and natural language processing.

In humanity, neutrosophic sets have been applied in the areas of social science, philosophy, and psychology, for instance in social science to analyze complex social phenomena such as human behavior, social norms, and cultural diversity. In philosophy, neutrosophic sets have been used to analyze and resolve paradoxes and contradictions, and in psychology, neutrosophic sets have been used to analyze the uncertainty and ambiguity inherent in the interpretation of human thought and behavior. In education, neutrosophic sets have been applied in the areas of educational psychology, curriculum development, and assessment.

For instance, in educational psychology, neutrosophic sets have been used to analyze and model the uncertainty and ambiguity in the learning process. In curriculum development, they have been used to design and implement educational programs that can handle diverse student needs and preferences. In assessment, neutrosophic sets have been used to develop evaluation methods that can handle complex and uncertain information about student performance and learning outcomes.

Overall, the application of neutrosophic sets in science, the humanities, and education has opened up new opportunities to deal with complex and uncertain information more accurately and efficiently. The ability of neutrosophic sets to handle the indeterminacy, inconsistency, and incompleteness of information makes them an ideal tool to deal with real-world problems that are inherently uncertain and complex.

Moreover, the use of neutrosophic sets can also enhance the quality of decision-making and problem-solving in various domains, enabling researchers and practitioners to make informed and effective decisions based on reliable and accurate information. The increasing interest in neutrosophic set theory in different fields is a testament to its potential to revolutionize the way we approach complex and uncertain problems. In conclusion, applications of neutrosophic sets are diverse and wide-ranging and offer a promising avenue for further research and development in the fields discussed in this chapter. The potential of neutrosophic set theory to transform the way we approach complex and uncertain information holds great promise for the future of research and practice in these domains.

REFERENCES

Akbulut, Y., & Kocak, O. (2020). An application of neutrosophic sets in assessing the effectiveness of teaching strategies in computer programming courses. Journal of Intelligent & Fuzzy Systems, 38(1), 133–142.

Alkhazali, K. M., & Smarandache, F. (2021). Neutrosophic set-based infrared spectral data analysis: An overview. Symmetry, 13(7), 1241.

Chakraborty, A., & Chattopadhyay, A. (2018). Attitude towards corruption: A neutrosophic approach. Journal of Intelligent & Fuzzy Systems, 35(4), 4263–4273.

Coffield, F., Moseley, D., Hall, E., & Ecclestone, K. (2004). Learning styles and pedagogy in post-16 learning: A systematic and critical review. Learning and Skills Research Centre.

Deshpande, S. S., & Kodag, P. B. (2020). Analysis of inconsistency in survey data using neutrosophic sets. Journal of Statistics Applications & Probability, 9(1), 1–12.

Felder, R. M., & Silverman, L. K. (1988). Learning and teaching styles in engineering education. Engineering Education, 78(7), 674–681.

Fleming, N. D., & Mills, C. (1992). Not another inventory, but rather a catalyst for reflection. To Improve the Academy, 11, 137–155.

Gao, J., & Smarandache, F. (2003). Neutrosophic sets and systems in ethics. Apeiron, 10(2), 81–102.

Karim, R., & Smarandache, F. (2014). Neutrosophic approach to population aging in Bangladesh. Journal of Applied Sciences, 14(19), 2244–2250.

Kaur, H., & Kumar, S. (2017). Diagnosis of autism spectrum disorder using neutrosophic sets. Procedia Computer Science, 122, 237–242.

Khan, M. A., & Smarandache, F. (2018). Neutrosophic sets and systems for ethical decision-making in business organizations. Journal of Business Ethics, 151(2), 489–503.

Khoshnevisan, L., & Vahidi, S. (2014). Neutrosophic sets and systems for biological sciences: A literature review. Journal of Biological Systems, 22(1), 1–23.

Li, Y., & Xu, Z. (2021). Neutrosophic set-based analysis of college students' motivation to participate in online learning during the COVID-19 pandemic. Journal of Educational Computing Research, 59(2), 509–525.

Majumder, S., Chattopadhyay, S., & Roy, P. K. (2018). Neutrosophic logic for modeling crime and poverty relationship in India. Journal of Interdisciplinary Mathematics, 21(1), 113–126.

Saaty, T. L., Vargas, L. G., & Abdel-Maguid, M. (2015). Uncertainty and rank order in the analytic hierarchy process. Annals of Operations Research, 242(1–2), 223–233.

Smarandache, F. (1998). Neutrosophy and neutrosophic logic, set, probability, and statistics. American Research Press.

Smarandache, F. (2003). Neutrosophic quantum theory. Progress in Physics, 1, 37–42.
Smarandache, F. (Ed.). (2018). Neutrosophic sets in biology and biostatistics. Neutrosophic Science International.
Smarandache, S. (2009). Neutrosophic set and its applications in physics. Infinite Study, 1, 19–24.
Wang, H., & Wei, X. (2020). Neutrosophic sets in the curriculum design of artificial intelligence. Journal of Ambient Intelligence and Humanized Computing, 11(4), 1545–1554.
Wang, L., & Smarandache, F. (2019). An application of neutrosophic sets in chemical bonding. Neutrosophic Sets and Systems, 27, 14–20.
Wei, Z., Zhang, B., & Xia, Y. (2021). Neutrosophic sets and their applications in dark matter and dark energy. Physics Letters B, 815, 136172.
Xu, Z., Xiao, M., Wang, X., Li, H., & Hu, X. (2019). A new method for diagnosis of major depressive disorder using neutrosophic sets. IEEE Access, 7, 98884–98891.
Yang, X., & Liu, J. (2015). Neutrosophic sets and systems in psychology. Psychology, 6(7), 900–910.
Zahra, S. A., & Smarandache, F. (2021). A survey on the applications of neutrosophic sets in bioinformatics and biomedicine. SN Applied Sciences, 3(2), 1–25.
Zhang, J., Liu, Y., & Pedrycz, W. (2018). A neutrosophic sets-based approach for molecular diversity analysis. IEEE Transactions on NanoBioscience, 17(3), 244–251.
Zhang, S., & Smarandache, F. (2015). The application of neutrosophic sets in evaluating the quality of online learning resources.
Zhang, Y. (2017). A survey of neutrosophic set theory and its applications. Complexity, 2017, 1–15.

Chapter 2

A First-Order Nonhomogeneous Ordinary Differential Equation with Initial Value Condition in a Pentagonal Neutrosophic Number Environment

Baisakhi Banik, Avishek Chakraborty, and Shariful Alam

2.1 INTRODUCTION

Neutrosophic sets are an extended version of Prof. Zadeh's fuzzy set [1], Prof. Attanossov's intuitionistic fuzzy set [2] and so forth which highlight the nature of neutralities in disjunctive fields. This legerdemain logic was first manifested by Prof. F. Smarandache [3], who accounted for the truth, indeterminacy or falsity of general data; accounting for these three components makes performance more logical and significant than that of established fuzzy logic. Recently, researchers developed the concepts of pentagonal [4], hexagonal [5] and heptagonal [6] fuzzy numbers and applied them in different areas, and Wang et al. [7] established the concept of single-valued neutrosophic set.

Chakraborty et al. [8, 9] manifested the idea of triangular and trapezoidal neutrosophic set theory and their classification based on dependency and independency of membership components, and Smarandache [10] in his article very elaborately portrayed the perception of neutrosophic measure, integral and probability. Further, several researchers have applied the knowledge of neutrosophic logic in distinct realistic problems like multicriterial decision-making (MCDM) [11], multicriteria group decision-making (MCGDM) [12], graph theory [13], operation research [14], etc. Recently, Chakraborty [15] introduced an amazing conception of pentagonal neutrosophic number (PNN), which is a congenial mixture of pentagonal fuzzy number and single-valued neutrosophic number. Chakraborty [16] also established the de-neutrosophication skill of PNN and applied the idea to solve an operatorbased MCGDM problem [17]. Additionally, modified score and accuracy function were developed in a PNN environment, and the concept was utilized in graph theory [18, 19] problem.

Several works have been published on neutrosophic logic. Basset et al. [20] ignited the idea of type 2 neutrosophic numbers, Bosc and Pivert [21] developed the bipolar neutrosophy number, Deli et al. [22] structured a bipolar-based

DOI: 10.1201/9781003487104-2

MCGDM problem, Broumi et al. [23] focused on a bipolar graph theory problem, Chaio [24] introduced an MCDM problem using type 2 linguistic fuzzy number, Wibowo [25] explored MCDM in administration systems, Nabeeh et al. [26] harnessed the neutrosophic analytic hierarchy process of the Internet of Things to estimate the significant factors, Haque et al. [27] manifested generalized spherical number and Chakraborty [28, 29] germinated a cylindrical neutrosophic number and applied it in distinct fields like MCGDM, graph theory and networking.

The conception of derivatives plays a fundamental role in mathematical modeling, engineering and medical diagnoses, and problems are generated with uncertain parameters and hesitation. To comprehend the concept, researchers developed fuzzy differential equation models [30–36], and with ongoing research, researchers focused on intuitionistic fuzzy differential equations [37–42] and membership and non-membership functions; however, these two established numbers do not contain the dilemma terms. Thus, researchers focused on neutrosophic differential equations to tackle the indeterminacy. Prof. F. Smarandache [43] discussed the neutrosophic exponential function, neutrosophic logarithmic function, neutrosophic inverse function and neutrosophic calculus. Only a few articles have been published in the area of neutrosophic differential equations [44, 45], and we are hopeful that researchers will fill the gap of this area after establishing neutro-logic articles.

In this article, we study a pentagonal neutrosophic number from different viewpoints and solve a first-order nonhomogeneous ordinary differential equation in a PNN environment with an assigned initial value condition. Apart from establishing the solution, we distinguish the strong versus weak solutions in the PNN environment. Lastly, we construct a numerical physical heat-transfer-based problem to illustrate the pertinence of our projected theory.

2.1.1 Motivation

As uncertainty is extensive in research on mathematical modeling, engineering and variousf computational difficulties, researchers are conducting ongoing research to elegantly expand neutrosophic theory. Because there are so few research articles on ordinary differential equations in the neutrosophic domain, our main motivation with this work was to ground the structure of ordinary differential equations in a higher dimension. Considering all of these objectives, we posed the following questions: What is the basic structure of an ordinary differential equation in a PNN environment? What is the configuration of the ordered differential equation when the conditions are nonhomogenous? Above all, are there any practical real-world applications of neutrosophic differential equations?

2.1.2 Novelties

As part of the progression of research on neutrosophic logic in PNN environments, we add some noteworthy aspects in this article. We construct a nonhomogeneous first-order ordinary differential equation in a PNN field with a specified initial value condition; we define the solution and we establish a numerical physical problem to strengthen our proposed theory.

2.2 PRELIMINARIES

Definition 2.2.1: Neutrosophic set: [3] A set $\tilde{X}_N$ is the universe of discourse of X most commonly defined as x and is named as a neutrosophic set if $\tilde{X}_N = \left\{\left\langle x;\left[\varphi_{\widetilde{X_N}}(x), 1_{\widetilde{X_N}}(x), \gamma_{\widetilde{X_N}}(x)\right]\right\rangle : x \in X\right\}$, where $\varphi_{\widetilde{X_N}}(x): X \to]-0,1+[$ measures confidence, $1_{\widetilde{X_N}}(x): X \to]-0,1+[$ stands for measures impreciseness and $\gamma_{\widetilde{X_N}}(x): X \to]-0,1+[$ measures falseness in the decision-making choice, and $\left[\varphi_{\widetilde{X_N}}(x), 1_{\widetilde{X_N}}(x), \gamma_{\widetilde{X_N}}(x)\right]$ satisfies the inequality

$$-0 \le Sup\{\varphi_{\widetilde{X_N}}(x)\} + Sup\left\{1_{\widetilde{X_N}}(x)\right\} + Sup\left\{\gamma_{\widetilde{X_N}}(x)\right\} \le 3+$$

Definition 2.2.2: Single-valued neutrosophic set: [7] The neutrosophic set explained in definition 2.2.1 is $\tilde{X}_N$ is designated as a single-valued neutrosophic set $\left(\tilde{X}_{SinN}\right)$ if x is a single-valued self-determining variable. $\tilde{X}_{SinN} = \left\{\left\langle x;\left[\varepsilon_{\tilde{X}_{Neu}}(x), \vartheta_{\tilde{X}_{Neu}}(x), \mu_{\tilde{X}_{Neu}}(x)\right]\right\rangle : x \in X\right\}$, where $\varepsilon_{\tilde{T}_{Neu}}(x), \vartheta_{\tilde{T}_{Neu}}(x)$ & $\omega_{\tilde{T}_{Neu}}(x)$ indicate the accuracy, impreciseness and falsity, respectively. $\widetilde{X_{NCon}}$ is named as neut-convex, which shows $\widetilde{X_{NCon}}$ is a subset of R by satisfying the following norms:

i. $\varepsilon_{\tilde{X}_{Neu}}\left\langle \varnothing l_1 + (1-\varnothing) l_2 \right\rangle \ge min\left\langle \varepsilon_{\tilde{X}_{Neu}}(l_1), \varepsilon_{\tilde{X}_{Neu}}(l_2)\right\rangle$
ii. $\vartheta_{\tilde{X}_{Neu}}\left\langle \varnothing l_1 + (1-\varnothing) l_2 \right\rangle \le max\left\langle \vartheta_{\tilde{X}_{Neu}}(l_1), \vartheta_{\tilde{X}_{Neu}}(l_2)\right\rangle$
iii. $\mu_{\tilde{X}_{Neu}}\left\langle \varnothing l_1 + (1-\varnothing) l_2 \right\rangle \le max\left\langle \mu_{\tilde{X}_{Neu}}(l_1), \mu_{\tilde{X}_{Neu}}(l_2)\right\rangle$

Where $l_1 \,\&\, l_2 \in \mathbb{R} and \varnothing \in [0,1]$.

Definition 2.2.3: Pentagonal neutrosophic number: [15] A pentagonal neutrosophic number $P_n = <\left(p^1, p^2, p^3, p^4, p^5\right); t, i, f>$ is a subset of a neutrosophic number in R determined according to the following truth, indeterminacy and falsity functions:

$$Tr(x) = \begin{cases} R\dfrac{(x-p_1)}{(p_2-p_1)} & p_1 \le x \le p_2 \\ t-(t-R)\dfrac{(x-p_1)}{(p_3-p_2)} & p_2 \le x < p_3 \\ t & x = p_3 \\ t-(t-R)\dfrac{(p_4-x)}{(p_4-p_3)} & p_3 < x \le p_4 \\ R\dfrac{(p_5-x)}{(p_5-p_4)} & p_4 \le x \le p_5 \\ 0 & otherwise \end{cases}, \quad In(x) = \begin{cases} \dfrac{R(p_2-x)}{(p_2-p_1)} & p_1 \le x \le p_2 \\ i-(i-R)\dfrac{(p_3-x)}{(p_3-p_2)} & p_2 \le x < p_3 \\ i & x = p_3 \\ i-(i-R)\dfrac{(x-p_3)}{(p_4-p_3)} & p_3 < x \le p_4 \\ R\dfrac{(x-p_4)}{(p_5-p_4)} & p_4 \le x \le p_5 \\ 0 & otherwise \end{cases}$$

$$Fa(x)=\begin{cases} \frac{R(p_2-x)}{(p_2-p_1)} & p_1\le x\le p_2 \\ f-(f-R)\frac{(p_3-x)}{(p_3-p_2)} & p_2\le x<p_3 \\ f & x=p_3 \\ f-(f-R)\frac{(x-p_3)}{(p_4-p_3)} & p_3<x\le p_4 \\ R\frac{(x-p_4)}{(p_5-p_4)} & p_4\le x\le p_5 \\ 0 & otherwise \end{cases}$$

Definition 2.2.4: Single-valued pentagonal neutrosophic number: [17] $\left(\widetilde{P_{SinN}}\right)$ is defined as $\widetilde{P_{SinN}}=\langle[(k_1,k_2,k_3,k_4,k_5);\&],[(k_1,k_2,k_3,k_4,k_5);\text{З}],$ $[(k_1,k_2,k_3,k_4,k_5);\upsilon]\rangle$, where $\&,\text{З},\vartheta\in[o,1]$. The accuracy membership function $(\aleph_{\tilde{X}}):\mathbb{R}\to[0,\&]$, the impreciseness membership function $(\theta_{\tilde{X}}):\mathbb{R}\to[\text{З},1]$ and the falsity membership function $(\beth_{\tilde{X}}):\mathbb{R}\to[o,1]$ are defined as follows:

$$\aleph_{\tilde{X}}(x)=\begin{cases} \frac{\&(x-k_1)}{(k_2-k_1)} & k_1\le x\le k_2 \\ \frac{\&(x-k_2)}{(k_3-k_2)} & k_2\le x<k_3 \\ \& & x=h_3 \\ \frac{\&(k_4-x)}{(k_4-k_3)} & k_3<x\le k_4 \\ \frac{\&(k_4-x)}{(k_5-k_4)} & k_4\le x\le k_5 \\ 0 & otherwise \end{cases},\quad \theta_{\tilde{X}}(x)=\begin{cases} \frac{k_2-x+\text{З}(x-k_1)}{(h_2-h_1)} & k_1\le x\le k_2 \\ \frac{k_3-x+\text{З}(x-k_2)}{(k_3-k_2)} & k_2\le x<k_3 \\ \text{З} & x=k_3 \\ \frac{x-k_3+\text{З}(k_4-x)}{(k_4-k_3)} & k_3<x\le k_4 \\ \frac{x-k_4+\text{З}(k_5-x)}{(k_5-k_4)} & k_4\le x\le k_5 \\ 1 & otherwise \end{cases}$$

$$\beth_{\tilde{X}}(x)=\begin{cases}\dfrac{k_2-x+o(x-k_1)}{(h_2-h_1)} & k_1\le x\le k_2\\ \dfrac{k_3-x+o(x-k_2)}{(k_3-k_2)} & k_2\le x<k_3\\ o & x=k_3\\ \dfrac{x-k_3+o(k_4-x)}{(k_4-k_3)} & k_3<x\le k_4\\ \dfrac{x-k_4+o(k_5-x)}{(k_5-k_4)} & k_4\le x\le k_5\\ 1 & otherwise\end{cases}$$

Definition 2.2.5: (α,β,γ) cut: [20] The (α,β,γ) cut of a neutrosophic set is designated by $C_{(\alpha,\beta,\gamma)}$, where $\alpha,\beta,\gamma\in[0,1]$, and it satisfies the relationship $\alpha+\beta+\gamma\le 3$, defined by $C_{(\alpha,\beta,\gamma)}=\{<T(x),I(x),F(x)>:$ where $x\in X, T(x)\ge\alpha, I(x)\le\beta, F(x)\le\gamma\}$

Definition 2.2.6: (α,β,γ) cut of a pentagonal neutrosophic number: [15] The (α,β,γ) cut of a pentagonal neutrosophic number $\widetilde{m_{Pen}}=<(m^1,m^2,m^3,m^4,m^5);$ $T,I,F>$ is defined as follows: $\tilde{m}_{(\alpha,\beta,\gamma)}$ = $\{[\tilde{m}^1(\alpha),\tilde{m}^2(\alpha),\tilde{m}^3(\alpha),\tilde{m}^4(\alpha)];$ $[\tilde{m}^{1'}(\beta),\tilde{m}^{2'}(\beta),\tilde{m}^{3'}(\beta),\tilde{m}^{4'}(\beta)];[\tilde{m}^{1''}(\gamma),\tilde{m}^{2''}(\gamma),\tilde{m}^{3''}(\gamma),\tilde{m}^{4''}(\gamma)]\}$

$$0\le\alpha+\beta+\gamma\le 3,\text{ where}$$

$$\widetilde{m}^1(\alpha)=m^1+\frac{\alpha}{R}(m^2-m^1),\ \alpha\epsilon[0,R],\ \widetilde{m}^2(\alpha)=m^2+\frac{T-\alpha}{T-R}(m^3-m^2),\ \alpha\epsilon[R,T]$$

$$\widetilde{m}^3(\alpha)=m^4-\frac{T-\alpha}{T-R}(m^4-m^3),\alpha\epsilon[R,T],\ \widetilde{m}^4(\alpha)=m^5-\frac{\alpha}{R}(D^5-m^4),\ \alpha\epsilon[0,R]$$

$$\widetilde{m}^{1'}(\beta)=m^2-\frac{\beta}{R}(m^2-m^1),\ \beta\epsilon[R,1],\ \widetilde{m}^{2'}(\beta)=m^3-\frac{I-\beta}{I-R}(m^3-m^2),\ \beta\epsilon[R,I]$$

$$\widetilde{m}^{3'}(\beta)=m^3+\frac{I-\beta}{I-R}(m^4-m^3),\ \beta\epsilon[R,I],\ \widetilde{m}^{4'}(\beta)=m^4+\frac{\beta}{0.5}(m^5-m^4),\ \beta\epsilon[R,1]$$

$$\widetilde{m}^{1''}(\gamma)=m^2-\frac{\gamma}{R}(m^2-m^1),\ \gamma\epsilon[R,1],\ \widetilde{m}^{2''}(\gamma)=m^3-\frac{F-\gamma}{F-R}(m^3-m^2),\ \gamma\epsilon[R,F]$$

$$\widetilde{m}^{3''}(\gamma)=m^3+\frac{F-\gamma}{F-R}(m^4-m^3),\ \gamma\epsilon[R,F],\ \widetilde{m}^{4'}(\beta)=m^4+\frac{\gamma}{R}(m^5-m^4),\ \gamma\epsilon[R,1]$$

Definition 2.2.7: Derivative of a fuzzy valued function: [30] The derivative of a fuzzy valued function $f(c,d)\to R$ at the point $x=x_0$ is defined as, $f'(x_0)=\lim\limits_{h\to 0}\dfrac{f(x_0+h)-f(x_0)}{h}$ and $f'(x_0)$ is D^1 differentiable at $x=x_0$ if, $[f'(x_0)]_\alpha=[f_1'(x_0,\alpha),f_2'(x_0,\alpha)]$ and $f'(x_0)$ is D^2 differentiable at $x=x_0$ if, $[f'(x_0)]_\alpha=[f_2'(x_0,\alpha),f_1'(x_0,\alpha)]$ for all $\alpha\in[0,1]$.

Definition 2.2.8: Strong solution and weak solution: Suppose the solution of the neutrosophic differential equation is y(x) and its corresponding (α,β,γ) cut is $y(x,\widetilde{\alpha,\beta},\gamma)=\left(\left[y^{1}(x,\alpha),y^{2}(x,\alpha),y^{3}(x,\alpha),y^{4}(x,\alpha)\right],\left[y^{1'}(x,\beta),y^{2'}(x,\beta),y^{3'}(x,\beta),y^{4'}(x,\beta)\right],\left[y^{1''}(x,\gamma),y^{2''}(x,\gamma),y^{3''}(x,\gamma),y^{4''}(x,\gamma)\right]\right)$ The solution is strong if

1. $$\frac{dy^{1}(x,\alpha)}{d\alpha}>0,\frac{dy^{4}(x,\alpha)}{d\alpha}>0 \text{ and } \frac{dy^{2}(x,\alpha)}{d\alpha}<0,$$
$$\frac{dy^{3}(x,\alpha)}{d\alpha}<0\,\forall\alpha\epsilon\,[0,1],y^{1}(x,1)\le y^{2}(x,1)$$

and 2.

$$\frac{dy^{1'}(x,\beta)}{d\beta}>0,\frac{dy^{4'}(x,\beta)}{d\beta}>0 \text{ and } \frac{dy^{2'}(x,\beta)}{d\beta}<0,$$
$$\frac{dy^{3'}(x,\beta)}{d\beta}<0\,\forall\beta\epsilon\,[0,1],\ y^{1'}(x,0)\le y^{2'}(x,0)$$

$$\frac{dy^{1''}(x,\gamma)}{d\gamma}>0,\frac{dy^{4''}(x,\gamma)}{d\gamma}>0 \text{ and } \frac{dy^{2''}(x,\gamma)}{d\gamma}<0,$$
$$\frac{dy^{3''}(x,\gamma)}{d\gamma}<0\,\forall\gamma\epsilon\,[0,1],y^{1''}(x,0)\le y^{2''}(x,0)$$

Otherwise the solution is weak.

2.3 SOLVING A FIRST-ORDER NONHOMOGENEOUS NEUTROSOPHIC DIFFERENTIAL EQUATION

Consider the nonhomogeneous first-order differential equation $\frac{dy(x)}{dx}=my(x)+\omega$ . . . (2.1) with the initial condition $y(0)=\tilde{m}$ where $\tilde{m}$ is a pentagonal neutrosophic number.

Let $\tilde{m}=<\left(m^{1},m^{2},m^{3},m^{4},m^{5}\right);T,I,F>$

Case 1: When $m>0$

Taking the (α,β,γ) cut of equation (2.1) gives

$$\frac{d\begin{pmatrix}\left[y^{1}(x,\alpha),D^{2}(x,\alpha),y^{3}(x,\alpha),y^{4}(x,\alpha)\right],\\ \left[y^{1'}(x,\beta),y^{2'}(x,\beta),y^{3'}(x,\beta),y^{4'}(x,\beta)\right],\\ \left[y^{1''}(x,\gamma),y^{2''}(x,\gamma),y^{3''}(x,\gamma),y^{4''}(x,\gamma)\right]\end{pmatrix}}{dx}$$

$$= m\begin{pmatrix} \left[y^{1}(x,\alpha), y^{2}(x,\alpha), y^{3}(x,\alpha), y^{4}(x,\alpha)\right], \\ \left[y^{1'}(x,\beta), y^{2'}(x,\beta), y^{3'}(x,\beta), y^{4'}(x,\beta)\right], \\ \left[y^{1''}(x,\gamma), y^{2''}(x,\gamma), y^{3''}(x,\gamma), y^{4''}(x,\gamma)\right] \end{pmatrix} + \omega$$

where $0 \le \alpha + \beta + \gamma \le 3$ and $\alpha, \beta, \gamma \epsilon [0,1]$

Thus we have the following:

$$\frac{dy^{1}(x,\alpha)}{dx} = my^{1}(x,\alpha) + \omega; \quad \frac{dy^{2}(x,\alpha)}{dx} = my^{2}(x,\alpha) + \omega;$$
$$\frac{dy^{3}(x,\alpha)}{dx} = my^{3}(x,\alpha) + \omega;$$

$$\frac{dy^{4}(x,\alpha)}{dx} = my^{4}(x,\alpha) + \omega; \quad \frac{dy^{1'}(x,\beta)}{dx} = my^{1'}(x,\beta) + \omega;$$
$$\frac{dy^{2'}(x,\beta)}{dx} = my^{2'}(x,\beta) + \omega;$$

$$\frac{dy^{3'}(x,\beta)}{dx} = my^{3'}(x,\beta) + \omega; \quad \frac{dy^{4'}(x,\beta)}{dx} = my^{4'}(x,\beta) + \omega;$$
$$\frac{dy^{1''}(x,\gamma)}{dx} = my^{1''}(x,\gamma) + \omega;$$

$$\frac{dy^{2''}(x,\gamma)}{dx} = my^{2''}(x,\gamma) + \omega; \quad \frac{dy^{3''}(x,\gamma)}{dx} = my^{3''}(x,\gamma) + \omega;$$
$$\frac{dy^{4''}(x,\gamma)}{dx} = my^{4''}(x,\gamma) + \omega.$$

with the initial conditions

$$y^{1}(0,\alpha) = m^{1} + \frac{\alpha}{0.5}\left(m^{2} - m^{1}\right), \; y^{2}(0,\alpha) = m^{2} + \frac{T-\alpha}{T-0.5}\left(m^{3} - m^{2}\right)$$
$$y^{3}(0,\alpha) = m^{4} - \frac{T-\alpha}{T-0.5}\left(m^{4} - m^{3}\right), \; y^{4}(0,\alpha) = m^{5} - \frac{\alpha}{0.5}\left(m^{5} - m^{4}\right)$$
$$y^{1'}(0,\beta) = m^{2} - \frac{\beta}{0.5}\left(m^{2} - m^{1}\right), \; y^{2'}(0,\beta) = m^{3} - \frac{I-\beta}{I-0.5}\left(m^{3} - m^{2}\right)$$
$$y^{3'}(0,\beta) = m^{3} + \frac{I-\beta}{I-0.5}\left(m^{4} - m^{3}\right), \; y^{4'}(0,\beta) = m^{4} + \frac{\beta}{0.5}\left(m^{5} - m^{4}\right)$$
$$y^{1''}(0,\gamma) = m^{2} - \frac{\gamma}{0.5}\left(m^{2} - m^{1}\right), \; y^{2''}(0,\gamma) = m^{3} - \frac{F-\gamma}{F-0.5}\left(m^{3} - m^{2}\right)$$

$$y^{3''}(0,\gamma)=m^3+\frac{F-\gamma}{F-0.5}\left(m^4-m^3\right),\ y^{4''}(0,\gamma)=m^4+\frac{\gamma}{0.5}\left(m^5-m^4\right).$$

Note: We assume R to be 0.5 for the level of symmetry.

Solution

The solutions of these equations are given as follows:

$$y^1(x,\alpha)=\left[m^1+\frac{\alpha}{0.5}\left(m^2-m^1\right)+\frac{\omega}{m}\right]e^{mx}-\frac{\omega}{m},$$

$$y^2(x,\alpha)=\left[m^2+\frac{T-\alpha}{T-0.5}\left(m^3-m^2\right)+\frac{\omega}{m}\right]e^{mx}-\frac{\omega}{m}$$

$$y^3(x,\alpha)=[m^4-\frac{T-\alpha}{T-0.5}\left(m^4-m^3\right)+\frac{\omega}{m}]e^{mx}-\frac{\omega}{m},$$

$$y^4(x,\alpha)=[m^5-\frac{\alpha}{0.5}\left(m^5-m^4\right)+\frac{\omega}{m}]e^{mx}-\frac{\omega}{m},$$

$$y^{1'}(x,\beta)=[m^2-\frac{\beta}{0.5}\left(m^2-m^1\right)+\frac{\omega}{m}]e^{mx}-\frac{\omega}{m},$$

$$y^{2'}(x,\beta)=[m^3-\frac{I-\beta}{I-0.5}\left(m^3-m^2\right)+\frac{\omega}{m}]e^{mx}-\frac{\omega}{m},$$

$$y^{3'}(x,\beta)=[m^3+\frac{I-\beta}{I-0.5}\left(m^4-m^3\right)+\frac{\omega}{m}]e^{mx}-\frac{\omega}{m},$$

$$y^{4'}(x,\beta)=[m^4+\frac{\beta}{0.5}\left(m^5-m^4\right)+\frac{\omega}{m}]e^{mx}-\frac{\omega}{m},$$

$$y^{1''}(x,\gamma)=[m^2-\frac{\gamma}{0.5}\left(m^2-m^1\right)+\frac{\omega}{m}]e^{mx}-\frac{\omega}{m},$$

$$y^{2''}(x,\gamma)=[m^3-\frac{F-\gamma}{F-0.5}\left(m^3-m^2\right)+\frac{\omega}{m}]e^{mx}-\frac{\omega}{m},$$

$$y^{3''}(x,\gamma)=[m^3+\frac{F-\gamma}{F-0.5}\left(m^4-m^3\right)+\frac{\omega}{m}]e^{mx}-\frac{\omega}{m},$$

$$y^{4''}(x,\gamma)=[m^4+\frac{\gamma}{0.5}\left(m^5-m^4\right)+\frac{\omega}{m}]e^{mx}-\frac{\omega}{m}.$$

Case 2: When m is a negative constant, that is, $m<0,\ let\, m=-n\, and\, n>0$, taking the (α,β,γ) cut of equation (2.1) gives

$$\frac{d\begin{pmatrix}\left[y^{1}(x,\alpha),y^{2}(x,\alpha),y^{3}(x,\alpha),y^{4}(x,\alpha)\right],\\ \left[y^{1'}(x,\beta),y^{2'}(x,\beta),y^{3'}(x,\beta),y^{4'}(x,\beta)\right],\\ \left[y^{1''}(x,\gamma),y^{2''}(x,\gamma),y^{3''}(x,\gamma),y^{4''}(x,\gamma)\right]\end{pmatrix}}{dx}$$

$$=-n\begin{pmatrix}\left[y^{1}(x,\alpha),y^{2}(x,\alpha),y^{3}(x,\alpha),y^{4}(x,\alpha)\right],\\ \left[y^{1'}(x,\beta),y^{2'}(x,\beta),y^{3'}(x,\beta),y^{4'}(x,\beta)\right],\\ \left[y^{1''}(x,\gamma),y^{2''}(x,\gamma),y^{3''}(x,\gamma),y^{4''}(x,\gamma)\right]\end{pmatrix}+\omega$$

where $0\le \alpha+\beta+\gamma\le 3$ and $\alpha,\beta,\gamma\epsilon\,[0,1]$.

Thus, we have the following:

$$\frac{dy^{1}(x,\alpha)}{dx}=-ny^{1}(x,\alpha)+\omega;\frac{dy^{2}(x,\alpha)}{dx}=-ny^{2}(x,\alpha)+\omega;$$

$$\frac{dy^{3}(x,\alpha)}{dx}=-ny^{3}(x,\alpha)+\omega$$

$$\frac{dy^{4}(x,\alpha)}{dx}=-ny^{4}(x,\alpha)+\omega;\frac{dy^{1'}(x,\beta)}{dx}=-ny^{1'}(x,\beta)+\omega;$$

$$\frac{dy^{2'}(x,\beta)}{dx}=-ny^{2'}(x,\beta)+\omega$$

$$\frac{dy^{3'}(x,\beta)}{dx}=-ny^{3'}(x,\beta)+\omega;\frac{dy^{4'}(x,\beta)}{dx}=-ny^{4'}(x,\beta)+\omega;$$

$$\frac{dy^{1''}(x,\gamma)}{dx}=-ny^{1''}(x,\gamma)+\omega$$

$$\frac{dy^{2''}(x,\gamma)}{dx}=-ny^{2''}(x,\gamma)+\omega;\frac{dy^{3''}(x,\gamma)}{dx}=-ny^{3''}(x,\gamma)+\omega;$$

$$\frac{dy^{4''}(x,\gamma)}{dx}=-ny^{4''}(x,\gamma)+\omega.$$

with the same initial conditions.

Solution

The solutions to these equations are given as follows:

$$y^{1}(x,\alpha)=\frac{1}{2}\left[\left(m^{1}+\frac{\alpha}{0.5}\left(m^{2}-m^{1}\right)\right)-\left(m^{2}+\frac{T-\alpha}{T-0.5}\left(m^{3}-m^{2}\right)\right)\right]e^{nx}$$
$$+\frac{1}{2}\left[\left(m^{1}+\frac{\alpha}{0.5}\left(m^{2}-m^{1}\right)\right)+\left(m^{2}+\frac{T-\alpha}{T-0.5}\left(m^{3}-m^{2}\right)\right)-\frac{2\omega}{n}\right]e^{-nx}+\frac{\omega}{n}$$

$$y^2(x,\alpha)=-\frac{1}{2}\left[\left(m^1+\frac{\alpha}{0.5}\left(m^2-m^1\right)\right)-\left(m^2+\frac{T-\alpha}{T-0.5}\left(m^3-m^2\right)\right)\right]e^{nx}$$
$$+\frac{1}{2}\left[\left(m^1+\frac{\alpha}{0.5}\left(m^2-m^1\right)\right)+\left(m^2+\frac{T-\alpha}{T-0.5}\left(m^3-m^2\right)\right)-\frac{2\omega}{n}\right]e^{-nx}+\frac{\omega}{n}$$

$$y^3(x,\alpha)=\frac{1}{2}\left[\left(m^4-\frac{T-\alpha}{T-0.5}\left(m^4-m^3\right)\right)-\left(m^5-\frac{\alpha}{0.5}\left(m^5-m^4\right)\right)\right]e^{nx}$$
$$+\frac{1}{2}\left[\left(m^4-\frac{T-\alpha}{T-0.5}\left(m^4-m^3\right)\right)+\left(m^5-\frac{\alpha}{0.5}\left(m^5-m^4\right)\right)-\frac{2\omega}{n}\right]e^{-nx}+\frac{\omega}{n}$$

$$y^4(x,\alpha)=-\frac{1}{2}\left[\left(m^4-\frac{T-\alpha}{T-0.5}\left(m^4-m^3\right)\right)-\left(m^5-\frac{\alpha}{0.5}\left(m^5-m^4\right)\right)\right]e^{nx}$$
$$+\frac{1}{2}\left[\left(m^4-\frac{T-\alpha}{T-0.5}\left(m^4-m^3\right)\right)+\left(m^5-\frac{\alpha}{0.5}\left(m^5-m^4\right)\right)-\frac{2\omega}{n}\right]e^{-nx}+\frac{\omega}{n}$$

$$y^{1'}(x,\beta)=\frac{1}{2}\left[\left(m^2-\frac{\beta}{0.5}\left(m^2-m^1\right)\right)-\left(m^3-\frac{I-\beta}{I-0.5}\left(m^3-m^2\right)\right)\right]e^{nx}$$
$$+\frac{1}{2}\left[\left(m^2-\frac{\beta}{0.5}\left(m^2-m^1\right)\right)+\left(m^3-\frac{I-\beta}{I-0.5}\left(m^3-m^2\right)\right)-\frac{2\omega}{n}\right]e^{-nx}+\frac{\omega}{n}$$

$$y^{2'}(x,\beta)=-\frac{1}{2}\left[\left(m^2-\frac{\beta}{0.5}\left(m^2-m^1\right)\right)-\left(m^3-\frac{I-\beta}{I-0.5}\left(m^3-m^2\right)\right)\right]e^{nx}$$
$$+\frac{1}{2}\left[\left(m^2-\frac{\beta}{0.5}\left(m^2-m^1\right)\right)+\left(m^3-\frac{I-\beta}{I-0.5}\left(m^3-m^2\right)\right)-\frac{2\omega}{n}\right]e^{-nx}+\frac{\omega}{n}$$

$$y^{3'}(x,\beta)=\frac{1}{2}\left[\left(m^3+\frac{I-\beta}{I-0.5}\left(m^4-m^3\right)\right)-\left(m^4+\frac{\beta}{0.5}\left(m^5-m^4\right)\right)\right]e^{nx}$$
$$+\frac{1}{2}\left[\left(m^3+\frac{I-\beta}{I-0.5}\left(m^4-m^3\right)\right)+\left(m^4+\frac{\beta}{0.5}\left(m^5-m^4\right)\right)-\frac{2\omega}{n}\right]e^{-nx}+\frac{\omega}{n}$$

$$y^{4'}(x,\beta)=-\frac{1}{2}\left[\left(m^3+\frac{I-\beta}{I-0.5}\left(m^4-m^3\right)\right)-\left(m^4+\frac{\beta}{0.5}\left(m^5-m^4\right)\right)\right]e^{nx}$$
$$+\frac{1}{2}\left[\left(m^3+\frac{I-\beta}{I-0.5}\left(m^4-m^3\right)\right)+\left(m^4+\frac{\beta}{0.5}\left(m^5-m^4\right)\right)-\frac{2\omega}{n}\right]e^{-nx}+\frac{\omega}{n}$$

$$y^{1''}(x,\gamma)=\frac{1}{2}\left[\left(m^2-\frac{\gamma}{0.5}\left(m^2-m^1\right)\right)-\left(m^3-\frac{F-\gamma}{F-0.5}\left(m^3-m^2\right)\right)\right]e^{nx}$$
$$+\frac{1}{2}\left[\left(m^2-\frac{\gamma}{0.5}\left(m^2-m^1\right)\right)+\left(m^3-\frac{F-\gamma}{F-0.5}\left(m^3-m^2\right)\right)-\frac{2\omega}{n}\right]e^{-nx}+\frac{\omega}{n}$$

$$y^{2''}(x,\gamma)=-\frac{1}{2}\left[\left(m^2-\frac{\gamma}{0.5}\left(m^2-m^1\right)\right)-\left(m^3-\frac{F-\gamma}{F-0.5}\left(m^3-m^2\right)\right)\right]e^{nx}$$
$$+\frac{1}{2}\left[\left(m^2-\frac{\gamma}{0.5}\left(m^2-m^1\right)\right)+\left(m^3-\frac{F-\gamma}{F-0.5}\left(m^3-m^2\right)\right)-\frac{2\omega}{n}\right]e^{-nx}+\frac{\omega}{n}$$

$$y^{3''}(x,\gamma)=\frac{1}{2}\left[\left(m^3+\frac{F-\gamma}{F-0.5}\left(m^4-m^3\right)\right)-\left(m^4+\frac{\gamma}{0.5}\left(m^5-m^4\right)\right)\right]e^{nx}$$
$$+\frac{1}{2}\left[\left(m^3+\frac{F-\gamma}{F-0.5}\left(m^4-m^3\right)\right)+\left(m^4+\frac{\gamma}{0.5}\left(m^5-m^4\right)\right)-\frac{2\omega}{n}\right]e^{-nx}+\frac{\omega}{n}$$

$$y^{4''}(x,\gamma)=-\frac{1}{2}\left[\left(m^3+\frac{F-\gamma}{F-0.5}\left(m^4-m^3\right)\right)-\left(m^4+\frac{\gamma}{0.5}\left(m^5-m^4\right)\right)\right]e^{nx}$$
$$+\frac{1}{2}\left[\left(m^3+\frac{F-\gamma}{F-0.5}\left(m^4-m^3\right)\right)+\left(m^4+\frac{\gamma}{0.5}\left(m^5-m^4\right)\right)-\frac{2\omega}{n}\right]e^{-nx}+\frac{\omega}{n}$$

***Case* 3:** When m is a zero constant, that is, $m=0$,
taking the (α,β,γ) cut of equation (2.1) gives

$$\frac{d\begin{pmatrix}\left[y^1(x,\alpha),y^2(x,\alpha),y^3(x,\alpha),y^4(x,\alpha)\right],\\ \left[y^{1'}(x,\beta),y^{2'}(x,\beta),y^{3'}(x,\beta),y^{4'}(x,\beta)\right],\\ \left[y^{1''}(x,\gamma),y^{2''}(x,\gamma),y^{3''}(x,\gamma),y^{4''}(x,\gamma)\right]\end{pmatrix}}{dx}=\omega$$

where $0\leq\alpha+\beta+\gamma\leq 3$ and $\alpha,\beta,\gamma\epsilon\left[0,1\right]$
Thus, we have the following:

$$\frac{dy^1(x,\alpha)}{dx}=\omega;\frac{dy^2(x,\alpha)}{dx}=\omega;\frac{dy^3(x,\alpha)}{dx}=\omega;$$
$$\frac{dy^1(x,\alpha)}{dx}=\omega;\frac{dy^{1'}(x,\beta)}{dx}=\omega;\frac{dy^{2'}(x,\beta)}{dx}=\omega;$$

$$\frac{dy^{3'}(x,\beta)}{dx}=\omega;\frac{dy^{4'}(x,\beta)}{dx}=\omega;\frac{dy^{1''}(x,\gamma)}{dx}=\omega;$$
$$\frac{dy^{2''}(x,\gamma)}{dx}=\omega;\frac{dy^{3''}(x,\gamma)}{dx}=\omega;\frac{dy^{4''}(x,\gamma)}{dx}=\omega.$$

with the same initial conditions.

Solution

The solutions to these equations are given as follows:

$$y^{1}(x,\alpha)=m^{1}+\frac{\alpha}{0.5}(m^{2}-m^{1})+\omega x,\ y^{2}(x,\alpha)=m^{2}+\frac{T-\alpha}{T-0.5}(m^{3}-m^{2})+\omega x,$$

$$y^{3}(x,\alpha)=m^{4}-\frac{T-\alpha}{T-0.5}(m^{4}-m^{3})+\omega x,\ y^{4}(x,\alpha)=m^{5}-\frac{\alpha}{0.5}(m^{5}-m^{4})+\omega x,$$

$$y^{1'}(x,\beta)=m^{2}-\frac{\beta}{0.5}(m^{2}-m^{1})+\omega x,\ y^{2'}(x,\beta)=m^{3}-\frac{I-\beta}{I-0.5}(m^{3}-m^{2})+\omega x,$$

$$y^{3'}(x,\beta)=m^{3}+\frac{I-\beta}{I-0.5}(m^{4}-m^{3})+\omega x,\ y^{4'}(x,\beta)=m^{4}+\frac{\beta}{0.5}(m^{5}-m^{4})+\omega x,$$

$$y^{1''}(x,\gamma)=m^{2}-\frac{\gamma}{0.5}(m^{2}-m^{1})+\omega x, y^{2''}(x,\gamma)=m^{3}-\frac{F-\gamma}{F-0.5}(m^{3}-m^{2})+\omega x,$$

$$y^{3''}(x,\gamma)=m^{3}+\frac{F-\gamma}{F-0.5}(m^{4}-m^{3})+\omega x,\ y^{4''}(x,\gamma)=m^{4}+\frac{\gamma}{0.5}(m^{5}-m^{4})+\omega x.$$

2.3.1 The Analytic Problem

Consider the following problem, which applies a first-order differential equation (mathematical model for heat transfer in a solid submerged in fluid) under neutrosophic uncertainty:

$$\frac{dT}{dt}=-\vartheta.A.T(t)+\vartheta.A.T_f, T(0)=T_0,$$

$$\text{Here, } \vartheta=\frac{h}{\rho cv}$$

where
T = solid temperature
h = heat transfer coefficient between the solid and the bulk fluid
ρ = mass density of the solid
c = specific heat of solid
v = volume of the solid
A =area of the contacting surface between solid and bulk fluid
T_f = bulk fluid temperature, with initial condition
$\widetilde{T_0} =< (0.3, 0.4, 0.7, 0.5, 0.1); 0.7, 0.3, 0.4 >$

Here, $\vartheta > 0$.

2.3.2 The Solution

Note that here, the solid is small, so the surface temperature and the solid temperature are the same. Indeed, this model obtained from the first law of thermodynamics:

$$T^{1}(t,\alpha)=\frac{1}{2}\Big[(0.3+0.2\alpha)-(0.4+1.5(0.7-\alpha)\Big]e^{\vartheta At}$$
$$+\frac{1}{2}\Big[(0.3+0.2\alpha)+(0.4+1.5(0.7-\alpha)-2T_f\Big]e^{-\vartheta At}+T_f$$

$$T^{2}(t,\alpha)=-\frac{1}{2}\Big[(0.3+0.2\alpha)-((0.4+1.5(0.7-\alpha))\Big]e^{\vartheta At}$$
$$+\frac{1}{2}\Big[(0.3+0.2\alpha)+((0.4+1.5(0.7-\alpha))-2T_f\Big]e^{-\vartheta At}+T_f$$

$$T^{3}(t,\alpha)=\frac{1}{2}\Big[(0.5+(0.7-\alpha))-(0.1+0.8\alpha)\Big]e^{\vartheta At}$$
$$+\frac{1}{2}\Big[(0.5+(0.7-\alpha))+(0.1+0.8\alpha)-2T_f\Big]e^{-\vartheta At}+T_f$$

$$T^{4}(t,\alpha)=-\frac{1}{2}\Big[(0.5+(0.7-\alpha))-(0.1+0.8\alpha)\Big]e^{\vartheta At}$$
$$+\frac{1}{2}\Big[(0.5+(0.7-\alpha))+(0.1+0.8\alpha)-2T_f\Big]e^{-\vartheta At}+T_f$$

$$T^{1'}(t,\beta)=\frac{1}{2}\Big[(0.4-0.2\beta)-(0.7+1.5(0.3-\beta))\Big]e^{\vartheta At}$$
$$+\frac{1}{2}\Big[(0.4-0.2\beta)+(0.7+1.5(0.3-\beta))-2T_f\Big]e^{-\vartheta At}+T_f$$

$$T^{2'}(t,\beta)=-\frac{1}{2}\Big[(0.4-0.2\beta)-(0.7+1.5(0.3-\beta))\Big]e^{\vartheta At}$$
$$+\frac{1}{2}\Big[(0.4-0.2\beta)+(0.7+1.5(0.3-\beta))-2T_f\Big]e^{-\vartheta At}+T_f$$

$$T^{3'}(t,\beta)=\frac{1}{2}\Big[(0.7+(0.3-\beta))-(0.5-0.8\beta))\Big]e^{\vartheta At}$$
$$+\frac{1}{2}\Big[(0.7+(0.3-\beta))+(0.5-0.8\beta))-2T_f\Big]e^{-\vartheta At}+T_f$$

$$T^{4'}(t,\beta)=-\frac{1}{2}\Big[((0.7+(0.3-\beta))-(0.5-0.8\beta))\Big]e^{\vartheta At}$$
$$+\frac{1}{2}\Big[((0.7+(0.3-\beta))+(0.5-0.8\beta))-2T_f\Big]e^{-\vartheta At}+T_f$$

$$T^{1''}(t,\gamma)=\frac{1}{2}\Big[(0.4-0.2\gamma)-(0.7+3(0.4-\gamma))\Big]e^{\vartheta At}$$
$$+\frac{1}{2}\Big[(0.4-0.2\gamma)+(0.7+3(0.4-\gamma))-2T_f\Big]e^{-\vartheta At}+T_f$$

$$T^{2''}(t,\gamma) = -\frac{1}{2}\left[(0.4-0.2\gamma)-(0.7+3(0.4-\gamma))\right]e^{\vartheta At} + \frac{1}{2}\left[(0.4-0.2\gamma)+(0.7+3(0.4-\gamma))-2T_f\right]e^{-\vartheta At} + T_f$$

$$T^{3''}(t,\gamma) = \frac{1}{2}\left[(0.7+2(0.4-\gamma))-(0.5-0.8\gamma)\right]e^{\vartheta At} + \frac{1}{2}\left[(0.7+2(0.4-\gamma))+(0.5-0.8\gamma)-2T_f\right]e^{-\vartheta At} + T_f$$

$$T^{4''}(t,\gamma) = -\frac{1}{2}\left[(0.7+2(0.4-\gamma))-(0.5-0.8\gamma)\right]e^{\vartheta At} + \frac{1}{2}\left[(0.7+2(0.4-\gamma))+(0.5-0.8\gamma)-2T_f\right]e^{-\vartheta At} + T_f$$

2.4.1.1 Numerical Solution

Here we construct tables for this solution with the parametric values of α, β, γ varying within the range [0, 1] with a 0.1 step difference.

Figures 2.1 to 2.3 give graphical interpretations of the Tables 2.1 to 2.3 solutions.

Table 2.1 Solution for $t = 1, T_f = 1, \vartheta A = 1$

α	$T^1(t,\alpha)$	$T^2(t,\alpha)$	$T^3(t,\alpha)$	$T^4(t,\alpha)$
0.0	–0.6090	2.5170	2.3663	–0.6239
0.1	–0.4019	2.2621	2.1180	–0.3828
0.2	–0.1947	2.0071	1.8697	–0.1419
0.3	0.0124	1.7521	1.6213	0.0991
0.4	0.2196	1.4972	1.3730	0.3400
0.5	0.4268	1.2422	1.1247	0.5811
0.6	0.6338	0.9872	0.8764	0.8220
0.7	0.8410	0.4322	0.8999	0.7911
0.8	1.0481	0.4773	0.3797	1.3039
0.9	1.2553	0.2223	0.1314	1.5448
1.0	1.4624	–0.0326	–0.1169	1.7859

Note 1: The table and the graphical presentation show that $T^1(t,\alpha)$, $T^4(t,\alpha)$ are increasing functions whereas $T^2(t,\alpha)$, $T^3(t,\alpha)$ are decreasing functions. Hence, the solution is strong.

Table 2.2 Solution for $t = 1, T_f = 1, \vartheta A = 1$

β	$T^{1'}(t,\beta)$	$T^{2'}(t,\beta)$	$T^{3'}(t,\beta)$	$T^{4'}(t,\beta)$
0.0	−0.1022	1.9366	1.5876	0.2284
0.1	0.0433	1.7287	1.5273	0.2225
0.2	0.1887	1.5207	1.4670	0.226
0.3	0.4972	1.1496	1.4067	0.2107
0.4	0.4796	1.048	1.3464	0.2848
0.5	0.6250	0.8968	1.2862	0.3288
0.6	0.7704	0.6888	1.2259	0.4229
0.7	0.9158	0.4808	1.256	0.5870
0.8	1.0612	0.2730	1.1053	0.7852
0.9	1.2066	0.0650	1.0450	0.9652
1.0	1.3520	−0.1430	0.9846	1.292

Note 2: The table and the graphical presentation show that $T^{1'}(t,\beta)$, $T^{4'}(t,\beta)$ are increasing functions whereas $T^{2'}(t,\beta)$, $T^{3'}(t,\beta)$ are decreasing functions. Hence, the solution is strong.

Table 2.3 Solution for $t = 1\ T_f = 1, \vartheta A = 1$

γ	$T^{1'}(t,\beta)$	$T^{2'}(t,\beta)$	$T^{3'}(t,\gamma)$	$T^{4'}(t,\gamma)$
0.0	−0.9835	3.0939	2.3591	−0.3591
0.1	−0.6618	2.6544	2.1445	−0.2475
0.2	−0.3401	2.2151	1.9299	−0.1359
0.3	−0.0184	1.7756	1.7153	−0.0243
0.4	0.3032	1.3362	1.5007	0.0873
0.5	0.6250	0.8968	1.2862	0.1988
0.6	0.9466	0.4574	1.072	0.3104
0.7	1.2684	0.0180	0.9563	0.5213
0.8	1.5901	−0.4215	0.6424	0.5336
0.9	1.9117	−0.8609	0.4278	0.6452
1.0	2.8507	−1.7703	0.2132	0.7568

Note 3: The table and the graphical presentation show that $T^{1'}(t,\beta)$, $T^{4'}(t,\gamma)$ are increasing functions whereas $T^{2'}(t,\beta)$, $T^{3'}(t,\gamma)$ are decreasing functions. Hence, the solution is strong.

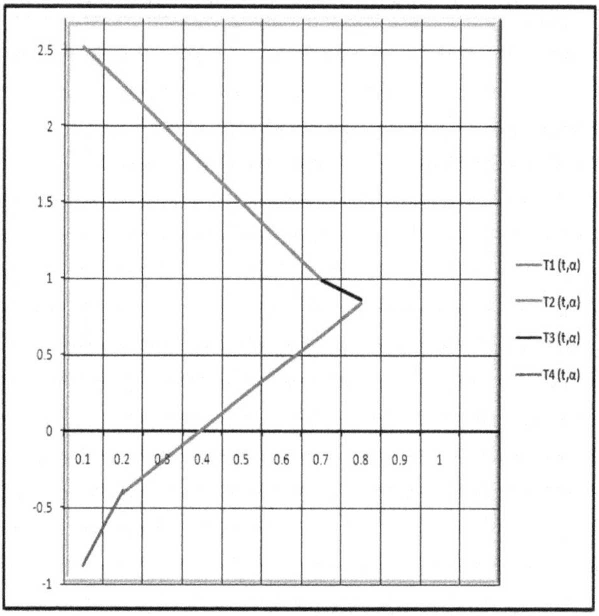

Figure 2.1 Graphical representation of the solution with respect to parameter α.

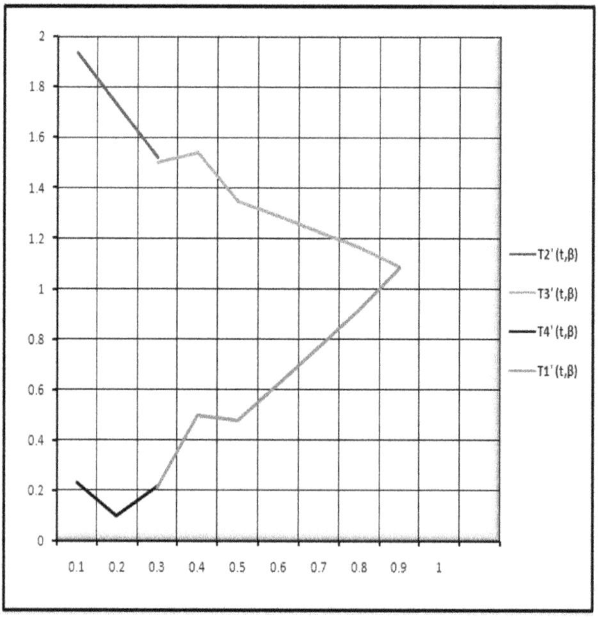

Figure 2.2 Graphical representation of the solution with respect to parameter β.

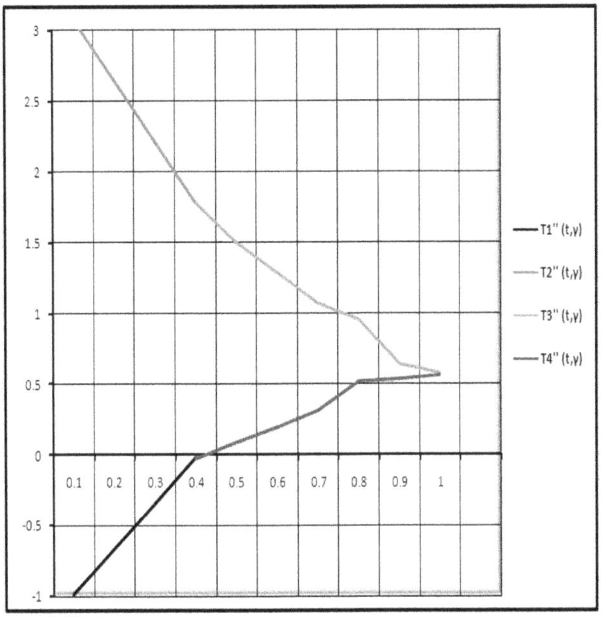

Figure 2.3 Graphical representation of the solution with respect to parameter γ.

2.5 CONCLUSION

In this chapter, we constructed a nonhomogeneous first-order differential equation with an initial value condition against a pentagonal neutrosophic number background. We erected a physical problem of heat transfer in a neutrosophic differential equation environment to intensify our suggested strategy. As this approach is reasonably innovative in the research field, researchers can fruitfully progress with the topic to solve higher-order linear and nonlinear differential equations, simultaneous differential equations and so on.

Conflicts of Interest: The authors have no conflicts of interest in this research article.

Data Availability: Not applicable.

Acknowledgement: The authors would like to give thanks to the editor-in-chief, the editors and the concerned reviewers for their valuable comments for the improvement of the study.

REFERENCES

[1] L. A. Zadeh. (1965). Fuzzy sets. Information and Control, 8(5), 338–353.
[2] K. Atanassov. (1986). Intuitionistic fuzzy sets. Fuzzy Sets and Systems, 20, 87–96.

[3] F. Smarandache. (1998). A unifying field in logics neutrosophy: Neutrosophic probability, set and logic. Rehoboth: American Research Press.

[4] A. Chakraborty, S. P. Mondal, A. Ahmadian, N. Senu, D. Dey, S. Alam, & S. Salahshour. (2019). The pentagonal fuzzy number: Its different representations, properties, ranking, defuzzification and application in game problem. Symmetry, 11(2), 248. doi:10.3390/sym11020248.

[5] A. Chakraborty, S. Maity, S. Jain, S. P. Mondal, & S. Alam. (2020). Hexagonal fuzzy number and its distinctive representation, ranking, defuzzification technique and application in production inventory management problem. Granular Computing. doi:10.1007/s41066-020-00212-8.

[6] S. Maity, A. Chakraborty, S. K. De, S. P. Mondal, & S. Alam. (2019). A comprehensive study of a backlogging EOQ model with nonlinear heptagonal dense fuzzy environment. Rairo Operations Research. doi:10.1051/ro/2018114.

[7] H. Wang, F. Smarandache, Q. Zhang, & R. Sunderraman. (2010). Single valued neutrosophic sets. Multispace and Multistructure, 4, 410–413.

[8] A. Chakraborty, S. P. Mondal, A. Ahmadian, N. Senu, S. Alam, & S. Salahshour. (2018). Different forms of triangular neutrosophic numbers, de-neutrosophication techniques, and their applications. Symmetry, 10, 327.

[9] A. Chakraborty, S. P. Mondal, S. Alam, & A. Mahata. (2019). Different linear and non-linear form of trapezoidal neutrosophic numbers, de-neutrosophication techniques and its application in time-cost optimization technique, sequencing problem. Rairo Operations Research. doi:10.1051/ro/2019090.

[10] F. Smarandache. (2013). Introduction to neutrosophic measure, neutrosophic integral, and neutrosophic probability (p. 140). Craiova: Sitech Educational.

[11] İ. Deli. (2019). Some operators with IVGSV TrN-numbers and their applications to multiple criteria group decision making. Neutrosophic Sets and Systems, 25, 33–53.

[12] H. Garg. (2019). Novel neutrality aggregation operators-based multi attribute group decision making method for single-valued neutrosophic numbers. Soft Computing. doi:10.1007/s00500-019-04535-w.

[13] S. Broumi, M. Talea, A. Bakali, & F. Smarandache. (2016). Single valued neutrosophic graphs. Journal of New Theory, 10, 86–101.

[14] S. Pal & A. Chakraborty. (2020). Triangular neutrosophic-based EOQ model for non instantaneous deteriorating item under shortages. American Journal of Business and Operations Research, 1(1), 28–35.

[15] A. Chakraborty, S. Broumi, & P. K. Singh. (2019). Some properties of pentagonal neutrosophic numbers and its applications in transportation problem environment. Neutrosophic Sets and Systems, 28, 200–215.

[16] A. Chakraborty, S. Mondal, & S. Broumi. (2019). De-neutrosophication technique of pentagonal neutrosophic number and application in minimal spanning tree. Neutrosophic Sets and Systems, 29, 1–18. doi:10.5281/zenodo.3514383.

[17] A. Chakraborty, B. Banik, S. P. Mondal, & S. Alam. (2020). Arithmetic and geometric operators of pentagonal neutrosophic number and its application in mobile communication service based MCGDM problem. Neutrosophic Sets and Systems, 32, 61–79.

[18] A. Chakraborty. (2020). A new score function of pentagonal neutrosophic number and its application in networking problem. International Journal of Neutrosophic Science (IJNS), 1(1).

[19] A. Chakraborty. (2020). Application of pentagonal neutrosophic number in shortest path problem. International Journal of Neutrosophic Science (IJNS), 3(1), 21–28.

[20] M. Abdel-Basset, M. Saleh, A. Gamal, & F. Smarandache. (2019). An approach of TOPSIS technique for developing supplier selection with group decision making under type-2 neutrosophic number. Applied Soft Computing, 77, 438–452.

[21] P. Bosc, & O. Pivert. (2013). On a fuzzy bipolar relational algebra. Information Sciences, 219, 1–2.

[22] I. Deli, M. Ali, & F. Smarandache. (2015). Bipolar neutrosophic sets and their application based on multi-criteria decision making problems. In: Proceedings of the 2015 InternationalConference on Advanced Mechatronic Systems, Beijing.

[23] S. Broumi, A. Bakali, M. Talea, F. Smarandache, & L. Vladareanu. (2016). Applying Dijkstra algorithm for solving neutrosophic shortest path problem. In: Proceedings on the International Conference on Advanced Mechatronic Systems, Melbourne.

[24] K.-P. Chiao. (2016). The multi-criteria group decision making methodology using type 2 fuzzy linguistic judgments. Applied Soft Computing, 49, 189–211.

[25] S. Wibowo, S. Grandhi, & H. Deng. (2016). Multicriteria group decision making for selecting human resources management information systems projects. IEEE, 1405–1410.

[26] N. A. Nabeeh, M. Abdel-Basset, H. A. El-Ghareeb, & A. Aboelfetouh. (2019). Neutrosophic multi-criteria decision making approach for iot-based enterprises. IEEE Access, 7, 59559–59574.

[27] T. Haque, A. Chakraborty, S. P. Mondal, & S. Alam. (2020). A new approach to solve multicriteria group decision making problems by exponential operational law in generalised spherical fuzzy environment. CAAI Transactions on Intelligence Technology, 5(2). doi:10.1049/trit.2019.0078.

[28] A. Chakraborty, S. P. Mondal, S. Alam, & A. Mahata. (2020). Cylindrical neutrosophic single-valued numberand its application in networking problem, multi criterion decision making problem and graph theory. CAAI Transactions on Intelligence Technology, 5(2). doi:10.1049/trit.2019.0083.

[29] A. Chakraborty. (2019). Minimal spanning tree in cylindrical single-valued neutrosophic arena. Neutrosophic Graph Theory and Algorithm. doi:10.4018/978-1-7998-1313-2.ch009.

[30] U. M. Pirzada, & D. C. Vakaskar. (2017). Existence of Hukuhara differentiability of fuzzy-valued functions. Journal of the Indian Mathematical Society, 84(3–4). doi:10.18311/jims/2017/5824.

[31] S. Seikkala. (1987). On the fuzzy initial value problem. Fuzzy Sets and Systems, 24(3), 319–330.

[32] O. Kaleva. (1987). Fuzzy differential equations. Fuzzy Sets and Systems, 24(3), 301–317.

[33] M. T. Malinowski. (2012). Random fuzzy differential equations under generalized Lipschitz condition. Nonlinear Analysis: Real World Applications, 13(2), 860–881.

[34] A. Kandel, & W. J. Byatt. (1978). Fuzzy differential equations. In: Proceedings of International Conference Cybernetics and Society, Tokyo, pp. 1213–1222.

[35] A. Khastana, & R. Rodriguez-Lopezb. (2016). On the solutions to first order linear fuzzy differential equations. Fuzzy Sets Syst, 295, 114–135.

[36] L. Stefanini, & B. Bede. (2009). Generalized Hukuhara differentiability of interval-valued functions and interval differential equations. Nonlinear Analysis, 71, 1311–1328.

[37] B. Ben Amma, S. Melliani, & L. S. Chadli. (2018). The Cauchy problem of intuitionistic fuzzy differential equations. Notes on Intuitionistic Fuzzy Sets, 24(1), 37–47.

[38] B. Ben Amma, S. Melliani, & L. S. Chadli. (2016). Numerical solution of intuitionistic fuzzy differential equations by Adams' three order predictor-corrector method. Notes on Intuitionistic Fuzzy Sets, 22(3), 47–69.

[39] S. Melliani, H. Atti, B. Ben Amma, & L. S. Chadli. (2018). Solution of n-th order intuitionistic fuzzy differential equation by variational iteration method. Notes on Intuitionistic Fuzzy Sets, 24(3), 92–105.

[40] M. Kumar, S. P. Yadav, & S. Kumar. (2011). A new approach for analyzing the fuzzy system reliability using intuitionistic fuzzy number. International Journal of Industrial and Systems Engineering, 8(2), 135–156.

[41] J. Wang, R. Nie, H. Zhang, & X. Chen. (2013). New operators on triangular intuitionistic fuzzy numbers and their applications in system fault analysis. Information Sciences, 251, 79–95.

[42] J. J. Buckley, & T. Feuring. (1999). Introduction to fuzzy partial differential equations. Fuzzy Sets and Systems, 105(2), 241–248.

[43] F. Smarandache. (2015). Neutrosophicprecalculus and neutrosophic calculus. Brussels: Sitech Educational, EuropaNova.

[44] I. R. Sumathi, & V. Mohana Priya. (2018). A new perspective on neutrosophic differential equation. International Journal of Engineering & Technology, 7(4.10), 422–425.

[45] I. R. Sumathi, & C. Antony Crispin Sweety. (2019). New approach on differential equation via trapezoidal neutrosophic number. Complex & Intelligent Systems, 5, 417–424.

Chapter 3

Analyzing an M/M/c Queue with Linguistic Single-Valued Neutrosophic Logic Using Parametric Nonlinear Programming

B. Vennila and C. Antony Crispin Sweety

3.1 INTRODUCTION

Queueing theory plays an important role in our daily life. It is important because it helps to describe features of the queue, like average wait time, and provides the tools for optimizing queues. The origin of queuing theory can be traced to the early 1900s in a study of the Copenhagen telephone exchange by Agner Krarup Erlang, a Danish engineer, statistician, and mathematician. His extensive studies of wait time in automated telephone services and his proposals for more efficient networks were widely adopted by telephone companies.

Zadeh (1965) set forth the concept of fuzzy sets, which makes an element belong to a set partially using the membership function that takes a value in the range [0,1]. The M/M/c model is applied in decision making to reduce customer waiting times. This concept has been applied in many real-world problems to administer unreliability.

Atanassov (2016) introduced the intuitionistic fuzzy set which is an extension of Lotfi Zadeh's notion of fuzzy set, where the elements in intuitionistic fuzzy set have degrees of membership and non-membership.

Atanassov (1986) generated an intuitionistic fuzzy set, a generalized fuzzy set that considered both membership and nonmembership functions (truth and falsity) with $0 \leq t(x) + f(x) \leq 1$, where $1 - t(x) + f(x)$ is an indeterminacy. F. Smarandache (1995) introduced neutrosophic logic, which combines fuzzy and intuitionistic fuzzy logic. If the parameters of the queueing system are in neutrosophic numbers, then the queueing system is neutrosophic. In a neutrosophic M/M/c queue, the arrival and departure rates are single-valued neutrosophic numbers (SVNNs) with T as truth, I as indeterminacy and F as falsity.

Smarandache (1998) later developed the concept of a neutrosophic set that annexed I(x) for indeterminacy, independent of t(x) and f(x). The applications of neutrosophic probability in queueing theory were first defined, and the author defined a neutrosophic event-based queueing model. Smarandache (2016) also

DOI: 10.1201/9781003487104-3

proposed the concept of subtracting and dividing neutrosophic SVNNs and the restrictions for these operations including for over, under, and off numbers.

Sahin et al. (2017) presented a similarity measure of properties based on falsity between SVNNs and properties based on falsity between single-valued neutrosophic sets. Zeina (2020a, 2020b) introduced the general definition of neutrosophic random variables and their properties and applications in various domains including quality control, stochastic modeling, reliability theory, queueing theory, decision making, electrical engineering, etc. He also derived the concept of a linguistic single-valued neutrosophic M/M/1 queue assuming that both arrival rate and departure rate are SVNNs: $\tilde{A}$= (T,I,F).

3.2 CRISP M/M/C QUEUE

Like M/M/1 queue, an M/M/c queue can be modeled as a birthdeath process since the arrival rate is constant at $\lambda_n = \lambda$ regardless of the number of customers in the system. If there is c or more customers in the system, then all c servers must be busy. Since each server processes customers with rate μ, the combined service-completion rate for the system is cμ.

The average number of customers in a queue is calculated as follows:

$$L_q = \left(\frac{r^c \rho}{C!(1-\rho)^2} \right) P_0, r = \frac{\lambda}{\mu}; \rho = \frac{\lambda}{c\mu} \quad (3.1).$$

The average number of customers in the system is

$$L_S = r + \left(\frac{r^c \rho}{c!(1-\rho)^2} \right) P_0, r = \frac{\lambda}{\mu}; \rho = \frac{\lambda}{c\mu} \quad (3.2).$$

The mean waiting time in the queue is

$$W_q = \left(\frac{r^c \rho}{C!(C\mu)(1-\rho)^2} \right) P_0, r = \frac{\lambda}{\mu}; \rho = \frac{\lambda}{c\mu} \quad (3.3).$$

The mean waiting time in the system is

$$W_s = \frac{1}{\mu_1} + \left(\frac{r^c}{C!(C\mu)(1-\rho)^2} \right) P_0, r = \frac{\lambda}{\mu}; \rho = \frac{\lambda}{c\mu} \quad (3.4).$$

Table 3.1 Linguistic Terms of SVNNs

Linguistic term	*SVN numbers*
Extremely good	(1,0,0)
Very very good	(0.95,0.05,0.10)
Very good	(0.85,0.15,0.20)
Good	(0.75,0.25,0.35)
Fairly good	(0.65,0.30,0.40)
Average	(0.55,0.45,0.50)
Moderately bad	(0.45,0.50,0.60)
Bad	(0.35,0.65,0.70)
Very bad	(0.25,0.75,0.80)
Very very bad	(0.15,0.90,0.90)
Extremely bad	(0.0,1.0,0.95)

3.3 NEUTROSOPHIC M/M/C QUEUE

Suppose that a queueing system is described by neutrosophic parameters $\lambda_N = (T_\lambda, I_\lambda, F_\lambda)$, $\mu_N = (T_\mu, I_\mu, F_\mu)$ such that both are SVNNs:

$$0 \le T_\lambda, I_\lambda, F_\lambda \le 1 \,\&\, 0 \le T_\mu, I_\mu, F_\mu \le 1$$

$$0 \le T_\lambda + I_\lambda + F_\lambda \le 3 \,\&\, 0 \le T_\mu + I_\mu + F_\mu \le 3.$$

This means we are not sure about the arrival rates and departure (serving) rates. We just know that these rates are extremely big, very very big, very big, big, fairly big, average, moderately small, small, very small, very very small, and extremely small.

3.3.1 Theorem

1. Average number of customers in the queue is

$$NL_q = \left[\left(\left(\frac{T_{\lambda_N}^{c+1}}{T_{\mu_N}^{c}\left(1-(1-T_{\mu_N})^{c}\right)}\right)\left(\left(\frac{T_{\lambda_N}}{T_{\mu_N}}\right)^{c} + \left(\sum_{n=0}^{c-1}\left(1-\left(1-\frac{T_{\lambda_N}^{n}}{T_{\mu_N}^{n}}\right)^{\frac{1}{n!}}\right) - \prod_{n=0}^{c-1}\left(1-\left(1-\frac{T_{\lambda_N}^{n}}{T_{\mu_N}^{n}}\right)^{\frac{1}{n!}}\right)\right)\right.\right.\right.$$

$$\left.\left.\left. - \left(\frac{T_{\lambda_N}}{T_{\mu_N}}\right)^{c}\left(\sum_{n=0}^{c-1}\left(1-\left(1-\frac{T_{\lambda_N}^{n}}{T_{\mu_N}^{n}}\right)^{\frac{1}{n!}}\right) - \prod_{n=0}^{c-1}\left(1-\left(1-\frac{T_{\lambda_N}^{n}}{T_{\mu_N}^{n}}\right)^{\frac{1}{n!}}\right)\right)\right)^{-1}\right)\right],$$

$$\Bigg(\bigg(1-\Big(\frac{1-I_{\lambda_N}}{1-I_{\mu_N}}\Big)^{c}\Big(\frac{1-I_{\lambda_N}}{1-I^{c}_{\mu_N}}\Big)\bigg)+\Bigg(1-\bigg(1-\bigg(\Big(1-\Big(\frac{1-I_{\lambda_N}}{1-I_{\mu_N}}\Big)^{c}\Big)\prod_{n=0}^{c-1}\Big(\Big(1-\Big(\frac{1-I^{n}_{\lambda_N}}{1-I^{n}_{\mu_N}}\Big)\Big)^{\frac{1}{n!}}\Big)\bigg)\bigg)\Bigg)^{-1}\Bigg)$$

$$-\bigg(1-\Big(\frac{1-I_{\lambda_N}}{1-I_{\mu_N}}\Big)^{c}\Big(\frac{1-I_{\lambda_N}}{1-I^{c}_{\mu_N}}\Big)\bigg)\Bigg(1-\bigg(1-\bigg(\Big(1-\Big(\frac{1-I_{\lambda_N}}{1-I_{\mu_N}}\Big)^{c}\Big)\prod_{n=0}^{c-1}\Big(\Big(1-\Big(\frac{1-I^{n}_{\lambda_N}}{1-I^{n}_{\mu_N}}\Big)\Big)^{\frac{1}{n!}}\Big)\bigg)\bigg)\Bigg)^{-1}\Bigg)\Bigg),$$

$$(3.5).$$

$$\Bigg(\bigg(1-\Big(\frac{1-F_{\lambda_N}}{1-F_{\mu_N}}\Big)^{c}\Big(\frac{1-F_{\lambda_N}}{1-F^{c}_{\mu_N}}\Big)\bigg)+\Bigg(1-\bigg(1-\bigg(\Big(1-\Big(\frac{1-F_{\lambda_N}}{1-F_{\mu_N}}\Big)^{c}\Big)\prod_{n=0}^{c-1}\Big(\Big(1-\Big(\frac{1-F^{n}_{\lambda_N}}{1-F^{n}_{\mu_N}}\Big)\Big)^{\frac{1}{n!}}\Big)\bigg)\bigg)\Bigg)^{-1}\Bigg)$$

$$-\bigg(1-\Big(\frac{1-F_{\lambda_N}}{1-F_{\mu_N}}\Big)^{c}\Big(\frac{1-F_{\lambda_N}}{1-F^{c}_{\mu_N}}\Big)\bigg)\Bigg(1-\bigg(1-\bigg(\Big(1-\Big(\frac{1-F_{\lambda_N}}{1-F_{\mu_N}}\Big)^{c}\Big)\prod_{n=0}^{c-1}\Big(\Big(1-\Big(\frac{1-F^{n}_{\lambda_N}}{1-F^{n}_{\mu_N}}\Big)\Big)^{\frac{1}{n!}}\Big)\bigg)\bigg)\Bigg)^{-1}\Bigg)\Bigg)\Bigg]$$

2. Average number of customers in the system is

$$NL_s=\Bigg[\Big(\frac{T_{\lambda_N}}{T_{\mu_N}}\Big)+\bigg(\Big(\frac{T^{c+1}_{\lambda_N}}{T^{c}_{\mu_N}\left(1-(1-T_{\mu_N})^{c}\right)}\Big)\Big(\Big(\frac{T_{\lambda_N}}{T_{\mu_N}}\Big)^{c}+\Big(\sum_{n=0}^{c-1}\Big(1-\Big(1-\frac{T^{n}_{\lambda_N}}{T^{n}_{\mu_N}}\Big)^{\frac{1}{n!}}\Big)-\prod_{n=0}^{c-1}\Big(1-\Big(1-\frac{T^{n}_{\lambda_N}}{T^{n}_{\mu_N}}\Big)^{\frac{1}{n!}}\Big)\Big)\Big)$$

$$-\Big(\frac{T_{\lambda_N}}{T_{\mu_N}}\Big)^{c}\Big(\sum_{n=0}^{c-1}\Big(1-\Big(1-\frac{T^{n}_{\lambda_N}}{T^{n}_{\mu_N}}\Big)^{\frac{1}{n!}}\Big)-\prod_{n=0}^{c-1}\Big(1-\Big(1-\frac{T^{n}_{\lambda_N}}{T^{n}_{\mu_N}}\Big)^{\frac{1}{n!}}\Big)\Big)\bigg)^{-1}\bigg)$$

$$-\Big(\frac{T_{\lambda_N}}{T_{\mu_N}}\Big)\bigg(\Big(\frac{T^{c+1}_{\lambda_N}}{T^{c}_{\mu_N}\left(1-(1-T_{\mu_N})^{c}\right)}\Big)\Big(\Big(\frac{T_{\lambda_N}}{T_{\mu_N}}\Big)^{c}+\Big(\sum_{n=0}^{c-1}\Big(1-\Big(1-\frac{T^{n}_{\lambda_N}}{T^{n}_{\mu_N}}\Big)^{\frac{1}{n!}}\Big)-\prod_{n=0}^{c-1}\Big(1-\Big(1-\frac{T^{n}_{\lambda_N}}{T^{n}_{\mu_N}}\Big)^{\frac{1}{n!}}\Big)\Big)\Big)$$

$$-\Big(\frac{T_{\lambda_N}}{T_{\mu_N}}\Big)^{c}\Big(\sum_{n=0}^{c-1}\Big(1-\Big(1-\frac{T^{n}_{\lambda_N}}{T^{n}_{\mu_N}}\Big)^{\frac{1}{n!}}\Big)-\prod_{n=0}^{c-1}\Big(1-\Big(1-\frac{T^{n}_{\lambda_N}}{T^{n}_{\mu_N}}\Big)^{\frac{1}{n!}}\Big)\Big)\bigg)^{-1}\bigg)\Bigg],$$

$$\Bigg[\Big(\frac{I_{\lambda_N}-I_{\mu_N}}{1-I_{\mu_N}}\Big)\Bigg(\bigg(1-\Big(\frac{1-I_{\lambda_N}}{1-I_{\mu_N}}\Big)^{c}\Big(\frac{1-I_{\lambda_N}}{1-I^{c}_{\mu_N}}\Big)\bigg)+\bigg(1-\bigg(1-\bigg(\Big(1-\Big(\frac{1-I_{\lambda_N}}{1-I_{\mu_N}}\Big)^{c}\Big)\prod_{n=0}^{c-1}\Big(\Big(1-\Big(\frac{1-I^{n}_{\lambda_N}}{1-I^{n}_{\mu_N}}\Big)\Big)^{\frac{1}{n!}}\Big)\bigg)\bigg)\bigg)^{-1}$$

$$-\left(1-\left(\frac{1-I_{\lambda_N}}{1-I_{\mu_N}}\right)^{c}\left(\frac{1-I_{\lambda_N}}{1-I_{\mu_N}^{c}}\right)\right)\left(1-\left(1-\left(\left(1-\left(\frac{1-I_{\lambda_N}}{1-I_{\mu_N}}\right)^{c}\right)\left(\prod_{n=0}^{c-1}\left(\left(1-\left(\frac{1-I_{\lambda_N}^{n}}{1-I_{\mu_N}^{n}}\right)\right)^{\frac{1}{n!}}\right)\right)\right)\right)^{-1}\right)\Bigg]\Bigg],$$

$$\Bigg[\left(\frac{F_{\lambda_N}-F_{\mu_N}}{1-F_{\mu_N}}\right)\left(\left(1-\left(\frac{1-F_{\lambda_N}}{1-F_{\mu_N}}\right)^{c}\left(\frac{1-F_{\lambda_N}}{1-F_{\mu_N}^{c}}\right)\right)+\left(1-\left(1-\left(\left(1-\left(\frac{1-F_{\lambda_N}}{1-F_{\mu_N}}\right)^{c}\right)\left(\prod_{n=0}^{c-1}\left(\left(1-\left(\frac{1-F_{\lambda_N}^{n}}{1-F_{\mu_N}^{n}}\right)\right)^{\frac{1}{n!}}\right)\right)\right)\right)\right)^{-1}\right) \quad (3.6).$$

$$-\left(1-\left(\frac{1-F_{\lambda_N}}{1-F_{\mu_N}}\right)^{c}\left(\frac{1-F_{\lambda_N}}{1-F_{\mu_N}^{c}}\right)\right)\left(1-\left(1-\left(\left(1-\left(\frac{1-F_{\lambda_N}}{1-F_{\mu_N}}\right)^{c}\right)\left(\prod_{n=0}^{c-1}\left(\left(1-\left(\frac{1-F_{\lambda_N}^{n}}{1-F_{\mu_N}^{n}}\right)\right)^{\frac{1}{n!}}\right)\right)\right)\right)^{-1}\right)\Bigg]\Bigg]$$

3. Mean waiting time in the queue is

$$NW_q=\Bigg[\Bigg[\left(\frac{\left(\frac{T_{\lambda_N}}{T_{\mu_N}}\right)^{c}}{\left(1-\left(1-T_{\mu_N}\right)^{c!c}\right)}\right)\left(\left(\frac{T_{\lambda_N}}{T_{\mu_N}}\right)^{c}+\left(\sum_{n=0}^{c-1}\left(1-\left(1-\frac{T_{\lambda_N}^{n}}{T_{\mu_N}^{n}}\right)^{\frac{1}{n!}}\right)-\prod_{n=0}^{c-1}\left(1-\left(1-\frac{T_{\lambda_N}^{n}}{T_{\mu_N}^{n}}\right)^{\frac{1}{n!}}\right)\right)\right.$$

$$\left.-\left(\frac{T_{\lambda_N}}{T_{\mu_N}}\right)^{c}\left(\sum_{n=0}^{c-1}\left(1-\left(1-\frac{T_{\lambda_N}^{n}}{T_{\mu_N}^{n}}\right)^{\frac{1}{n!}}\right)-\prod_{n=0}^{c-1}\left(1-\left(1-\frac{T_{\lambda_N}^{n}}{T_{\mu_N}^{n}}\right)^{\frac{1}{n!}}\right)\right)\right)^{-1}\Bigg],$$

$$\Bigg[\frac{\left(1-\left(\frac{1-I_{\lambda_N}}{1-I_{\mu_N}}\right)^{c}\right)-I_{\mu_N}^{c!c}}{1-I_{\mu_N}^{c!c}}+\left(1-\left(1-\left(\left(1-\left(\frac{1-I_{\lambda_N}}{1-I_{\mu_N}}\right)^{c}\right)\left(\prod_{n=0}^{c-1}\left(\left(1-\left(\frac{1-I_{\lambda_N}^{n}}{1-I_{\mu_N}^{n}}\right)\right)^{\frac{1}{n!}}\right)\right)\right)\right)\right)^{-1}$$

$$-\frac{\left(1-\left(\frac{1-I_{\lambda_N}}{1-I_{\mu_N}}\right)^{c}\right)-I_{\mu_N}^{c!c}}{1-I_{\mu_N}^{c!c}}\left(1-\left(1-\left(\left(1-\left(\frac{1-I_{\lambda_N}}{1-I_{\mu_N}}\right)^{c}\right)\left(\prod_{n=0}^{c-1}\left(\left(1-\left(\frac{1-I_{\lambda_N}^{n}}{1-I_{\mu_N}^{n}}\right)\right)^{\frac{1}{n!}}\right)\right)\right)\right)^{-1}\Bigg], \quad (3.7).$$

$$\Bigg[\frac{\left(1-\left(\frac{1-F_{\lambda_N}}{1-F_{\mu_N}}\right)^{c}\right)-F_{\mu_N}^{c!c}}{1-F_{\mu_N}^{c!c}}+\left(1-\left(1-\left(\left(1-\left(\frac{1-F_{\lambda_N}}{1-F_{\mu_N}}\right)^{c}\right)\left(\prod_{n=0}^{c-1}\left(\left(1-\left(\frac{1-F_{\lambda_N}^{n}}{1-F_{\mu_N}^{n}}\right)\right)^{\frac{1}{n!}}\right)\right)\right)\right)\right)^{-1}$$

$$-\frac{\left(1-\left(\frac{1-F_{\lambda_N}}{1-F_{\mu_N}}\right)^{c}\right)-F_{\mu_N}^{c!c}}{1-F_{\mu_N}^{c!c}}\left(1-\left(1-\left(\left(1-\left(\frac{1-F_{\lambda_N}}{1-F_{\mu_N}}\right)^{c}\right)\left(\prod_{n=0}^{c-1}\left(\left(1-\left(\frac{1-F_{\lambda_N}^{n}}{1-F_{\mu_N}^{n}}\right)\right)^{\frac{1}{n!}}\right)\right)\right)\right)^{-1}\Bigg]\Bigg]$$

4. Mean waiting time in the system is

$$
NW_s = \Bigg[\Bigg[\left(\frac{1}{T_{\mu_N}}\right) + \left[\left(\frac{\left(\frac{T_{\lambda_N}}{T_{\mu_N}}\right)^c}{\left(1-\left(1-T_{\mu_N}\right)^{c!c}\right)}\right)\left(\left(\frac{T_{\lambda_N}}{T_{\mu_N}}\right)^c + \sum_{n=0}^{c-1}\left(1-\left(1-\frac{T_{\lambda_N}^n}{T_{\mu_N}^n}\right)^{\frac{1}{n!}}\right) - \prod_{n=0}^{c-1}\left(1-\left(1-\frac{T_{\lambda_N}^n}{T_{\mu_N}^n}\right)^{\frac{1}{n!}}\right)\right)\right.
$$

$$
\left.\left. - \left(\frac{T_{\lambda_N}}{T_{\mu_N}}\right)^c - \left(\sum_{n=0}^{c-1}\left(1-\left(1-\frac{T_{\lambda_N}^n}{T_{\mu_N}^n}\right)^{\frac{1}{n!}}\right) - \prod_{n=0}^{c-1}\left(1-\left(1-\frac{T_{\lambda_N}^n}{T_{\mu_N}^n}\right)^{\frac{1}{n!}}\right)\right)\right)^{-1}\right]
$$

$$
-\left(\frac{1}{T_{\mu_N}}\right)\left[\left(\frac{\left(\frac{T_{\lambda_N}}{T_{\mu_N}}\right)^c}{\left(1-\left(1-T_{\mu_N}\right)^{c!c}\right)}\right)\left(\left(\frac{T_{\lambda_N}}{T_{\mu_N}}\right)^c + \sum_{n=0}^{c-1}\left(1-\left(1-\frac{T_{\lambda_N}^n}{T_{\mu_N}^n}\right)^{\frac{1}{n!}}\right) - \prod_{n=0}^{c-1}\left(1-\left(1-\frac{T_{\lambda_N}^n}{T_{\mu_N}^n}\right)^{\frac{1}{n!}}\right)\right)\right.
$$

$$
\left.\left. - \left(\frac{T_{\lambda_N}}{T_{\mu_N}}\right)^c\left(\sum_{n=0}^{c-1}\left(1-\left(1-\frac{T_{\lambda_N}^n}{T_{\mu_N}^n}\right)^{\frac{1}{n!}}\right) - \prod_{n=0}^{c-1}\left(1-\left(1-\frac{T_{\lambda_N}^n}{T_{\mu_N}^n}\right)^{\frac{1}{n!}}\right)\right)\right)^{-1}\right]\Bigg],
$$

$$
\Bigg[\left(\frac{-I_{\mu_N}}{1-I_{\mu_N}}\right)\left[\frac{\left(1-\left(\frac{1-I_{\lambda_N}}{1-I_{\mu_N}}\right)^c\right)-I_{\mu_N}^{c!c}}{1-I_{\mu_N}^{c!c}} + \left(1-\left(1-\left(\left(1-\left(\frac{1-I_{\lambda_N}}{1-I_{\mu_N}}\right)^c\right)\prod_{n=0}^{c-1}\left(\left(1-\left(\frac{1-I_{\lambda_N}^n}{1-I_{\mu_N}^n}\right)\right)^{\frac{1}{n!}}\right)\right)\right)\right)^{-1}\right. \quad (3.8).
$$

$$
\left. -\frac{\left(1-\left(\frac{1-I_{\lambda_N}}{1-I_{\mu_N}}\right)^c\right)-I_{\mu_N}^{c!c}}{1-I_{\mu_N}^{c!c}}\left(1-\left(1-\left(\left(1-\left(\frac{1-I_{\lambda_N}}{1-I_{\mu_N}}\right)^c\right)\prod_{n=0}^{c-1}\left(\left(1-\left(\frac{1-I_{\lambda_N}^n}{1-I_{\mu_N}^n}\right)\right)^{\frac{1}{n!}}\right)\right)\right)\right)^{-1}\right]\Bigg],
$$

$$
\Bigg[\left(\frac{-F_{\mu_N}}{1-F_{\mu_N}}\right)\left[\frac{\left(1-\left(\frac{1-F_{\lambda_N}}{1-F_{\mu_N}}\right)^c\right)-F_{\mu_N}^{c!c}}{1-F_{\mu_N}^{c!c}} + \left(1-\left(1-\left(\left(1-\left(\frac{1-F_{\lambda_N}}{1-F_{\mu_N}}\right)^c\right)\prod_{n=0}^{c-1}\left(\left(1-\left(\frac{1-F_{\lambda_N}^n}{1-F_{\mu_N}^n}\right)\right)^{\frac{1}{n!}}\right)\right)\right)\right)^{-1}\right.
$$

$$
\left. -\frac{\left(1-\left(\frac{1-F_{\lambda_N}}{1-F_{\mu_N}}\right)^c\right)-F_{\mu_N}^{c!c}}{1-F_{\mu_N}^{c!c}}\left(1-\left(1-\left(\left(1-\left(\frac{1-F_{\lambda_N}}{1-F_{\mu_N}}\right)^c\right)\prod_{n=0}^{c-1}\left(\left(1-\left(\frac{1-F_{\lambda_N}^n}{1-F_{\mu_N}^n}\right)\right)^{\frac{1}{n!}}\right)\right)\right)\right)^{-1}\right]\Bigg]\Bigg]
$$

Proof:

$$r_N = \frac{\lambda_N}{\mu_N} = \left(\frac{T_{\lambda_N}}{T_{\mu_N}}, \frac{I_{\lambda_N} - I_{\mu_N}}{1 - I_{\mu_N}}, \frac{F_{\lambda_N} - F_{\mu_N}}{1 - F_{\mu_N}}\right); T_{\mu_N} \neq 0, I_{\mu_N} \neq 1, F_{\mu_N} \neq 1 \quad (3.9)$$

$$C\mu_N = \left(1 - (1 - T_{\mu_N})^c, I_{\mu_N}^c, F_{\mu_N}^c\right); C > 0 \quad (3.10)$$

$$\rho_N = \frac{\lambda_N}{C\mu_N}$$

$$\rho_N = \left(\frac{T_{\lambda_N}}{1 - \left(1 - T_{\mu_N}\right)^c}, \frac{I_{\lambda_N} - I_{\mu_N}^c}{1 - I_{\mu_N}^c}, \frac{F_{\lambda_N} - F_{\mu_N}^c}{1 - F_{\mu_N}^c}\right), \left(1 - T_{\mu_N}\right)^c \neq 1, I_{\mu_N}^c \neq 1, F_{\mu_N}^c \neq 1 \quad (3.11)$$

$$r_N^c = \left(\frac{\lambda_N}{\mu_N}\right)^c$$

$$r_N^c = \left(\left(\frac{T_{\lambda_N}}{T_{\mu_N}}\right)^c, 1 - \left(\frac{I_{\lambda_N} - I_{\mu_N}}{1 - I_{\mu_N}}\right)^c, 1 - \left(\frac{F_{\lambda_N} - F_{\mu_N}}{1 - F_{\mu_N}}\right)^c\right) \quad (3.12).$$

Using (3.11) and (3.12), we get $r_N^c . \rho_N$:

$$r_N^c . \rho_N = \left(\frac{T_{\lambda_N}^{c+1}}{T_{\mu_N}^c - T_{\mu_N}^c (1 - T_{\mu_N})^c}\right), \left[1 - \left(\frac{1 - I_{\lambda_N}}{1 - I_{\mu_N}}\right)^c \left(\frac{(1 - I_{\lambda_N})}{1 - I_{\mu_N}^c}\right)\right], \left[1 - \left(\frac{1 - F_{\lambda_N}}{1 - F_{\mu_N}}\right)^c \left(\frac{1 - F_{\lambda_N}}{1 - F_{\mu_N}^c}\right)\right] \quad (3.13)$$

$$1 - \rho_N = (1,0,0) - \left(\frac{T_{\lambda_N}}{\left(1 - (1 - T_{\mu_N})^c\right)}, \frac{I_{\lambda_N} - I_{\mu_N}^c}{1 - I_{\mu_N}^c}, \frac{F_{\lambda_N} - F_{\mu_N}^c}{1 - F_{\mu_N}^c}\right)$$

$$= \left(\frac{1 - \frac{T_{\lambda_N}}{1 - (1 - T_{\mu_N}^c)}}{1 - \frac{T_{\lambda_N}}{1 - (1 - T_{\mu_N}^c)}}, 0, 0\right)$$

$$\therefore 1 - \rho_N = (1,0,0) \quad (3.14)$$

$$C!(1-\rho_N) = C!(1,0,0)$$

$$= (1-(1-1)^c, 0^{c!}, 0^{c!})$$

$$\therefore C!(1-\rho_N) = (1,0,0) \tag{3.15}$$

$$(1-\rho_N)^2 = (1,0,0)^2$$

$$= \left(1^2, 1-(1-0)^2, 1-(1-0)^2\right)$$

$$= (1,\ 1-1,\ 1-1) = (1,\ 0,\ 0) \tag{3.16}$$

$$C!(1-\rho_N)^2 = C!(1,0,0)$$

$$= \left(1-(1-1)^c, 0^{c!}, 0^{c!}\right)$$

$$= (1,\ 0,\ 0) \tag{3.17}$$

1. Average number of customers in the queue is

$$NL_q = \left(\frac{r_N^c \rho_N}{c!(1-\rho_N)^2}\right) NP_0$$

Using (3.13) and (3.17), we get $\left(\dfrac{r_N^c \rho_N}{c!(1-\rho_N)^2}\right)$

$$\frac{r_N^c \rho_N}{C!(1-\rho_N)^2} = \left(\frac{T_{\lambda_N}^{c+1}}{T_{\mu_N}^c - T_{\mu_N}^c (1-T_{\mu_N})^c}\right), \left[1-\left(\frac{1-I_{\lambda_N}}{1-I_{\mu_N}}\right)^c \left(\frac{1-I_{\lambda_N}}{1-I_{\mu_N}^c}\right)\right],$$

$$\left[1-\left(\frac{1-F_{\lambda_N}}{1-F_{\mu_N}}\right)^c \left(\frac{1-F_{\lambda_N}}{1-F_{\mu_N}^c}\right)\right] \tag{3.18}$$

To find NP_0:

$$NP_0 = \left(\frac{r_N^c}{C!(1-\rho_N)} + \sum_{n=0}^{c-1} \frac{r_N^n}{n!}\right)^{-1}$$

To find $\dfrac{r_N^c}{C!(1-\rho_N)}$

from (3.12) and (3.15), we get

$$\left(\frac{r_N^c}{C!\left(1-\rho_N\right)}\right)=\left[\left(\frac{T_{\lambda_N}}{T_{\mu_N}}\right)^c,\left(1-\left(\frac{1-I_{\lambda_N}}{1-I_{\mu_N}}\right)^c\right),\left(1-\left(\frac{1-F_{\lambda_N}}{1-F_{\mu_N}}\right)^c\right)\right] \quad (3.19).$$

To find $\sum_{n=0}^{c-1}\frac{r_N^n}{n!}$

from (3.12), we know r_N^c:

$$\frac{r_N^n}{n!}=\left[1-\left(1-\frac{T^n_{\lambda_N}}{T^n_{\mu_N}}\right)^{\frac{1}{n!}},\left(1-\left(\frac{1-I^n_{\lambda_N}}{1-I^n_{\mu_N}}\right)\right)^{\frac{1}{n!}},\left(1-\left(\frac{1-F^n_{\lambda_N}}{1-F^n_{\mu_N}}\right)\right)^{\frac{1}{n!}}\right]$$

$$\sum_{n=0}^{c-1}\frac{r_N^n}{n!}=\left[\left(\sum_{n=0}^{c-1}\left(1-\left(1-\frac{T^n_{\lambda_N}}{T^n_{\mu_N}}\right)^{\frac{1}{n!}}\right)-\prod_{n=0}^{c-1}\left(1-\left(1-\frac{T^n_{\lambda_N}}{T^n_{\mu_N}}\right)^{\frac{1}{n!}}\right)\right),\right.$$
$$\left.\left(\prod_{n=0}^{c-1}\left(\left(1-\left(\frac{1-I^n_{\lambda_N}}{1-I^n_{\mu_N}}\right)\right)^{\frac{1}{n!}}\right)\right),\left(\prod_{n=0}^{c-1}\left(\left(1-\left(\frac{1-F^n_{\lambda_N}}{1-F^n_{\mu_N}}\right)\right)^{\frac{1}{n!}}\right)\right)\right] \quad (3.20).$$

From (3.19) and (3.20), we get

$$NP_0=\left[\left(\left(\frac{T_{\lambda_N}}{T_{\mu_N}}\right)^c+\left(\sum_{n=0}^{c-1}\left(1-\left(1-\frac{T^n_{\lambda_N}}{T^n_{\mu_N}}\right)^{\frac{1}{n!}}\right)-\prod_{n=0}^{c-1}\left(1-\left(1-\frac{T^n_{\lambda_N}}{T^n_{\mu_N}}\right)^{\frac{1}{n!}}\right)\right)\right.\right.$$
$$\left.-\left(\frac{T_{\lambda_N}}{T_{\mu_N}}\right)^c\left(\sum_{n=0}^{c-1}\left(1-\left(1-\frac{T^n_{\lambda_N}}{T^n_{\mu_N}}\right)^{\frac{1}{n!}}\right)-\prod_{n=0}^{c-1}\left(1-\left(1-\frac{T^n_{\lambda_N}}{T^n_{\mu_N}}\right)^{\frac{1}{n!}}\right)\right)\right),$$

$$\left(\left(1-\left(\frac{1-I_{\lambda_N}}{1-I_{\mu_N}}\right)^c\right)\left(\prod_{n=0}^{c-1}\left(\left(1-\left(\frac{1-I^n_{\lambda_N}}{1-I^n_{\mu_N}}\right)\right)^{\frac{1}{n!}}\right)\right)\right),$$

$$\left.\left(\left(1-\left(\frac{1-F_{\lambda_N}}{1-F_{\mu_N}}\right)^c\right)\left(\prod_{n=0}^{c-1}\left(\left(1-\left(\frac{1-F^n_{\lambda_N}}{1-F^n_{\mu_N}}\right)\right)^{\frac{1}{n!}}\right)\right)\right)\right]^{-1}$$

$$NP_0=\left[\left(\left(\frac{T_{\lambda_N}}{T_{\mu_N}}\right)^c+\left(\sum_{n=0}^{c-1}\left(1-\left(1-\frac{T^n_{\lambda_N}}{T^n_{\mu_N}}\right)^{\frac{1}{n!}}\right)-\prod_{n=0}^{c-1}\left(1-\left(1-\frac{T^n_{\lambda_N}}{T^n_{\mu_N}}\right)^{\frac{1}{n!}}\right)\right)\right.\right.$$

$$\left.\left.-\left(\frac{T_{\lambda_N}}{T_{\mu_N}}\right)^c\left(\sum_{n=0}^{c-1}\left(1-\left(1-\frac{T^n_{\lambda_N}}{T^n_{\mu_N}}\right)^{\frac{1}{n!}}\right)-\prod_{n=0}^{c-1}\left(1-\left(1-\frac{T^n_{\lambda_N}}{T^n_{\mu_N}}\right)^{\frac{1}{n!}}\right)\right)\right)^{-1},\right.$$

$$\left(1-\left(1-\left(\left(1-\left(\frac{1-I_{\lambda_N}}{1-I_{\mu_N}}\right)^c\right)\left(\prod_{n=0}^{c-1}\left(\left(1-\left(\frac{1-I^n_{\lambda_N}}{1-I^n_{\mu_N}}\right)\right)^{\frac{1}{n!}}\right)\right)\right)\right)^{-1}\right),\tag{3.21}$$

$$\left.\left(1-\left(1-\left(\left(1-\left(\frac{1-F_{\lambda_N}}{1-F_{\mu_N}}\right)^c\right)\left(\prod_{n=0}^{c-1}\left(\left(1-\left(\frac{1-F^n_{\lambda_N}}{1-F^n_{\mu_N}}\right)\right)^{\frac{1}{n!}}\right)\right)\right)\right)^{-1}\right)\right]$$

From (3.18) and (3.21), we get

$$NL_q=\left[\left(\left(\frac{T^{c+1}_{\lambda_N}}{T^c_{\mu_N}\left(1-(1-T_{\mu_N})^c\right)}\right)\left(\left(\frac{T_{\lambda_N}}{T_{\mu_N}}\right)^c+\left(\sum_{n=0}^{c-1}\left(1-\left(1-\frac{T^n_{\lambda_N}}{T^n_{\mu_N}}\right)^{\frac{1}{n!}}\right)-\prod_{n=0}^{c-1}\left(1-\left(1-\frac{T^n_{\lambda_N}}{T^n_{\mu_N}}\right)^{\frac{1}{n!}}\right)\right)\right.\right.\right.$$

$$\left.\left.\left.-\left(\frac{T_{\lambda_N}}{T_{\mu_N}}\right)^c\left(\sum_{n=0}^{c-1}\left(1-\left(1-\frac{T^n_{\lambda_N}}{T^n_{\mu_N}}\right)^{\frac{1}{n!}}\right)-\prod_{n=0}^{c-1}\left(1-\left(1-\frac{T^n_{\lambda_N}}{T^n_{\mu_N}}\right)^{\frac{1}{n!}}\right)\right)\right)^{-1}\right),\right.$$

$$\left(\left(1-\left(\frac{1-I_{\lambda_N}}{1-I_{\mu_N}}\right)^{c}\left(\frac{1-I_{\lambda_N}}{1-I^{c}_{\mu_N}}\right)\right)+\left(1-\left(1-\left(\left(1-\left(\frac{1-I_{\lambda_N}}{1-I_{\mu_N}}\right)^{c}\right)\left(\prod_{n=0}^{c-1}\left(\left(1-\left(\frac{1-I^{n}_{\lambda_N}}{1-I^{n}_{\mu_N}}\right)\right)^{\frac{1}{n!}}\right)\right)\right)\right)\right)^{-1}\right)$$

$$-\left(1-\left(\frac{1-I_{\lambda_N}}{1-I_{\mu_N}}\right)^{c}\left(\frac{1-I_{\lambda_N}}{1-I^{c}_{\mu_N}}\right)\right)\left(1-\left(1-\left(\left(1-\left(\frac{1-I_{\lambda_N}}{1-I_{\mu_N}}\right)^{c}\right)\left(\prod_{n=0}^{c-1}\left(\left(1-\left(\frac{1-I^{n}_{\lambda_N}}{1-I^{n}_{\mu_N}}\right)\right)^{\frac{1}{n!}}\right)\right)\right)\right)^{-1}\right)\Bigg),$$

$$\left(\left(1-\left(\frac{1-F_{\lambda_N}}{1-F_{\mu_N}}\right)^{c}\left(\frac{1-F_{\lambda_N}}{1-F^{c}_{\mu_N}}\right)\right)+\left(1-\left(1-\left(\left(1-\left(\frac{1-F_{\lambda_N}}{1-F_{\mu_N}}\right)^{c}\right)\left(\prod_{n=0}^{c-1}\left(\left(1-\left(\frac{1-F^{n}_{\lambda_N}}{1-F^{n}_{\mu_N}}\right)\right)^{\frac{1}{n!}}\right)\right)\right)\right)\right)^{-1}\right)$$

$$-\left(1-\left(\frac{1-F_{\lambda_N}}{1-F_{\mu_N}}\right)^{c}\left(\frac{1-F_{\lambda_N}}{1-F^{c}_{\mu_N}}\right)\right)\left(1-\left(1-\left(\left(1-\left(\frac{1-F_{\lambda_N}}{1-F_{\mu_N}}\right)^{c}\right)\left(\prod_{n=0}^{c-1}\left(\left(1-\left(\frac{1-F^{n}_{\lambda_N}}{1-F^{n}_{\mu_N}}\right)\right)^{\frac{1}{n!}}\right)\right)\right)\right)^{-1}\right)\Bigg)\Bigg]$$

2. Average number of customers in the system is (using (3.1), (3.5), and (3.9))

$$NL_s = r_N + \left(\frac{r_N^c \rho_N}{c!(1-\rho_N)^2}\right)NP_0$$

$$NL_s = \left[\left(\frac{T_{\lambda_N}}{T_{\mu_N}}\right)+\left(\left(\frac{T^{c+1}_{\lambda_N}}{T^{c}_{\mu_N}\left(1-(1-T_{\mu_N})^{c}\right)}\right)\left(\left(\frac{T_{\lambda_N}}{T_{\mu_N}}\right)^{c}+\sum_{n=0}^{c-1}\left(1-\left(1-\frac{T^{n}_{\lambda_N}}{T^{n}_{\mu_N}}\right)^{\frac{1}{n!}}\right)-\prod_{n=0}^{c-1}\left(1-\left(1-\frac{T^{n}_{\lambda_N}}{T^{n}_{\mu_N}}\right)^{\frac{1}{n!}}\right)\right)\right.\right.$$

$$\left.-\left(\frac{T_{\lambda_N}}{T_{\mu_N}}\right)^{c}\left(\sum_{n=0}^{c-1}\left(1-\left(1-\frac{T^{n}_{\lambda_N}}{T^{n}_{\mu_N}}\right)^{\frac{1}{n!}}\right)-\prod_{n=0}^{c-1}\left(1-\left(1-\frac{T^{n}_{\lambda_N}}{T^{n}_{\mu_N}}\right)^{\frac{1}{n!}}\right)\right)\right)^{-1}\Bigg)$$

$$-\left(\frac{T_{\lambda_N}}{T_{\mu_N}}\right)\left(\left(\frac{T^{c+1}_{\lambda_N}}{T^{c}_{\mu_N}\left(1-(1-T_{\mu_N})^{c}\right)}\right)\left(\left(\frac{T_{\lambda_N}}{T_{\mu_N}}\right)^{c}+\sum_{n=0}^{c-1}\left(1-\left(1-\frac{T^{n}_{\lambda_N}}{T^{n}_{\mu_N}}\right)^{\frac{1}{n!}}\right)-\prod_{n=0}^{c-1}\left(1-\left(1-\frac{T^{n}_{\lambda_N}}{T^{n}_{\mu_N}}\right)^{\frac{1}{n!}}\right)\right)\right.$$

$$\left.\left.-\left(\frac{T_{\lambda_N}}{T_{\mu_N}}\right)^{c}\left(\sum_{n=0}^{c-1}\left(1-\left(1-\frac{T^{n}_{\lambda_N}}{T^{n}_{\mu_N}}\right)^{\frac{1}{n!}}\right)-\prod_{n=0}^{c-1}\left(1-\left(1-\frac{T^{n}_{\lambda_N}}{T^{n}_{\mu_N}}\right)^{\frac{1}{n!}}\right)\right)\right)^{-1}\right)\right],$$

$$\begin{aligned}&\Bigg[\Bigg(\frac{I_{\lambda_N}-I_{\mu_N}}{1-I_{\mu_N}}\Bigg)\Bigg(\Bigg(1-\Bigg(\frac{1-I_{\lambda_N}}{1-I_{\mu_N}}\Bigg)^c\Bigg(\frac{1-I_{\lambda_N}}{1-I^c_{\mu_N}}\Bigg)\Bigg)+\Bigg(1-\Bigg(1-\Bigg(\Bigg(1-\Bigg(\frac{1-I_{\lambda_N}}{1-I_{\mu_N}}\Bigg)^c\Bigg)\Bigg(\prod_{n=0}^{c-1}\Bigg(\Bigg(1-\Bigg(\frac{1-I^n_{\lambda_N}}{1-I^n_{\mu_N}}\Bigg)\Bigg)^{\frac{1}{n!}}\Bigg)\Bigg)\Bigg)\Bigg)\Bigg)^{-1}\Bigg)\\&-\Bigg(1-\Bigg(\frac{1-I_{\lambda_N}}{1-I_{\mu_N}}\Bigg)^c\Bigg(\frac{1-I_{\lambda_N}}{1-I^c_{\mu_N}}\Bigg)\Bigg)\Bigg(1-\Bigg(1-\Bigg(\Bigg(1-\Bigg(\frac{1-I_{\lambda_N}}{1-I_{\mu_N}}\Bigg)^c\Bigg)\Bigg(\prod_{n=0}^{c-1}\Bigg(\Bigg(1-\Bigg(\frac{1-I^n_{\lambda_N}}{1-I^n_{\mu_N}}\Bigg)\Bigg)^{\frac{1}{n!}}\Bigg)\Bigg)\Bigg)\Bigg)\Bigg)^{-1}\Bigg)\Bigg)\Bigg],\end{aligned}$$

$$\begin{aligned}&\Bigg[\Bigg(\frac{F_{\lambda_N}-F_{\mu_N}}{1-F_{\mu_N}}\Bigg)\Bigg(\Bigg(1-\Bigg(\frac{1-F_{\lambda_N}}{1-F_{\mu_N}}\Bigg)^c\Bigg(\frac{1-F_{\lambda_N}}{1-F^c_{\mu_N}}\Bigg)\Bigg)+\Bigg(1-\Bigg(1-\Bigg(\Bigg(1-\Bigg(\frac{1-F_{\lambda_N}}{1-F_{\mu_N}}\Bigg)^c\Bigg)\Bigg(\prod_{n=0}^{c-1}\Bigg(\Bigg(1-\Bigg(\frac{1-F^n_{\lambda_N}}{1-F^n_{\mu_N}}\Bigg)\Bigg)^{\frac{1}{n!}}\Bigg)\Bigg)\Bigg)\Bigg)\Bigg)^{-1}\Bigg)\\&-\Bigg(1-\Bigg(\frac{1-F_{\lambda_N}}{1-F_{\mu_N}}\Bigg)^c\Bigg(\frac{1-F_{\lambda_N}}{1-F^c_{\mu_N}}\Bigg)\Bigg)\Bigg(1-\Bigg(1-\Bigg(\Bigg(1-\Bigg(\frac{1-F_{\lambda_N}}{1-F_{\mu_N}}\Bigg)^c\Bigg)\Bigg(\prod_{n=0}^{c-1}\Bigg(\Bigg(1-\Bigg(\frac{1-F^n_{\lambda_N}}{1-F^n_{\mu_N}}\Bigg)\Bigg)^{\frac{1}{n!}}\Bigg)\Bigg)\Bigg)\Bigg)\Bigg)^{-1}\Bigg)\Bigg)\Bigg]\end{aligned}$$

3. Mean waiting time in the queue is

$$NW_q=\left(\frac{r_N^c}{C!(C\mu_N)(1-\rho_N)^2}\right)NP_0.$$

From (3.5) and (3.10), we get

$$C!C\mu_N=\left(\left(1-\left(1-T_{\mu_N}\right)\right)^{c!c},I_{\mu_N}^{c!c},F_{\mu_N}^{c!c}\right);c>0.$$

Using (3.16), we get

$$\begin{aligned}(C!C\mu_N)(1-\rho_N)^2&=\left(\left(1-\left(1-T_{\mu_N}\right)\right)^{c!c}(1),I_{\mu_N}^{c!c}+0-I_{\mu_N}^{c!c}(0),F_{\mu_N}^{c!c}+0-F_{\mu_N}^{c!c}(0)\right)\\&=\left(\left(1-\left(1-T_{\mu_N}\right)\right)^{c!c},I_{\mu_N}^{c!c},F_{\mu_N}^{c!c}\right)\qquad(3.22).\end{aligned}$$

Using (3.12) and (3.22), we get

$$\frac{r_N^c}{C!(C\mu_N)(1-\rho_N)^2}=\left(\frac{\left(\frac{T_{\lambda_N}}{T_{\mu_N}}\right)^c}{\left(1-(1-T\mu_N)^{c!c}\right)},\frac{1-\left(\frac{1-I_{\lambda_N}}{1-I_{\mu_N}}\right)^c-I_{\mu_N}^{c!c}}{1-I_{\mu_N}^{c!c}},\frac{1-\left(\frac{1-F_{\lambda_N}}{1-F_{\mu_N}}\right)^c-F_{\mu_N}^{c!c}}{1-F_{\mu_N}^{c!c}}\right)\qquad(3.23)$$

Multiplying (3.21) and (3.23), we get

$$NW_q = \left[\left[\left(\frac{\left(\frac{T_{\lambda_N}}{T_{\mu_N}}\right)^c}{\left(1-\left(1-T_{\mu_N}\right)^{c!c}\right)}\right)\left(\left(\frac{T_{\lambda_N}}{T_{\mu_N}}\right)^c + \left(\sum_{n=0}^{c-1}\left(1-\left(1-\frac{T^n_{\lambda_N}}{T^n_{\mu_N}}\right)^{\frac{1}{n!}}\right) - \prod_{n=0}^{c-1}\left(1-\left(1-\frac{T^n_{\lambda_N}}{T^n_{\mu_N}}\right)^{\frac{1}{n!}}\right)\right)\right.\right.\right.$$

$$\left.\left.\left. - \left(\frac{T_{\lambda_N}}{T_{\mu_N}}\right)^c\left(\sum_{n=0}^{c-1}\left(1-\left(1-\frac{T^n_{\lambda_N}}{T^n_{\mu_N}}\right)^{\frac{1}{n!}}\right) - \prod_{n=0}^{c-1}\left(1-\left(1-\frac{T^n_{\lambda_N}}{T^n_{\mu_N}}\right)^{\frac{1}{n!}}\right)\right)\right)^{-1}\right],$$

$$-\left[\left(\frac{\left(1-\left(\frac{1-I_{\lambda_N}}{1-I_{\mu_N}}\right)^c\right) - I^{c!c}_{\mu_N}}{1-I^{c!c}_{\mu_N}} + \left(1-\left(1-\left(\left(1-\left(\frac{1-I_{\lambda_N}}{1-I_{\mu_N}}\right)^c\right)\prod_{n=0}^{c-1}\left(\left(1-\left(\frac{1-I^n_{\lambda_N}}{1-I^n_{\mu_N}}\right)\right)^{\frac{1}{n!}}\right)\right)\right)\right)^{-1}\right)\right.$$

$$\left.\frac{\left(1-\left(\frac{1-I_{\lambda_N}}{1-I_{\mu_N}}\right)^c\right) - I^{c!c}_{\mu_N}}{1-I^{c!c}_{\mu_N}}\left(1-\left(1-\left(\left(1-\left(\frac{1-I_{\lambda_N}}{1-I_{\mu_N}}\right)^c\right)\prod_{n=0}^{c-1}\left(\left(1-\left(\frac{1-I^n_{\lambda_N}}{1-I^n_{\mu_N}}\right)\right)^{\frac{1}{n!}}\right)\right)\right)\right)^{-1}\right],$$

$$\left[\left(\frac{\left(1-\left(\frac{1-F_{\lambda_N}}{1-F_{\mu_N}}\right)^c\right) - F^{c!c}_{\mu_N}}{1-F^{c!c}_{\mu_N}} + \left(1-\left(1-\left(\left(1-\left(\frac{1-F_{\lambda_N}}{1-F_{\mu_N}}\right)^c\right)\prod_{n=0}^{c-1}\left(\left(1-\left(\frac{1-F^n_{\lambda_N}}{1-F^n_{\mu_N}}\right)\right)^{\frac{1}{n!}}\right)\right)\right)\right)^{-1}\right)\right.$$

$$\left.\left. -\frac{\left(1-\left(\frac{1-F_{\lambda_N}}{1-F_{\mu_N}}\right)^c\right) - F^{c!c}_{\mu_N}}{1-F^{c!c}_{\mu_N}}\left(1-\left(1-\left(\left(1-\left(\frac{1-F_{\lambda_N}}{1-F_{\mu_N}}\right)^c\right)\prod_{n=0}^{c-1}\left(\left(1-\left(\frac{1-F^n_{\lambda_N}}{1-F^n_{\mu}}\right)\right)^{\frac{1}{n!}}\right)\right)\right)\right)^{-1}\right]\right]$$

4. Mean waiting time in the system is

$$NW_s = \frac{1}{\mu_N} + \left(\frac{r_N^c}{C!(C\mu_N)\left(1-\rho_N\right)^2}\right)p_0$$

$$\mu_N^{-1} = \left(T^{-1}_{\mu_N}, 1-\left(1-I_{\mu_N}\right)^{-1}, 1-\left(1-F_{\mu_N}\right)^{-1}\right)$$

$$= \left(\frac{1}{T_{\mu_N}}, 1-\frac{1}{1-I_{\mu_N}}, 1-\frac{1}{1-F_{\mu_N}}\right)$$

$$\mu_N^{-1} = \left(\frac{1}{T_{\mu_N}}, \frac{-I_{\mu_N}}{1-I_{\mu_N}}, \frac{-F_{\mu_N}}{1-F_{\mu_N}}\right) \quad (3.24)$$

By adding (3.7) and (3.24), we get

$$NW_s = \left[\left[\left(\frac{1}{T_{\mu_N}}\right) + \left(\frac{\left(\frac{T_{\lambda_N}}{T_{\mu_N}}\right)^c}{\left(1-\left(1-T_{\mu_N}\right)^{c!c}\right)}\right)\right]\left(\left(\frac{T_{\lambda_N}}{T_{\mu_N}}\right)^c + \left(\sum_{n=0}^{c-1}\left(1-\left(1-\frac{T^n_{\lambda_N}}{T^n_{\mu_N}}\right)^{\frac{1}{n!}}\right) - \prod_{n=0}^{c-1}\left(1-\left(1-\frac{T^n_{\lambda_N}}{T^n_{\mu_N}}\right)^{\frac{1}{n!}}\right)\right)\right.\right.$$

$$\left.\left. - \left(\frac{T_{\lambda_N}}{T_{\mu_N}}\right)^c \left(\sum_{n=0}^{c-1}\left(1-\left(1-\frac{T^n_{\lambda_N}}{T^n_{\mu_N}}\right)^{\frac{1}{n!}}\right) - \prod_{n=0}^{c-1}\left(1-\left(1-\frac{T^n_{\lambda_N}}{T^n_{\mu_N}}\right)^{\frac{1}{n!}}\right)\right)\right)^{-1}\right]$$

$$- \left(\frac{1}{T_{\mu_N}}\right)\left[\left[\left(\frac{\left(\frac{T_{\lambda_N}}{T_{\mu_N}}\right)^c}{\left(1-\left(1-T_{\mu_N}\right)^{c!c}\right)}\right)\right]\left(\left(\frac{T_{\lambda_N}}{T_{\mu_N}}\right)^c + \left(\sum_{n=0}^{c-1}\left(1-\left(1-\frac{T^n_{\lambda_N}}{T^n_{\mu_N}}\right)^{\frac{1}{n!}}\right) - \prod_{n=0}^{c-1}\left(1-\left(1-\frac{T^n_{\lambda_N}}{T^n_{\mu_N}}\right)^{\frac{1}{n!}}\right)\right)\right.\right.$$

$$\left.\left. - \left(\frac{T_{\lambda_N}}{T_{\mu_N}}\right)^c \left(\sum_{n=0}^{c-1}\left(1-\left(1-\frac{T^n_{\lambda_N}}{T^n_{\mu_N}}\right)^{\frac{1}{n!}}\right) - \prod_{n=0}^{c-1}\left(1-\left(1-\frac{T^n_{\lambda_N}}{T^n_{\mu_N}}\right)^{\frac{1}{n!}}\right)\right)\right)^{-1}\right]\right],$$

$$\left[\left(\frac{-I_{\mu_N}}{1-I_{\mu_N}}\right)\left[\frac{\left(1-\left(\frac{1-I_{\lambda_N}}{1-I_{\mu_N}}\right)^c\right) - I^{c!c}_{\mu_N}}{1-I^{c!c}_{\mu_N}} + \left(1-\left(1-\left(\left(1-\left(\frac{1-I_{\lambda_N}}{1-I_{\mu_N}}\right)^c\right)\prod_{n=0}^{c-1}\left(\left(1-\left(\frac{1-I^n_{\lambda_N}}{1-I^n_{\mu_N}}\right)\right)^{\frac{1}{n!}}\right)\right)\right)\right)^{-1}\right.\right.$$

$$\left.\left. - \frac{\left(1-\left(\frac{1-I_{\lambda_N}}{1-I_{\mu_N}}\right)^c\right) - I^{c!c}_{\mu_N}}{1-I^{c!c}_{\mu_N}}\left(1-\left(1-\left(\left(1-\left(\frac{1-I_{\lambda_N}}{1-I_{\mu_N}}\right)^c\right)\prod_{n=0}^{c-1}\left(\left(1-\left(\frac{1-I^n_{\lambda_N}}{1-I^n_{\mu_N}}\right)\right)^{\frac{1}{n!}}\right)\right)\right)\right)^{-1}\right]\right],$$

$$\left[\left(\frac{-F_{\mu_N}}{1-F_{\mu_N}}\right)\left[\frac{\left(1-\left(\frac{1-F_{\lambda_N}}{1-F_{\mu_N}}\right)^c\right) - F^{c!c}_{\mu_N}}{1-F^{c!c}_{\mu_N}} + \left(1-\left(1-\left(\left(1-\left(\frac{1-F_{\lambda_N}}{1-F_{\mu_N}}\right)^c\right)\prod_{n=0}^{c-1}\left(\left(1-\left(\frac{1-F^n_{\lambda_N}}{1-F^n_{\mu_N}}\right)\right)^{\frac{1}{n!}}\right)\right)\right)\right)^{-1}\right.\right.$$

$$\left.\left. - \frac{\left(1-\left(\frac{1-F_{\lambda_N}}{1-F_{\mu_N}}\right)^c\right) - F^{c!c}_{\mu_N}}{1-F^{c!c}_{\mu_N}}\left(1-\left(1-\left(\left(1-\left(\frac{1-F_{\lambda_N}}{1-F_{\mu_N}}\right)^c\right)\prod_{n=0}^{c-1}\left(\left(1-\left(\frac{1-F^n_{\lambda_N}}{1-F^n_{\mu_N}}\right)\right)^{\frac{1}{n!}}\right)\right)\right)\right)^{-1}\right]\right]$$

3.4 NUMERIC EXAMPLE

$$r_N = \frac{\lambda_N}{\mu_N} = \left(\frac{T_{\lambda_N}}{T_{\mu_N}}, \frac{I_{\lambda_N} - I_{\mu_N}}{1 - I_{\mu_N}}, \frac{F_{\lambda_N} - F_{\mu_N}}{1 - F_{\mu_N}}\right)$$

$$= \left(\frac{0.7}{0.8}, \frac{0.6 - 0.4}{1 - 0.4}, \frac{0.6 - 0.5}{1 - 0.5}\right) = (0.875, 0.333, 0.111)$$

$$\rho_N = \left(\frac{T_{\lambda_N}}{1 - \left(1 - T_{\mu_N}\right)^c}, \frac{I_{\lambda_N} - I^c_{\mu_N}}{1 - I^c_{\mu_N}}, \frac{F_{\lambda_N} - F^c_{\mu_N}}{1 - F^c_{\mu_N}}\right), \left(1 - T_{\mu_N}\right)^c \neq 1, I^c_{\mu_N} \neq 1, F^c_{\mu_N} \neq 1$$

$$= \left(\left(\frac{0.7}{1 - (1 - 0.8)^2}\right), \left(\frac{0.6 - (0.4)^2}{1 - (0.4)^2}\right), \left(\frac{0.6 - (0.5)^2}{1 - (0.5)^2}\right)\right)$$

$$= (0.729, 0.523, 0.466)$$

1. Average number of customers in the queue is

$$NL_q = \Bigg[\Bigg(\left(\frac{T^{c+1}_{\lambda_N}}{T^c_{\mu_N}\left(1 - (1 - T_{\mu_N})^c\right)}\right)\Bigg(\left(\frac{T_{\lambda_N}}{T_{\mu_N}}\right)^c + \Bigg(\sum_{n=0}^{c-1}\Bigg(1 - \left(1 - \frac{T^n_{\lambda_N}}{T^n_{\mu_N}}\right)^{\frac{1}{n!}}\Bigg) - \prod_{n=0}^{c-1}\Bigg(1 - \left(1 - \frac{T^n_{\lambda_N}}{T^n_{\mu_N}}\right)^{\frac{1}{n!}}\Bigg)\Bigg)\Bigg)$$

$$- \left(\frac{T_{\lambda_N}}{T_{\mu_N}}\right)^c \Bigg(\sum_{n=0}^{c-1}\Bigg(1 - \left(1 - \frac{T^n_{\lambda_N}}{T^n_{\mu_N}}\right)^{\frac{1}{n!}}\Bigg) - \prod_{n=0}^{c-1}\Bigg(1 - \left(1 - \frac{T^n_{\lambda_N}}{T^n_{\mu_N}}\right)^{\frac{1}{n!}}\Bigg)\Bigg)\Bigg)^{-1}\Bigg),$$

$$- \Bigg(\Bigg(1 - \left(\frac{1 - I_{\lambda_N}}{1 - I_{\mu_N}}\right)^c \left(\frac{1 - I_{\lambda_N}}{1 - I^c_{\mu_N}}\right)\Bigg) + \Bigg(1 - \Bigg(1 - \Bigg(\Bigg(1 - \left(\frac{1 - I_{\lambda_N}}{1 - I_{\mu_N}}\right)^c\Bigg)\Bigg(\prod_{n=0}^{c-1}\Bigg(\Bigg(1 - \left(\frac{1 - I^n_{\lambda_N}}{1 - I^n_{\mu_N}}\right)\Bigg)^{\frac{1}{n!}}\Bigg)\Bigg)\Bigg)\Bigg)\Bigg)^{-1}\Bigg)$$

$$\Bigg(1 - \left(\frac{1 - I_{\lambda_N}}{1 - I_{\mu_N}}\right)^c \left(\frac{1 - I_{\lambda_N}}{1 - I^c_{\mu_N}}\right)\Bigg)\Bigg(1 - \Bigg(1 - \Bigg(\Bigg(1 - \left(\frac{1 - I_{\lambda_N}}{1 - I_{\mu_N}}\right)^c\Bigg)\Bigg(\prod_{n=0}^{c-1}\Bigg(\Bigg(1 - \left(\frac{1 - I^n_{\lambda_N}}{1 - I^n_{\mu_N}}\right)\Bigg)^{\frac{1}{n!}}\Bigg)\Bigg)\Bigg)\Bigg)^{-1}\Bigg)\Bigg),$$

$$\Bigg(\Bigg(1 - \left(\frac{1 - F_{\lambda_N}}{1 - F_{\mu_N}}\right)^c \left(\frac{1 - F_{\lambda_N}}{1 - F^c_{\mu_N}}\right)\Bigg) + \Bigg(1 - \Bigg(1 - \Bigg(\Bigg(1 - \left(\frac{1 - F_{\lambda_N}}{1 - F_{\mu_N}}\right)^c\Bigg)\Bigg(\prod_{n=0}^{c-1}\Bigg(\Bigg(1 - \left(\frac{1 - F^n_{\lambda_N}}{1 - F^n_{\mu_N}}\right)\Bigg)^{\frac{1}{n!}}\Bigg)\Bigg)\Bigg)\Bigg)\Bigg)^{-1}\Bigg)$$

$$- \Bigg(1 - \left(\frac{1 - F_{\lambda_N}}{1 - F_{\mu_N}}\right)^c \left(\frac{1 - F_{\lambda_N}}{1 - F^c_{\mu_N}}\right)\Bigg)\Bigg(1 - \Bigg(1 - \Bigg(\Bigg(1 - \left(\frac{1 - F_{\lambda_N}}{1 - F_{\mu_N}}\right)^c\Bigg)\Bigg(\prod_{n=0}^{c-1}\Bigg(\Bigg(1 - \left(\frac{1 - F^n_{\lambda_N}}{1 - F^n_{\mu_N}}\right)\Bigg)^{\frac{1}{n!}}\Bigg)\Bigg)\Bigg)\Bigg)^{-1}\Bigg)\Bigg)\Bigg]$$

$$= (0.558, -0.428, -1.666)$$

Table 3.2 Linguistic Terms NL_q

Linguistic terms	*SVN numbers*	*Hausdorff distance*
Extremely good	**(1,0,0)**	1.667
Very very good	(0.95,0.05,0.10)	**1.767**
Very good	(0.85,0.15,0.20)	**1.867**
Good	(0.75,0.25,0.35)	**2.017**
Fairly good	(0.65,0.30,0.40)	**2.067**
Average	(0.55,0.45,0.50)	**2.167**
Moderately bad	(0.45,0.50,0.60)	**2.267**
Bad	(0.35,0.65,0.70)	**2.367**
Very bad	(0.25,0.75,0.80)	**2.467**
Very very bad	(0.15,0.90,0.90)	**2.567**
Extremely bad	(0.0,1.0,0.95)	**2.617**

Table 3.2 displays the findings for calculating the Hausdorff distance using linguistic terms in NL_q.

These findings imply that the average number of customers in the queue is going to be extremely good because it has the smallest distance.

2. Average number of customers in the system is

$$NL_s = \Bigg[\left(\frac{T_{\lambda_\mathrm{N}}}{T_{\mu_\mathrm{N}}}\right) + \Bigg(\left(\frac{T_{\lambda_\mathrm{N}}^{c+1}}{T_{\mu_\mathrm{N}}^{c}\left(1-(1-T_{\mu_\mathrm{N}})^{c}\right)}\right) \Bigg(\left(\frac{T_{\lambda_\mathrm{N}}}{T_{\mu_\mathrm{N}}}\right)^{c} + \Bigg(\sum_{n=0}^{c-1} \Bigg(1-\left(1-\frac{T_{\lambda_\mathrm{N}}^{n}}{T_{\mu_\mathrm{N}}^{n}}\right)^{\frac{1}{n!}}\Bigg) - \prod_{n=0}^{c-1} \Bigg(1-\left(1-\frac{T_{\lambda_\mathrm{N}}^{n}}{T_{\mu_\mathrm{N}}^{n}}\right)^{\frac{1}{n!}}\Bigg)\Bigg)\Bigg)$$

$$-\left(\frac{T_{\lambda_\mathrm{N}}}{T_{\mu_\mathrm{N}}}\right)^{c} \Bigg(\sum_{n=0}^{c-1} \Bigg(1-\left(1-\frac{T_{\lambda_\mathrm{N}}^{n}}{T_{\mu_\mathrm{N}}^{n}}\right)^{\frac{1}{n!}}\Bigg) - \prod_{n=0}^{c-1} \Bigg(1-\left(1-\frac{T_{\lambda_\mathrm{N}}^{n}}{T_{\mu_\mathrm{N}}^{n}}\right)^{\frac{1}{n!}}\Bigg)\Bigg)\Bigg)\Bigg)^{-1}\Bigg)$$

$$-\left(\frac{T_{\lambda_\mathrm{N}}}{T_{\mu_\mathrm{N}}}\right) \Bigg(\left(\frac{T_{\lambda_\mathrm{N}}^{c+1}}{T_{\mu_\mathrm{N}}^{c}\left(1-(1-T_{\mu_\mathrm{N}})^{c}\right)}\right) \Bigg(\left(\frac{T_{\lambda_\mathrm{N}}}{T_{\mu_\mathrm{N}}}\right)^{c} + \Bigg(\sum_{n=0}^{c-1} \Bigg(1-\left(1-\frac{T_{\lambda_\mathrm{N}}^{n}}{T_{\mu_\mathrm{N}}^{n}}\right)^{\frac{1}{n!}}\Bigg) - \prod_{n=0}^{c-1} \Bigg(1-\left(1-\frac{T_{\lambda_\mathrm{N}}^{n}}{T_{\mu_\mathrm{N}}^{n}}\right)^{\frac{1}{n!}}\Bigg)\Bigg)\Bigg)$$

$$-\left(\frac{T_{\lambda_\mathrm{N}}}{T_{\mu_\mathrm{N}}}\right)^{c} \Bigg(\sum_{n=0}^{c-1} \Bigg(1-\left(1-\frac{T_{\lambda_\mathrm{N}}^{n}}{T_{\mu_\mathrm{N}}^{n}}\right)^{\frac{1}{n!}}\Bigg) - \prod_{n=0}^{c-1} \Bigg(1-\left(1-\frac{T_{\lambda_\mathrm{N}}^{n}}{T_{\mu_\mathrm{N}}^{n}}\right)^{\frac{1}{n!}}\Bigg)\Bigg)\Bigg)\Bigg)^{-1}\Bigg)\Bigg],$$

$$\Bigg[\left(\frac{I_{\lambda_\mathrm{N}}-I_{\mu_\mathrm{N}}}{1-I_{\mu_\mathrm{N}}}\right) \Bigg(\left(1-\left(\frac{1-I_{\lambda_\mathrm{N}}}{1-I_{\mu_\mathrm{N}}}\right)^{c}\left(\frac{1-I_{\lambda_\mathrm{N}}}{1-I_{\mu_\mathrm{N}}^{c}}\right)\right) + \Bigg(1-\Bigg(1-\Bigg(\left(1-\left(\frac{1-I_{\lambda_\mathrm{N}}}{1-I_{\mu_\mathrm{N}}}\right)^{c}\right) \Bigg(\prod_{n=0}^{c-1} \Bigg(\left(1-\left(\frac{1-I_{\lambda_\mathrm{N}}^{n}}{1-I_{\mu_\mathrm{N}}^{n}}\right)\right)^{\frac{1}{n!}}\Bigg)\Bigg)\Bigg)\Bigg)\Bigg)^{-1}\Bigg)$$

$$-\left(1-\left(\frac{1-I_{\lambda_N}}{1-I_{\mu_N}}\right)^{c}\left(\frac{1-I_{\lambda_N}}{1-I^{c}_{\mu_N}}\right)\right)\Bigg(1-\Bigg(1-\Bigg(\Bigg(1-\left(\frac{1-I_{\lambda_N}}{1-I_{\mu_N}}\right)^{c}\Bigg)\prod_{n=0}^{c-1}\Bigg(\left(1-\left(\frac{1-I^{n}_{\lambda_N}}{1-I^{n}_{\mu_N}}\right)\right)^{\frac{1}{n!}}\Bigg)\Bigg)\Bigg)^{-1}\Bigg)\Bigg]\Bigg],$$

$$\Bigg[\left(\frac{F_{\lambda_N}-F_{\mu_N}}{1-F_{\mu_N}}\right)\Bigg(\left(1-\left(\frac{1-F_{\lambda_N}}{1-F_{\mu_N}}\right)^{c}\left(\frac{1-F_{\lambda_N}}{1-F^{c}_{\mu_N}}\right)\right)+1-\Bigg(1-\Bigg(1-\left(\frac{1-F_{\lambda_N}}{1-F_{\mu_N}}\right)^{c}\Bigg)\prod_{n=0}^{c-1}\Bigg(\left(1-\left(\frac{1-F^{n}_{\lambda_N}}{1-F^{n}_{\mu_N}}\right)\right)^{\frac{1}{n!}}\Bigg)\Bigg)^{-1}\Bigg)$$

$$-\left(1-\left(\frac{1-F_{\lambda_N}}{1-F_{\mu_N}}\right)^{c}\left(\frac{1-F_{\lambda_N}}{1-F^{c}_{\mu_N}}\right)\right)\Bigg(1-\Bigg(1-\Bigg(\Bigg(1-\left(\frac{1-F_{\lambda_N}}{1-F_{\mu_N}}\right)^{c}\Bigg)\prod_{n=0}^{c-1}\Bigg(\left(1-\left(\frac{1-F^{n}_{\lambda_N}}{1-F^{n}_{\mu_N}}\right)\right)^{\frac{1}{n!}}\Bigg)\Bigg)\Bigg)^{-1}\Bigg)\Bigg]\Bigg]$$

$$=(0.944,-0.142,-0.333)$$

Table 3.3 displays the findings for calculating the Hausdorff distance using linguistic terms in NL_S.
which means that the average number of customers in the system is going to be extremely good because it has the smallest distance.

3. Mean waiting time in the queue is

$$NW_q=\Bigg[\Bigg[\Bigg(\frac{\left(\frac{T_{\lambda_N}}{T_{\mu_N}}\right)^{c}}{\left(1-\left(1-T_{\mu_N}\right)^{c!c}\right)}\Bigg)\Bigg(\left(\frac{T_{\lambda_N}}{T_{\mu_N}}\right)^{c}+\sum_{n=0}^{c-1}\Bigg(1-\left(1-\frac{T^{n}_{\lambda_N}}{T^{n}_{\mu_N}}\right)^{\frac{1}{n!}}\Bigg)-\prod_{n=0}^{c-1}\Bigg(1-\left(1-\frac{T^{n}_{\lambda_N}}{T^{n}_{\mu_N}}\right)^{\frac{1}{n!}}\Bigg)\Bigg)$$

$$-\left(\frac{T_{\lambda_N}}{T_{\mu_N}}\right)^{c}\Bigg(\sum_{n=0}^{c-1}\Bigg(1-\left(1-\frac{T^{n}_{\lambda_N}}{T^{n}_{\mu_N}}\right)^{\frac{1}{n!}}\Bigg)-\prod_{n=0}^{c-1}\Bigg(1-\left(1-\frac{T^{n}_{\lambda_N}}{T^{n}_{\mu_N}}\right)^{\frac{1}{n!}}\Bigg)\Bigg)^{-1}\Bigg],$$

Table 3.3 Linguistic Terms NL_s

Linguistic terms	*SVN numbers*	*Hausdorff distance*
Extremely good	**(1,0,0)**	0.333
Very very good	(0.95,0.05,0.10)	**0.433**
Very good	(0.85,0.15,0.20)	**0.533**
Good	(0.75,0.25,0.35)	**0.683**
Fairly good	(0.65,0.30,0.40)	**0.733**
Average	(0.55,0.45,0.50)	**0.833**
Moderately bad	(0.45,0.50,0.60)	**0.933**
Bad	(0.35,0.65,0.70)	**1.033**
Very bad	(0.25,0.75,0.80)	**1.173**
Very very bad	(0.15,0.90,0.90)	**1.233**
Extremely bad	(0.0,1.0,0.95)	**1.283**

$$\left[\frac{\left(1-\left(\frac{1-I_{\lambda_N}}{1-I_{\mu_N}}\right)^{c}\right)-I_{\mu_N}^{c!c}}{1-I_{\mu_N}^{c!c}}+\left(1-\left(1-\left(\left(1-\left(\frac{1-I_{\lambda_N}}{1-I_{\mu_N}}\right)^{c}\right)\left(\prod_{n=0}^{c-1}\left(1-\left(\frac{1-I^{n}_{\lambda_N}}{1-I^{n}_{\mu_N}}\right)\right)^{\frac{1}{n!}}\right)\right)\right)^{-1}\right)\right.$$

$$\left.-\frac{\left(1-\left(\frac{1-I_{\lambda_N}}{1-I_{\mu_N}}\right)^{c}\right)-I_{\mu_N}^{c!c}}{1-I_{\mu_N}^{c!c}}\left(1-\left(1-\left(\left(1-\left(\frac{1-I_{\lambda_N}}{1-I_{\mu_N}}\right)^{c}\right)\left(\prod_{n=0}^{c-1}\left(1-\left(\frac{1-I^{n}_{\lambda_N}}{1-I^{n}_{\mu_N}}\right)\right)^{\frac{1}{n!}}\right)\right)\right)^{-1}\right)\right],$$

$$\left[\frac{\left(1-\left(\frac{1-F_{\lambda_N}}{1-F_{\mu_N}}\right)^{c}\right)-F_{\mu_N}^{c!c}}{1-F_{\mu_N}^{c!c}}+\left(1-\left(1-\left(\left(1-\left(\frac{1-F_{\lambda_N}}{1-F_{\mu_N}}\right)^{c}\right)\left(\prod_{n=0}^{c-1}\left(1-\left(\frac{1-F^{n}_{\lambda_N}}{1-F^{n}_{\mu_N}}\right)\right)^{\frac{1}{n!}}\right)\right)\right)^{-1}\right)\right.$$

$$\left.-\frac{\left(1-\left(\frac{1-F_{\lambda_N}}{1-F_{\mu_N}}\right)^{c}\right)-F_{\mu_N}^{c!c}}{1-F_{\mu_N}^{c!c}}\left(1-\left(1-\left(\left(1-\left(\frac{1-F_{\lambda_N}}{1-F_{\mu_N}}\right)^{c}\right)\left(\prod_{n=0}^{c-1}\left(1-\left(\frac{1-F^{n}_{\lambda_N}}{1-F^{n}_{\mu}}\right)\right)^{\frac{1}{n!}}\right)\right)\right)^{-1}\right)\right]\Bigg]$$

$$=(0.766,-2.256,-4.333)$$

Table 3.4 displays the findings for calculating the Hausdorff distance using linguistic terms in NW_q.

Table 3.4 Linguistic Terms NW_q

Linguistic terms	*SVN numbers*	*Hausdorff distance*
Extremely good	**(1,0,0)**	4.333
Very very good	(0.95,0.05,0.10)	**4.433**
Very good	(0.85,0.15,0.20)	**4.533**
Good	(0.75,0.25,0.35)	**4.683**
Fairly good	(0.65,0.30,0.40)	**4.733**
Average	(0.55,0.45,0.50)	**4.833**
Moderately bad	(0.45,0.50,0.60)	**4.933**
Bad	(0.35,0.65,0.70)	**5.033**
Very bad	(0.25,0.75,0.80)	**5.133**
Very very bad	(0.15,0.90,0.90)	**5.233**
Extremely bad	(0.0,1.0,0.95)	**5.283**

which means that the mean waiting time in the queue is going to be extremely good because it has the smallest distance.

4. Mean waiting time in the system is

$$NL_s = \Bigg[\left(\frac{T_{\lambda_N}}{T_{\mu_N}}\right) + \Bigg(\left(\frac{T_{\lambda_N}^{c+1}}{T_{\mu_N}^{c}\left(1-(1-T_{\mu_N})^{c}\right)}\right) \Bigg(\left(\frac{T_{\lambda_N}}{T_{\mu_N}}\right)^{c} + \sum_{n=0}^{c-1}\left(1-\left(1-\frac{T_{\lambda_N}^{n}}{T_{\mu_N}^{n}}\right)^{\frac{1}{n!}}\right) - \prod_{n=0}^{c-1}\left(1-\left(1-\frac{T_{\lambda_N}^{n}}{T_{\mu_N}^{n}}\right)^{\frac{1}{n!}}\right)$$

$$- \left(\frac{T_{\lambda_N}}{T_{\mu_N}}\right)^{c} \left(\sum_{n=0}^{c-1}\left(1-\left(1-\frac{T_{\lambda_N}^{n}}{T_{\mu_N}^{n}}\right)^{\frac{1}{n!}}\right) - \prod_{n=0}^{c-1}\left(1-\left(1-\frac{T_{\lambda_N}^{n}}{T_{\mu_N}^{n}}\right)^{\frac{1}{n!}}\right)\right) \Bigg)^{-1} \Bigg)$$

$$- \left(\frac{T_{\lambda_N}}{T_{\mu_N}}\right) \Bigg(\left(\frac{T_{\lambda_N}^{c+1}}{T_{\mu_N}^{c}\left(1-(1-T_{\mu_N})^{c}\right)}\right) \Bigg(\left(\frac{T_{\lambda_N}}{T_{\mu_N}}\right)^{c} + \sum_{n=0}^{c-1}\left(1-\left(1-\frac{T_{\lambda_N}^{n}}{T_{\mu_N}^{n}}\right)^{\frac{1}{n!}}\right) - \prod_{n=0}^{c-1}\left(1-\left(1-\frac{T_{\lambda_N}^{n}}{T_{\mu_N}^{n}}\right)^{\frac{1}{n!}}\right)$$

$$- \left(\frac{T_{\lambda_N}}{T_{\mu_N}}\right)^{c} \left(\sum_{n=0}^{c-1}\left(1-\left(1-\frac{T_{\lambda_N}^{n}}{T_{\mu_N}^{n}}\right)^{\frac{1}{n!}}\right) - \prod_{n=0}^{c-1}\left(1-\left(1-\frac{T_{\lambda_N}^{n}}{T_{\mu_N}^{n}}\right)^{\frac{1}{n!}}\right)\right) \Bigg)^{-1} \Bigg) \Bigg],$$

$$\Bigg[\left(\frac{I_{\lambda_N}-I_{\mu_N}}{1-I_{\mu_N}}\right) \Bigg(\left(1-\left(\frac{1-I_{\lambda_N}}{1-I_{\mu_N}}\right)^{c}\left(\frac{1-I_{\lambda_N}}{1-I_{\mu_N}^{c}}\right)\right) + 1 - \left(1-\left(1-\left(\frac{1-I_{\lambda_N}}{1-I_{\mu_N}}\right)^{c}\right)\prod_{n=0}^{c-1}\left(1-\left(\frac{1-I_{\lambda_N}^{n}}{1-I_{\mu_N}^{n}}\right)^{\frac{1}{n!}}\right)\right)\Bigg)^{-1}$$

$$- \left(1-\left(\frac{1-I_{\lambda_N}}{1-I_{\mu_N}}\right)^{c}\left(\frac{1-I_{\lambda_N}}{1-I_{\mu_N}^{c}}\right)\right)\left(1-\left(1-\left(1-\left(\frac{1-I_{\lambda_N}}{1-I_{\mu_N}}\right)^{c}\right)\prod_{n=0}^{c-1}\left(1-\left(\frac{1-I_{\lambda_N}^{n}}{1-I_{\mu_N}^{n}}\right)^{\frac{1}{n!}}\right)\right)\right)^{-1} \Bigg],$$

$$\Bigg[\left(\frac{F_{\lambda_N}-F_{\mu_N}}{1-F_{\mu_N}}\right) \Bigg(\left(1-\left(\frac{1-F_{\lambda_N}}{1-F_{\mu_N}}\right)^{c}\left(\frac{1-F_{\lambda_N}}{1-F_{\mu_N}^{c}}\right)\right) + 1 - \left(1-\left(1-\left(\frac{1-F_{\lambda_N}}{1-F_{\mu_N}}\right)^{c}\right)\prod_{n=0}^{c-1}\left(1-\left(\frac{1-F_{\lambda_N}^{n}}{1-F_{\mu_N}^{n}}\right)^{\frac{1}{n!}}\right)\right)\Bigg)^{-1}$$

$$- \left(1-\left(\frac{1-F_{\lambda_N}}{1-F_{\mu_N}}\right)^{c}\left(\frac{1-F_{\lambda_N}}{1-F_{\mu_N}^{c}}\right)\right)\left(1-\left(1-\left(1-\left(\frac{1-F_{\lambda_N}}{1-F_{\mu_N}}\right)^{c}\right)\prod_{n=0}^{c-1}\left(1-\left(\frac{1-F_{\lambda_N}^{n}}{1-F_{\mu_N}^{n}}\right)^{\frac{1}{n!}}\right)\right)\right)^{-1} \Bigg]$$

$$= (1.058, 1.504, 4.333)$$

Table 3.5 displays the findings for calculating the Hausdorff distance using linguistic terms in NW_s.

Table 3.5 Linguistic Terms NW_s

Linguistic terms	*SVN numbers*	*Hausdorff Distance*
Extremely good	(1,0,0)	**4.333**
Very very good	(0.95,0.05,0.10)	**4.233**
Very good	(0.85,0.15,0.20)	**4.133**
Good	(0.75,0.25,0.35)	**3.983**
Fairly good	(0.65,0.30,0.40)	**3.933**
Average	(0.55,0.45,0.50)	**3.833**
Moderately bad	(0.45,0.50,0.60)	**3.733**
Bad	(0.35,0.65,0.70)	**3.633**
Very bad	(0.25,0.75,0.85)	**3.533**
Very very bad	(0.15,0.90,0.90)	**3.433**
Extremely bad	**(0.0,1.0,0.95)**	3.383

which means that the mean waiting time in the system is going to be extremely bad (because it has the smallest distance).

3.5 NEUTROSOPHIC EXTENSION PRINCIPLE

We use the neutrosophic extension principle to define the membership function of the system:

$$\text{Let } \mathrm{A} = \Big\{\big(s, (\mathrm{T}_{\mathrm{A}}^{1}(s), \mathrm{T}_{\mathrm{A}}^{2}(s), \mathrm{T}_{\mathrm{A}}^{3}(s), \ldots\ldots, \mathrm{T}_{\mathrm{A}}^{p}(s)\big), \\ \big(\mathrm{I}_{\mathrm{A}}^{1}(s), \mathrm{I}_{\mathrm{A}}^{2}(s), \mathrm{I}_{\mathrm{A}}^{3}(s), \ldots\ldots \mathrm{I}_{\mathrm{A}}^{p}(s)\big), \\ \big(F_{\mathrm{A}}^{1}(s), F_{\mathrm{A}}^{2}(s), F_{\mathrm{A}}^{3}(s), \ldots\ldots F_{\mathrm{A}}^{p}(s)\big)\Big\}; s \in S,$$

$\{\mathrm{T}_{\mathrm{A}}^{i}(s), \mathrm{T}_{\mathrm{A}}^{i}(s), \mathrm{T}_{\mathrm{A}}^{i}(s) \in [0.1]\}, (i \in \{1, 2, \ldots, p\})$ is any neutrosophic multi-sets on S. Then, extending the function $f: \mathrm{X} \to Y$, the neutrosophic multi-subset A of X is made to correspond to neutrosophic multi-subsets of $f(\mathrm{A}) = \{\mathrm{T}_{f(\mathrm{A})}^{i}, \mathrm{I}_{f(\mathrm{A})}^{i}, F_{f(\mathrm{A})}^{i}\}$ of Y in the following ways:

$$\mathrm{T}_{f(\mathrm{A})}^{i}(y) = \begin{cases} \vee\{\mathrm{T}_{\mathrm{A}}^{i}(x) : x \in f^{-1}(y)\}, if \ \mathrm{f}^{-1}(y) \neq \varphi, \\ 0, \quad \text{otherwise} \end{cases}$$

$$\mathrm{I}_{f(\mathrm{A})}^{i}(y) = \begin{cases} \vee\{\mathrm{I}_{\mathrm{A}}^{i}(x) : x \in f^{-1}(y)\}, if \ \mathrm{f}^{-1}(y) \neq \varphi, \\ 1, \quad \text{otherwise} \end{cases}$$

$$F_{f(\mathrm{A})}^{i}(y) = \begin{cases} \wedge\{F_{\mathrm{A}}^{i}(x) : x \in f^{-1}(y)\}, if \ \mathrm{f}^{-1}(y) \neq \varphi, \\ 1, \quad \text{otherwise} \end{cases}$$

For i=1, 2 . . ., p.

3.5.1 (α, β, γ) Cut Set of Neutrosophic Numbers

$$\text{Let } A = \{(s, (T_A^1(s), T_A^2(s), T_A^3(s), \ldots\ldots, T_A^p(s)),$$
$$(I_A^1(s), I_A^2(s), I_A^3(s), \ldots\ldots I_A^p(s)),$$
$$(F_A^1(s), F_A^2(s), F_A^3(s), \ldots\ldots F_A^p(s))\}; s \in S,$$

$\{T_A^i(s), T_A^i(s), T_A^i(s) \in [0.1]\}, (i \in \{1, 2, \ldots, p\})$ is any neutrosophic multi-sets on S. For any order, (α, β, γ) where $\alpha, \beta, \gamma \in [0,1]$, so $0 \le \alpha + \beta + \gamma \le 3$. Then, the (α, β, γ) cut set of neutrosophic multi-set Ais denoted by $A_{(\alpha,\beta,\gamma)}$:

$$A_{(\alpha,\beta,\gamma)} = \{s : T_A^i \ge \alpha, I_A^i \ge \beta, F_A^i \le \gamma, s \in S\};$$

$$\text{that is, } A_{(\alpha,\beta,\gamma)} = \{s : T_A^i(s) \wedge \alpha = \alpha, I_A^i(s) \wedge \beta = \beta, F_A^i(s) \vee \gamma = \gamma, s \in S\}.$$

The strong (α, β, γ) cut set of neutrosophic multi-set A is denoted by $A_{\overline{(\alpha,\beta,\gamma)}}$,

$$A_{\overline{(\alpha,\beta,\gamma)}} = \{s : T_A^i(s) > \alpha, I_A^i(s) > \beta, F_A^i(s) < \gamma, s \in S\}.$$

3.6 PARAMETRIC NONLINEAR PROGRAMMING

The membership function for the average number of customers in the system is defined as

$$\phi_{(N(x,y))} = \sup\min < T_{\hat{\lambda}}(x), T_{\hat{\mu}}(y) \mid \kappa = N(x,y) >,$$
$$\sup\min < I_{\hat{\lambda}}(x), I_{\hat{\mu}}(y) \mid \kappa = N(x,y) >,$$
$$\inf\max < F_{\hat{\lambda}}(x), F_{\hat{\mu}}(y) \mid \kappa = N(x,y) >$$

Therefore, the neutrosophic (α, β, γ) cut for the average number of customers in the system are $\lambda(\delta)$ and $\mu(\delta)$, which are actually deterministic sets. The neutrosophic queuing model NM/NM/c can be reduced to a family of deterministic queues of M/M/c with different (α, β, γ).

$$\lambda(\delta) = [x_\alpha^L, x_\alpha^U] = \left[\min\{x \mid T_{\hat{\lambda}}(x) \ge \alpha\}, \max\{x \mid T_{\hat{\lambda}}(x) \ge \alpha\}\right],$$
$$\lambda(\delta) = [x_\beta^L, x_\beta^U] = \left[\min\{x \mid T_{\hat{\lambda}}(x) \ge \beta\}, \max\{x \mid T_{\hat{\lambda}}(x) \ge \beta\}\right],$$
$$\lambda(\delta) = [x_\gamma^L, x_\gamma^U] = \left[\min\{x \mid I_{\hat{\lambda}}(x) \le \gamma\}, \max\{x \mid I_{\hat{\lambda}}(x) \le \gamma\}\right],$$
$$\mu(\delta) = [y_\alpha^L, y_\alpha^U] = \left[\min\{y \mid I_{\hat{\mu}}(y) \ge \alpha\}, \max\{y \mid I_{\hat{\mu}}(y) \ge \alpha\}\right],$$

$$\mu(\delta)=\left[y_\beta^L, y_\beta^U\right]=\left[\min\left\{y \mid F_{\hat{\mu}}(y)\geq \beta\right\}, \max\left\{y \mid F_{\hat{\mu}}(y)\geq \beta\right\}\right],$$

$$\mu(\delta)=\left[y_\gamma^L, y_\gamma^U\right]=\left[\min\left\{y \mid F_{\hat{\mu}}(y)\leq \gamma\right\}, \max\left\{y \mid F_{\hat{\mu}}(y)\leq \gamma\right\}\right],$$

Assuming that the considered neutrosophic numbers are convex and concave, the upper and lower bounds of these distances will be a function of (α, β, γ). The membership function $\phi_{(\mathrm{N}(x,y))}$ is also a function of (α, β, γ). As a result, $\phi_{(\mathrm{N}(x,y))}$ would be equal to the minimum $\left(\mathrm{T}_{\hat{\lambda}}(x), \mathrm{T}_{\hat{\mu}}(y)\right)$ and $\left(\mathrm{I}_{\hat{\lambda}}(x), \mathrm{I}_{\hat{\mu}}(y)\right)$ and the maximum $\left(F_{\hat{\lambda}}(x), \mathrm{F}_{\hat{\mu}}(y)\right)$.

Finding the membership function $\phi_{(\mathrm{N}(x,y))}$ needs at least one of the following, in which $\mathrm{N}(x, y)$ satisfies $\phi_{(\mathrm{N}(x,y))}$:

Case (1): $(\mathrm{T}_{\hat{\lambda}}(x)=\alpha, \mathrm{T}_{\hat{\mu}}(y)\geq\alpha)$

Case (2): $(\mathrm{T}_{\hat{\lambda}}(x)\geq\alpha, \mathrm{T}_{\hat{\mu}}(y)=\alpha)$

3.6.1 Upper and Lower Boundaries of the α Cuts in $\phi_{(\mathrm{N}(x,y))}$

The nonlinear programming technique for finding the upper and lower boundaries of the α cuts in $\phi_{(\mathrm{N}(x,y))}$ is given as follows
for case (1):

$$(L)_\alpha^{L_1} = \min[\mathrm{N(x,y)}]$$

$$(L)_\alpha^{U_1} = \max[\mathrm{N(x,y)}]$$

and as follows for case (2):

$$(L)_\alpha^{L_2} = \min[\mathrm{N(x,y)}]$$

$$(L)_\alpha^{U_2} = \min[\mathrm{N(x,y)}],$$

where the upper and lower bounds are given as

$$(L)_\alpha^{L} = \min\{(L)_\alpha^{L_1}, (L)_\alpha^{L_2}\}$$
$$(L)_\alpha^{U} = \min\{(L)_\alpha^{U_1}, (L)_\alpha^{U_2}\}.$$

Thus,

$$(L)_\alpha^{L} = \min[\mathrm{N}(x, y)]$$

such that

$$x_\alpha^L \le x \le x_\alpha^U$$
$$y_\alpha^L \le y \le y_\alpha^U$$

and $(L)_\alpha^U = \min[\mathrm{N}(x,y)]$
such that

$$x_\alpha^L \le x \le x_\alpha^U$$
$$y_\alpha^L \le y \le y_\alpha^U.$$

3.6.2 Upper and Lower Boundaries of the β Cuts in $\phi_{(\mathrm{N}(x,y))}$

Similarly, the nonlinear programming technique for finding the upper and lower boundaries of the β cuts in $\phi_{(\mathrm{N}(x,y))}$ is given as follows for $(\mathrm{I}_{\hat{\lambda}}(x) = \beta, \mathrm{I}_{\hat{\mu}}(y) \ge \beta)$:

$$(L)_\beta^{L_1} = \min[\mathrm{N(x,y)}]$$

$$(L)_\beta^{U_1} = \max[\mathrm{N(x,y)}].$$

For $(\mathrm{I}_{\hat{\lambda}}(x) \ge \beta, \mathrm{I}_{\hat{\mu}}(y) = \beta)$

$$(L)_\beta^{L_2} = \min[\mathrm{N(x,y)}]$$

$$(L)_\beta^{U_2} = \min[\mathrm{N(x,y)}]$$

where the upper and lower bounds are given as

$$(L)_\beta^L = \min\{(L)_\beta^{L_1}, (L)_\beta^{L_2}\}$$
$$(L)_\beta^U = \min\{(L)_\beta^{U_1}, (L)_\beta^{U_2}\};$$

Thus,

$$(L)_\beta^L = \min[\mathrm{N}(x,y)]$$

such that

$$x_\beta^L \le x \le x_\beta^U$$
$$y_\beta^L \le y \le y_\beta^U$$

and

$$(L)_\beta^U = \min[\mathrm{N}(x,y)]$$

such that

$$x_\beta^L \le x \le x_\beta^U$$
$$y_\beta^L \le y \le y_\beta^U.$$

Here, $(L)_\beta^L$ and $(L)_\beta^U$ are nonincreasing and nondecreasing with respect to β.

3.6.3 Upper and Lower Boundaries of the γ Cuts in $\phi_{(N(x,y))}$

Similarly, the nonlinear programming technique for finding the upper and lower boundaries of the γ cuts in $\phi_{(N(x,y))}$ is given as follows for $(F_{\hat{\lambda}}(x) = \gamma, F_{\hat{\mu}}(y) \le \gamma)$:

$$(L)_\gamma^{L_1} = \min[N(x,y)]$$

$$(L)_\gamma^{U_1} = \max[N(x,y)].$$

For $(F_{\hat{\lambda}}(x) \le \gamma, F_{\hat{\mu}}(y) = \gamma)$

$$(L)_\gamma^{L_2} = \max[N(x,y)]$$

$$(L)_\gamma^{U_2} = \max[N(x,y)]$$

where the upper and lower bounds are given as

$$(L)_\gamma^L = \max\{(L)_\gamma^{L_1}, (L)_\gamma^{L_2}\}$$
$$(L)_\gamma^U = \max\{(L)_\gamma^{U_1}, (L)_\gamma^{U_2}\}.$$

Thus,

$$(L)_\gamma^L = \max[N(x,y)]$$

such that

$$x_\gamma^L \le x \le x_\gamma^U$$
$$y_\gamma^L \le y \le y_\gamma^U$$

and $(L)_\gamma^U = \max[N(x,y)]$
such that

$$x_\gamma^L \le x \le x_\gamma^U$$
$$y_\gamma^L \le y \le y_\gamma^U.$$

Here, $(L)_\gamma^L$ and $(L)_\gamma^U$ are nonincreasing and nondecreasing with respect to γ.

3.7 NUMERIC EXAMPLE USING PARAMETRIC NONLINEAR PROGRAMMING

Next, we illustrate our numeric example with the combination of numbers and show our methods. In this section, we consider an NM/NM/c queueing system with two servers, where the arrival and service are hexagonal, heptagonal, and octagonal neutrosophic numbers, which we next describe.

Hexagonal neutrosophic number:

$$\hat{\lambda} = [8,9,10,11,12,13] \text{ and } \hat{\mu} = [15,16,17,18,19,20].$$

Thus,

$$[x_\alpha^L, x_\alpha^U] = [8+\alpha, 13-\alpha]$$
$$[y_\alpha^L, y_\alpha^U] = [15+\alpha, 20-\alpha]$$

$$[x_\beta^L, x_\beta^U] = [8+\beta, 13-\beta]$$
$$[y_\beta^L, y_\beta^U] = [15+\beta, 20-\beta]$$

$$[x_\gamma^L, x_\gamma^U] = [10-\gamma, 11+\gamma]$$
$$[y_\gamma^L, y_\gamma^U] = [17-\gamma, 18+\gamma].$$

Heptagonal neutrosophic number:

$$\hat{\lambda} = [25,26,27,28,29,30,31] \text{ and } \hat{\mu} = [33,34,35,36,37,38,39]$$

and thus,

$$[x_\alpha^L, x_\alpha^U] = [25+\alpha, 31-\alpha]$$
$$[y_\alpha^L, y_\alpha^U] = [33+\alpha, 39-\alpha]$$

$$[x_\beta^L, x_\beta^U] = [25+\beta, 31-\beta]$$
$$[y_\beta^L, y_\beta^U] = [33+\beta, 39-\beta]$$

$$[x_\gamma^L, x_\gamma^U] = [27.5 - \gamma, 28.5 + \gamma]$$
$$[y_\gamma^L, y_\gamma^U] = [35.5 - \gamma, 36.5 + \gamma]^{\cdot}$$

Octagonal neutrosophic number:

$$\hat{\lambda} = [43, 44, 45, 46, 47, 48, 49, 50] \text{ and } \hat{\mu} = [52, 53, 54, 55, 56, 57, 58, 59]$$

and thus,

$$[x_\alpha^L, x_\alpha^U] = [43 + \alpha, 50 - \alpha]$$
$$[y_\alpha^L, y_\alpha^U] = [52 + \alpha, 59 - \alpha]$$

$$[x_\beta^L, x_\beta^U] = [43 + \beta, 50 - \beta]$$
$$[y_\beta^L, y_\beta^U] = [52 + \beta, 59 - \beta]$$

$$[x_\gamma^L, x_\gamma^U] = [46 - \gamma, 47 + \gamma]$$
$$[y_\gamma^L, y_\gamma^U] = [55 - \gamma, 56 + \gamma]^{\cdot}$$

We applied parametric nonlinear programming techniques for each performance measure: Tables 3.6–3.8, hexagonal neutrosophic number; Tables 3.9 to 3.11, heptagonal neutrosophic number; Tables 3.12 to 3.14, octagonal neutrosophic number.

Table 3.6 The α Cut System Lengths at 11 Distinct α Values for the Hexagonal Neutrosophic Number

α	x_α^L	x_α^U	y_α^L	y_α^U	TL_α^L	TL_α^U
0.0	8	13	15	20	0.3978	0.8599
0.1	8.1	12.9	15.1	19.9	0.4048	0.8473
0.2	8.2	12.8	15.2	19.8	0.4118	0.8349
0.3	8.3	12.7	15.3	19.7	0.4188	0.8227
0.4	8.4	12.6	15.4	19.6	0.4260	0.8107
0.5	8.5	12.5	15.5	19.5	0.4332	0.7989
0.6	8.6	12.4	15.6	19.4	0.4405	0.7872
0.7	8.7	12.3	15.7	19.3	0.4478	0.7758
0.8	8.8	12.2	15.8	19.2	0.4553	0.7645
0.9	8.9	12.1	15.9	19.1	0.4628	0.7534
1.0	9	12	14	19	0.4703	0.7424

Table 3.7 The β Cut System Lengths at 11 Distinct β Values for the Hexagonal Neutrosophic Number

β	x^L_β	x^U_β	y^L_β	y^U_β	IL^U_β	IL^U_β
0.0	8	13	15	20	0.6140	0.0892
0.1	8.1	12.9	15.1	19.9	0.6063	0.1036
0.2	8.2	12.8	15.2	19.8	0.5986	0.1177
0.3	8.3	12.7	15.3	19.7	0.5907	0.1316
0.4	8.4	12.6	15.4	19.6	0.5828	0.1435
0.5	8.5	12.5	15.5	19.5	0.5748	0.1588
0.6	8.6	12.4	15.6	19.4	0.5667	0.1721
0.7	8.7	12.3	15.7	19.3	0.5585	0.1852
0.8	8.8	12.2	15.8	19.2	0.5502	0.1981
0.9	8.9	12.1	15.9	19.1	0.5418	0.2109
1.0	9	12	14	19	0.5333	0.2235

Table 3.8 The γ Cut System Lengths at 11 Distinct γ Values for the Hexagonal Neutrosophic Number

γ	x^L_γ	x^U_γ	y^L_γ	y^U_γ	FL^L_γ	FL^U_γ
0.0	10	11	17	18	0.4424	0.3402
0.1	9.9	11.1	16.9	18.1	0.4522	0.3292
0.2	9.8	11.2	16.8	18.2	0.4617	0.3181
0.3	9.7	11.3	16.7	18.3	0.4710	0.3068
0.4	9.6	11.4	16.6	18.4	0.4802	0.2954
0.5	9.5	11.5	16.5	18.5	0.4893	0.2838
0.6	9.4	11.6	16.4	18.6	0.4983	0.2721
0.7	9.3	11.7	16.3	18.7	0.5072	0.2602
0.8	9.2	11.8	16.2	18.8	0.5160	0.2481
0.9	9.1	11.9	16.1	18.9	0.5247	0.2359
1.0	9	12	16	19	0.5333	0.2235

Table 3.9 The α Cut System Lengths at 11 Distinct α Values for the Heptagonal Neutrosophic Number

α	x^L_α	x^U_α	y^L_α	y^U_α	TL^L_α	TL^U_α
0.0	25	31	33	39	0.6384	0.9377
0.1	25.1	30.9	33.1	38.9	0.6426	0.9317
0.2	25.2	30.8	33.2	38.8	0.6468	0.9258
0.3	25.3	30.7	33.3	38.7	0.6511	0.9199
0.4	25.4	30.6	33.4	38.6	0.6553	0.9141

(Continued)

Table 3.9 (Continued)

α	x_α^L	x_α^U	y_α^L	y_α^U	TL_α^L	TL_α^U
0.5	25.5	30.5	33.5	38.5	0.6596	0.9083
0.6	25.6	30.4	33.6	38.4	0.6639	0.9025
0.7	25.7	30.3	33.7	38.3	0.6682	0.8967
0.8	25.8	30.2	33.8	38.2	0.6726	0.8911
0.9	25.9	30.1	33.9	38.1	0.6770	0.8854
1.0	26	30	34	38	0.6814	0.8798

Table 3.10 The β Cut System Lengths At 11 Distinct β Values for the Heptagonal Neutrosophic Number

β	x_β^L	x_β^U	y_β^L	y_β^U	IL_β^L	IL_β^U
0.0	25	31	33	39	0.3526	0.3492
0.1	25.1	30.9	33.1	38.9	0.3481	0.0412
0.2	25.2	30.8	33.2	38.8	0.3437	0.0474
0.3	25.3	30.7	33.3	38.7	0.3392	0.0536
0.4	25.4	30.6	33.4	38.6	0.3346	0.0586
0.5	25.5	30.5	33.5	38.5	0.3301	0.0659
0.6	25.6	30.4	33.6	38.4	0.3255	0.0720
0.7	25.7	30.3	33.7	38.3	0.3209	0.0781
0.8	25.8	30.2	33.8	38.2	0.3163	0.0841
0.9	25.9	30.1	33.9	38.1	0.3116	0.0901
1.0	26	30	34	38	0.3069	0.0961

Table 3.11 The γ Cut System Lengths at 11 distinct γ values for the Heptagonal Neutrosophic Number

γ	x_γ^L	x_γ^U	y_γ^L	y_γ^U	FL_γ^L	FL_γ^U
0.0	27.5	28.5	35.5	36.5	0.2336	0.1810
0.1	27.4	28.6	35.4	36.6	0.2387	0.1756
0.2	27.3	28.7	35.3	36.7	0.2437	0.1701
0.3	27.2	28.8	35.2	36.8	0.2487	0.1646
0.4	27.1	28.9	35.1	36.9	0.2537	0.1591
0.5	27	29	35	37	0.2587	0.1535
0.6	26.9	29.1	34.9	37.1	0.2637	0.1480
0.7	26.8	29.2	34.8	37.2	0.2686	0.1423
0.8	26.7	29.3	34.7	37.3	0.2735	0.1367
0.9	26.6	29.4	34.6	37.4	0.2783	0.1310
1.0	26.5	29.5	34.5	37.5	0.2832	0.1252

Table 3.12 The α Cut System Lengths at 11 Distinct α Values for the Octagonal Neutrosophic Number

α	x_α^L	x_α^U	y_α^L	y_α^U	TL_α^L	TL_α^U
0.0	43	50	52	59	0.7269	0.9608
0.1	43.1	49.9	52.1	58.9	0.7299	0.9570
0.2	43.2	49.8	52.2	58.8	0.7328	0.9532
0.3	43.3	49.7	52.3	58.7	0.7357	0.9494
0.4	43.4	49.6	52.4	58.6	0.7387	0.9456
0.5	43.5	49.5	52.5	58.5	0.7417	0.9419
0.6	43.6	49.4	52.6	58.4	0.7447	0.9381
0.7	43.7	49.3	52.7	58.3	0.7476	0.9344
0.8	43.8	49.2	52.8	58.2	0.7507	0.9307
0.9	43.9	49.1	52.9	58.1	0.7573	0.9270
1.0	44	49	53	58	0.7567	0.9233

Table 3.13 The β Cut System Lengths at 11 Distinct β Values for the Octagonal Neutrosophic Number

β	x_β^L	x_β^U	y_β^L	y_β^U	IL_β^L	IL_β^U
0.0	43	50	52	59	0.2637	0.0210
0.1	43.1	49.9	52.1	58.9	0.2607	0.0250
0.2	43.2	49.8	52.2	58.8	0.2576	0.0289
0.3	43.3	49.7	52.3	58.7	0.2546	0.0328
0.4	43.4	49.6	52.4	58.6	0.2515	0.0367
0.5	43.5	49.5	52.5	58.5	0.2484	0.0406
0.6	43.6	49.4	52.6	58.4	0.2453	0.0445
0.7	43.7	49.3	52.7	58.3	0.2422	0.0483
0.8	43.8	49.2	52.8	58.2	0.2391	0.0522
0.9	43.9	49.1	52.9	58.1	0.2359	0.0560
1.0	44	49	53	58	0.2328	0.0598

Table 3.14 The γ Cut System Lengths at 11 Distinct γ Values for the Octagonal Neutrosophic Number

γ	x_γ^L	x_γ^U	y_γ^L	y_γ^U	FL_γ^L	FL_γ^U
0.0	46	47	55	56	0.1674	0.1329
0.1	45.9	47.1	54.9	56.1	0.1708	0.1294
0.2	45.8	47.2	54.8	56.2	0.1742	0.1258
0.3	45.7	47.3	54.7	56.3	0.1775	0.1222
0.4	45.6	47.4	54.6	56.4	0.1809	0.1187

(Continued)

Table 3.14 (Continued)

γ	x_γ^L	x_γ^U	y_γ^L	y_γ^U	FL_γ^L	FL_γ^U
0.5	45.5	47.5	54.5	56.5	0.1842	0.1151
0.6	45.4	47.6	54.4	56.6	0.1875	0.1115
0.7	45.3	47.7	54.3	56.7	0.1908	0.1079
0.8	45.2	47.8	54.2	56.8	0.1941	0.1043
0.9	45.1	47.9	54.1	56.9	0.1974	0.1007
1.0	45	48	54	57	0.2007	0.0970

3.8 CONCLUSION

In this chapter, we demonstrated simulating queues with various unknown parameters in a neutrosophic logic model using linguistic terms. Specifically, we discussed hexagonal, heptagonal, and octagonal neutrosophic numbers with (α, β, γ) cut operations with the help of parametric nonlinear programming. Furthermore, we depicted the performance measures of the neutrosophic queueing model: average number of customers in the queue (NL_q) and in the system (NL_s) and mean waiting time in the queue (NW_q) and in the system (NW_s) using a numeric example. As a future work, other important performance measures can be analyzed. Ranking techniques could be employed with this proposed work for analysing decision-making problems.

REFERENCES

Atanassov, K., (1986). "Intuitionistic fuzzy sets", Fuzzy Sets and Systems, Vol. 20, pp. 87–96.

Atanassov, K. T., (2016). "Intuitionistic fuzzy sets, VII ITKR session, Sofia, 20–23 June 1983 (Deposed in Centr. Sci.-Techn. Library of the Bulg. Acad. of Sci., 1697/84) (in Bulgarian)", Reprinted: International Journal Bioautomation, Vol. 20, No. S1, pp. S1–S6.

Sahin, M., Olgun, N., Uluçay, V., Kargın, A., and Smarandache, F., (2017). "A new similarity measure based on falsity value between single valued neutrosophic sets based on the centroid points of transformed single valued neutrosophic numbers with applications to pattern recognition", Neutrosophic Sets and Systems, Vol. 15, pp. 31–48.

Smarandache, F., (1995). Neutrosophic Logic and Set. http://fs.gallup.unm.edu/neutrosophy.htm

Smarandache, F., (1998). A Unifying Field in Logics. Neutrosophy: Neutrosophic Probability, Set and Logic. Rehoboth: American Research Press.

Smarandache, F., (2016). "Subtraction and division of neutrosophic numbers", Uncertainty, Vol. XIII, pp. 103–110.

Sumathi, I. R., and Antony Crispin Sweety, C., (2019). "New approach on differential equation via trapezoidal neutrosophic number", Complex & Intelligent Systems, Vol. 5, pp. 417–424.

Zadeh, L., (1965). "Fuzzy sets", Inform and Control, Vol. 8, pp. 338–353.

Zeina, M. B., (2020a)."Neutrosophic event-based queueing model", International Journal of Neutrosophic Science, Vol. 6, No. 1, pp. 48–55.

Zeina, M. B., (2020b). "Erlang service queueing model with neutrosophic parameters", International Journal of Neutrosophic Science, Vol. 6, No. 1, pp. 106–112.

Chapter 4

Cardinalities of Neutrosophic Sets and Neutrosophic Crisp Sets

Bhimraj Basumatary and Jili Basumatary

4.1 INTRODUCTION

In 1995, Smarandache [1] found that some objects have indeterminacy or neutrality rather than membership and non-membership and introduced the notion of neutrosophic sets by generalizing intuitionistic fuzzy sets [2]. Salama and Smarandache [3] introduced the concept of neutrosophic crisp sets, which have been defined along with descriptions of their operations [4, 5]. Smarandache [6] extended the neutrosophic set to neutrosophic oversets, undersets, and offsets.

A simple finite set has a small number of subsets, so enumerating them is a straightforward problem. The challenges appear only when that set is unlimited. Cantor [7] solved the problem of infiniteness by introducing the ideas of aleph and aleph-null. Yager [8] introduced the concept of a count-bag to represent the cardinality of a fuzzy set and studied some of its properties. However, in a fuzzy environment and a neutrosophic environment, the problem of counting becomes complicated due to the lack of sharp boundaries.

Many authors consider the counting problem in the case of fuzzy subsets [9–11], which becomes more complicated even if the set is finite. This is considered a challenging problem due to the uncertainty in membership values. Murali [12, 13] studied the number of k-level equivalence classes of fuzzy subsets of a finite set of n elements under a natural equivalence that was related to Stirling numbers. The author showed that there exists a bijection between all the equivalence classes of fuzzy subsets of X and all the chains in the power set of X. Chamorro-Martínez et al. [14] discussed different representations of the cardinality of a fuzzy set and their use in fuzzy quantification. Benoumhani [15] computed several results concerning chains in Y^X which is the lattice of mappings from a finite set X into a finite totally ordered set Y.

4.2 PRELIMINARIES

Definition 4.2.1 [1] A neutrosophic set A^{NT} on a universe of discourse X is defined as $A^{NT} = \left\langle \frac{x}{\left(T(x), I(x), F(x)\right)} : x \in X \right\rangle$, where $T, I, F : X \to]^{-}0, 1^{+}[$.

DOI: 10.1201/9781003487104-4

Note that $^-0 \leq T(x)+I(x)+F(x) \leq 3^+$; $T(x)$, $I(x)$, and $F(x)$ represent degree of membership function, degree of indeterminacy, and degree of non-membership function, respectively.

Definition 4.2.2 [16] The neutrosophic subsets 0^{NT} and 1^{NT} in X are as follows:

0^{NT} may be defined as

$$0^{NT} = \{\langle x,0,0,1\rangle : x \in X\},$$
$$0^{NT} = \{\langle x,0,1,1\rangle : x \in X\},$$
$$0^{NT} = \{\langle x,0,1,0\rangle : x \in X\},$$
$$0^{NT} = \{\langle x,0,0,0\rangle : x \in X\}.$$

1^{NT} may be defined as

$$1^{NT} = \{\langle x,1,0,0\rangle : x \in X\},$$
$$1^{NT} = \{\langle x,1,0,1\rangle : x \in X\},$$
$$1^{NT} = \{\langle x,1,1,0\rangle : x \in X\},$$
$$1^{NT} = \{\langle x,1,1,1\rangle : x \in X\}.$$

Definition 4.2.3 [16] Let X be a non-empty set and neutrosophic subsets A and B in the form $A = \langle \mu_A, \rho_A, \gamma_A \rangle$, $B = \langle \mu_B, \rho_B, \gamma_B \rangle$. Then, two possible definitions may be considered for subsets $(A \subseteq B)$:

$$A \subseteq B \Leftrightarrow \mu_A(x) \leq \mu_B(x), \rho_A(x) \geq \rho_B(x) \text{ and } \gamma_A(x) \leq \gamma_B(x) \text{ or}$$
$$A \subseteq B \Leftrightarrow \mu_A(x) \leq \mu_B(x), \rho_A(x) \geq \rho_B(x) \text{ and } \gamma_A(x) \geq \gamma_B(x).$$

Definition 4.2.4 [4] Let X be a non-empty fixed set. A neutrosophic crisp set (NCrS) A is an object with the form $A = \langle A_1, A_2, A_3 \rangle$, where A_1, A_2, and A_3 are subsets of X satisfying $A_1 \cap A_2 = \phi$, $A_1 \cap A_3 = \phi$, and $A_2 \cap A_3 = \phi$, respectively.

Remark 4.2.1 [4] A NCrS $A = \langle A_1, A_2, A_3 \rangle$ can be identified as an ordered triple $\langle A_1, A_2, A_3 \rangle$, where A_1, A_2, and A_3 are subsets of X.

Definition 4.2.5 [4] ϕ_N may be defined in many ways as a NCrS, as follows:

(i) $\phi_N = \langle \phi, \phi, X \rangle$, or
(ii) $\phi_N = \langle \phi, X, X \rangle$, or

(iii) $\phi_N = \langle \phi, X, \phi \rangle$, or
(iv) $\phi_N = \langle \phi, \phi, \phi \rangle$.

X_N may also be defined in many ways as a NCrS:

(i) $X_N = \langle X, \phi, \phi \rangle$,
(ii) $X_N = \langle X, X, \phi \rangle$,
(iii) $X_N = \langle X, X, X \rangle$.

Definition 4.2.6 [4] Let X be a non-empty set and NCrSs A and B in the form $A = \langle A_1, A_2, A_3 \rangle$, $B = \langle B_1, B_2, B_3 \rangle$. Then, the following two possible definitions may be considered for subsets $(A \subseteq B)$:

(i) $A \subseteq B \Leftrightarrow A_1 \subseteq B_1, A_2 \subseteq B_2$, and $A_3 \supseteq B_3$, or
(ii) $A \subseteq B \Leftrightarrow A_1 \subseteq B_1$, $A_2 \supseteq B_2$, and $A_3 \supseteq B_3$.

Lemma 4.2.1 [17] Let m and n be positive integers, then for any numbers $y_1, y_2, \ldots, y_m$, we have, $\sum_{(i_1, i_2, \ldots, i_n) \in \{1,2,\ldots,m\}^n} y_{i_1} y_{i_2} \cdots y_{i_n} = \left(\sum_{i=1}^{m} y_i \right)^n$.

Algorithm 4.2.1 [17] Let P be a finite ordered set and $c_k(P)$ or c_k denote the number of chains with k elements in the ordered set P. Also, for each $u \in P$, let $c_k(u)$ be the number of chains with k elements from P and with maximal element u. The numbers $c_1, c_2, \ldots, c_n$ are obtained recursively as follows:

(i) $c_1(u) = 1$, for each $u \in P$,
(ii) $c_k(u) = \sum_{v<u} c_{k-1}(v)$, $2 \le k \le n$, for each $u \in P$,
(iii) $c_k := c_k(P) = \sum_{u \in P} c_k(u)$, $1 \le k \le n$.

Definition 4.2.7 [18] The Stirling number of the second kind is the number of partitions of a finite set with n elements into k blocks. It is denoted by $S(n,k)$ or $S_{n,k}$ and its explicit formula is $S(n,k) = S_{n,k} = \frac{1}{k!} \sum_{j=0}^{k} (-1)^j \binom{k}{j} (k-j)^n$.

Lemma 4.2.2 [19] For $p \ge q \ge 0$, $\binom{p}{q} = \frac{p!}{q!(p-q)!}$ with the convention that $\binom{p}{q} = 0$ for any $q > p$.

Theorem 4.2.1 [19] For any integer $n \ge 0$, $(a+b)^n = \sum_{m=0}^{n} \binom{n}{m} a^m b^{n-m}$.

4.3 CARDINALITIES OF THE NEUTROSOPHIC SET

Let us assume that X and $\mathcal{M}$ are both finite, with $X = \{v_1, v_2, \ldots, v_n\}$, and $\mathcal{M} = \{t_0, t_1, t_2, \ldots, t_{m-1}\}$ is a totally ordered set such that $(0,1,1) = t_0 < t_1 = (T_1, I_1, F_1) < t_2 = (T_2, I_2, F_2) < \ldots < t_{m-2} = (T_{m-2}, I_{m-2}, F_{m-2}) < t_{m-1} = (1,0,0)$, where $t_i = (T_i, I_i, F_i) < t_j = (T_j, I_j, F_j)$ iff [$T_i \leq T_j, I_i \geq I_j$, $F_i \geq F_j$ and at least one of $T_i < T_j$ or $I_i > I_j$ or $F_i > F_j$] or [$T_i \leq T_j, I_i \geq I_j$, $F_i \leq F_j$ and at least one of $T_i < T_j$ or $I_i > I_j$ or $F_i < F_j$]. Also, let $\mathcal{N}_X$ be the collection of neutrosophic subsets (Nsubs) of X with neutrosophic values in $\mathcal{M}$.

$\mathcal{N}_X$ is partially ordered by

$A^{NT} \preceq B^{NT}$ if and only if $T_1(v_i) \leq T_2(v_i), I_1(v_i) \geq I_2(v_i), F_1(v_i) \geq F_2(v_i)$ or $T_1(v_i) \leq T_2(v_i), I_1(v_i) \geq I_2(v_i), F_1(v_i) \leq F_2(v_i)$ for each $i \in \{1, 2, \ldots, n\}$, where

$$A^{NT} = \left\langle \frac{x}{(T_1(x), I_1(x), F_1(x))} : x \in X \right\rangle \text{ and } B^{NT} = \left\langle \frac{x}{(T_2(x), I_2(x), F_2(x))} : x \in X \right\rangle.$$

We also have

$A^{NT} \prec B^{NT}$ iff $A^{NT} \preceq B^{NT}$ and [at least one of $T_1(v_i) < T_2(v_i)$ or $I_1(v_i) > I_2(v_i)$ or $F_1(v_i) > F_2(v_i)$] or [at least one of $T_1(v_i) < T_2(v_i)$ or $I_1(v_i) > I_2(v_i)$ or $F_1(v_i) < F_2(v_i)$] for some $i \in \{1, 2, \ldots, n\}$.

Then the first question that arises in our mind is "How many Nsubs are there in a non-empty finite set X of n elements?" The study is based on a non-empty finite set because a NSub of the empty set does not have a conventional meaning since there are no elements to talk of in this set.

Definition 4.3.1 The set of all NSubs of a non-empty finite set X whose neutrosophic values lie in $\mathcal{M}$ with $|\mathcal{M}| = m \geq 2$ is called the neutrosophic power set of X with neutrosophic values in $\mathcal{M}$. The notation for the neutrosophic power set of X whose neutrosophic values lie in $\mathcal{M}$ is $P_{\mathcal{M}}(X)$, and its cardinality is denoted by $|P_{\mathcal{M}}(X)|$.

Proposition 4.3.1 A non-empty finite set X with $|X| = n$ whose neutrosophic values lie in $\mathcal{M}$ with $|\mathcal{M}| = m \geq 2$ has m^n NSubs.

Proof: Each element of X has m choices for neutrosophic values as $|\mathcal{M}| = m$. Hence, the total number of NSubs of X whose neutrosophic values lie in $\mathcal{M}$ is $\underbrace{m.m\ldots m}_{n \text{ times}} = m^n$.

Proposition 4.3.2 If $|X| = n$ and $|\mathcal{M}| = m \geq 2$, then the cardinality of the power set of neutrosophic set of X whose neutrosophic values lie in $\mathcal{M}$ is $|P_{\mathcal{M}}(X)| = m^n$.

Proof: By Definition 4.3.1 and Proposition 4.3.1, the proof is straightforward.

Example 4.3.1 Let $X=\{u\}$ and $\mathcal{M}=\{(0,1,1),(T,I,F),(1,0,0)\}$. It is seen that $|X|=n=1,|\mathcal{M}|=m=3$. Then, the NSubs of X whose neutrosophic values lie i $\mathcal{M}$ are

$$0^{NT}=\left\langle\frac{u}{(0,1,1)}\right\rangle,\ 1^{NT}=\left\langle\frac{u}{(1,0,0)}\right\rangle,\ \text{and } A_1^{NT}=\left\langle\frac{u}{(T,I,F)}\right\rangle.$$

Therefore, $|P_{\mathcal{M}}(X)|=3=m^n$.

Example 4.3.2 Let $X=\{u,v\}$ and $\mathcal{M}=\{(0,1,1),(T,I,F),(1,0,0)\}$. It is seen that $|X|=n=2,|\mathcal{M}|=m=3$. Then, the NSubs of X whose neutrosophic values lie in $\mathcal{M}$ are

$$0^{NT}=\left\langle\frac{u}{(0,1,1)},\frac{v}{(0,1,1)}\right\rangle,\ 1^{NT}=\left\langle\frac{u}{(1,0,0)},\frac{v}{(1,0,0)}\right\rangle,\ A_1^{NT}=\left\langle\frac{u}{(0,0,1)},\frac{v}{(T,I,F)}\right\rangle,$$

$$A_2^{NT}=\left\langle\frac{u}{(0,0,1)},\frac{v}{(1,0,0)}\right\rangle,\ A_3^{NT}=\left\langle\frac{u}{(T,I,F)},\frac{v}{(0,1,1)}\right\rangle,$$

$$A_4^{NT}=\left\langle\frac{u}{(T,I,F)},\frac{v}{(T,I,F)}\right\rangle,\ \text{and}$$

$$A_5^{NT}=\left\langle\frac{u}{(T,I,F)},\frac{v}{(1,0,0)}\right\rangle,\ A_6^{NT}=\left\langle\frac{u}{(1,0,0)},\frac{v}{(0,1,1)}\right\rangle,$$

$$A_7^{NT}=\left\langle\frac{u}{(1,0,0)},\frac{v}{(T,I,F)}\right\rangle.$$

Therefore, $|P_{\mathcal{M}}(X)|=9=m^n$.

Proposition 4.3.3 The number of chains (NCs) of length one in the neutrosophic power set on X whose neutrosophic values lie in $\mathcal{M}$ is $C_N(n,m,1)=|P_{\mathcal{M}}(X)|=m^n$.

Proof: Every NSub of X whose neutrosophic values lie in $\mathcal{M}$ are taken as the chain of length one, where $|X|=n$ and $|\mathcal{M}|=m$. Then clearly, $C_N(n,m,1)=|P_{\mathcal{M}}(X)|=m^n$.

Example 4.3.3 Every NSub of X whose neutrosophic values lie in $\mathcal{M}$ is a chain of length one. If we consider example 4.3.2, then the chains of length one are

$$0^{NT}, 1^{NT}, A_1^{NT}, A_2^{NT}, A_3^{NT}, A_4^{NT}, A_5^{NT}, A_6^{NT}, A_7^{NT}.$$

Proposition 4.3.4 The NCs of length two in $\widehat{P}_{\mathcal{M}}(X)=P_{\mathcal{M}}(X)-\{0^{NT},\ 1^{NT}\}$ is $\binom{m+1}{2}^n - m^n + 3$.

Proof: The existence of a chain of length two in $\widehat{P}_{\mathcal{M}}(X)$ is first subject to the condition that there are enough levels to contain such chains. That is, we first assumed that $n(m-1)-1 \geq 2$, or equivalently $n(m-1)\geq 3$. To compute the NCs of length two in $\widehat{P}_{\mathcal{M}}(X)$, we used algorithm 4.2.1. That gave $c_1\left(t_{i_1},t_{i_2},\ldots,t_{i_n}\right)=1$ for each $\left(t_{i_1},t_{i_2},\ldots,t_{i_n}\right)\in \widehat{P}_{\mathcal{M}}(X)$. For each $\left(t_{j_1},t_{j_2},\ldots,t_{j_n}\right)$ that satisfies $\left(t_{j_1},t_{j_2},\ldots,t_{j_n}\right)\preceq\left(t_{i_1},t_{i_2},\ldots,t_{i_n}\right)$, we have $0\leq j_k\leq i_k$ for each k in $\{1,2,\ldots,n\}$. That is, there are exactly $(i_1+1)(i_2+1)\ldots(i_n+1)-2$ such $\left(t_{j_1},t_{j_2},\ldots,t_{j_n}\right)$s. Removing $(i_1,i_2,\ldots,i_n)$ and $(0,0,\ldots,0)$ from the list, we obtain that the NCs containing two elements in $\widehat{P}_{\mathcal{M}}(X)$ and ending with $\left(t_{i_1},t_{i_2},\ldots,t_{i_n}\right)$ is

$$c_2\left(t_{i_1},t_{i_2},\ldots,t_{i_n}\right)=(i_1+1)(i_2+1)\ldots(i_n+1)2.$$

Taking the sum of all elements of $P_{\mathcal{M}}(X)$ lying on levels 2 to $n(m-1)-1$, we obtained c_2 chains of length two in $\widehat{P}_{\mathcal{M}}(X)$, given by

$$\begin{aligned}
c_2\left(\widehat{P}_{\mathcal{M}}(X)\right) &= \sum_{2\leq i_1+i_2+\cdots+i_n\leq n(m-1)-1} c_2\left(t_{i_1},t_{i_2},\ldots,t_{i_n}\right)\\
&= \sum_{2\leq i_1+i_2+\cdots+i_n\leq n(m-1)-1}\left((i_1+1)(i_2+1)\ldots(i_n+1)-2\right)\\
&= \sum_{h=2}^{n(m-1)-1}\sum_{\substack{i_1+i_2+\cdots+i_n=h,\\ t_0\leq t_i\leq t_{m-1}}}\left((i_1+1)(i_2+1)\ldots(i_n+1)-2\right)\\
&= \sum_{h=2}^{n(m-1)-1}\left(\sum_{\substack{y_1+y_2+\ldots+y_n=n+h,\\ 1\leq y_i\leq m}}(y_1y_2\ldots y_n-2)\right);\ (\text{taking } y_k=i_k+1)\\
&= \sum_{k=n+2}^{nm-1}\left(\sum_{\substack{y_1+y_2+\ldots+y_n=k,\\ 1\leq y_i\leq m}}(y_1y_2\ldots y_n-2)\right)\\
&= -2\left(m^n-2-n\right)+\sum_{k=n+2}^{nm-1}\left(\sum_{\substack{y_1+y_2+\ldots+y_n=k,\\ 1\leq y_i\leq m}}(y_1y_2\ldots y_n)\right)
\end{aligned}$$

Letting $s=\sum_{\substack{k=n+2\\ 1\leq y_i\leq m}}^{nm-1}\left(\sum_{\substack{y_1+y_2+\ldots+y_n=k,\\ 1\leq y_i\leq m}}(y_1y_2\ldots y_n)\right)$

and using Lemma 4.2.1, we obtained:

$$\left(\frac{m(m+1)}{2}\right)^n = \sum_{k=n}^{nm}\left(\sum_{\substack{y_1+y_2+\ldots+y_n=k,\\ 1\le y_i\le m}}(y_1y_2\ldots y_n)\right)$$
$$= y_1^{\,n} + ny_1^{\,n-1}y_2 + s + y_m^n$$
$$= 1 + 2n + s + m^n,$$

from which we deduced that $s = \left(\frac{m(m+1)}{2}\right)^n - m^n - 2n - 1.$

Therefore, $c_2\left(\widehat{P}_{\mathcal{M}}(X)\right) = -2\left(m^n - 2 - n\right) + \left(\frac{m(m+1)}{2}\right)^n - m^n - 2n - 1$

$$= -3m^n + \left(\frac{m+1}{2}\right)^n + 3.$$

This proves the result in the case $n(m-1)\ge 3$.

Now, if n and m are positive integers such that $n(m-1)<3$ and taking that in account, we assumed that $n\ge 1$ and $m\ge 2$ we get either

(a) $m=2$ and $n=1$,
or (b) $m=2$ and $n=2$,
or (c) $m=3$ and $n=1$.

On one hand, these values substituted in the above formula give the value 0. On the other hand, these values result in ordered sets. It is easy to check that we have no chain with length two in any case, (a), (b), or (c), which is consistent with the obtained formula. This proves the proposition.

Corollary 4.3.1 In $P_{\mathcal{M}}(X)$, the NCs of length four having both 0^{NT} and 1^{NT} is the same as $c_2\left(\widehat{P}_{\mathcal{M}}(X)\right)$.

Proposition 4.3.5 The NCs of length two in the neutrosophic power set on X whose neutrosophic values lie in $\mathcal{M}$ is $C_N(n,m,2) = \left(\frac{m+1}{2}\right)^n - m^n$.

Proof: Let $|X| = n$ and $|\mathcal{M}| = m$, then $|P_{\mathcal{M}}(X)| = m^n$. To prove the result, we have the following cases:

Case 1 Chain of 0^{NT} with 1^{NT}: The NCs of length two of 0^{NT} with 1^{NT} is 1. i.e., $0^{NT} \subset 1^{NT}$.

Case 2 Chain of length one with 0^{NT} as subset: We know every neutrosophic subset is a chain of length one and 0^{NT} is a subset of every neutrosophic subset. Therefore, in this case, the NCs of length two is $|P_{\mathcal{M}}(X)|-1=m^n-1$ as $0^{NT} \subseteq 0^{NT}$, which is a chain of length one.

Case 3 Chain of length one with 1^{NT} as super-subset: Since 1^{NT} is a supersubset of every neutrosophic subset. Therefore, in this case, the NCs of length two is $|P_{\mathcal{M}}(X)|-1=m^n-1$ as $1^{NT} \subseteq 1^{NT}$, which is a chain of length one.

Case 4 Chain of length two from $\widehat{P}_{\mathcal{M}}(X)$: Following Proposition 4.3.4, the NCs of length two is

$$\binom{m+1}{2}^n - 3m^n + 3.$$

Since the chain from case 1, that is, $0^{NT} \subset 1^{NT}$ is present in both case 2 and case 3, the total NCs of length two in the neutrosophic power set on X whose neutrosophic values lie in $\mathcal{M}$ is

$$1+\left(m^n-1-1\right)+\left(m^n-1-1\right)+\binom{m+1}{2}^n-3m^n+3$$

$$\text{that is, } C_N(n,m,2)=\binom{m+1}{2}^n-m^n.$$

Example 4.3.4 If we consider example 4.3.2, then the NCs of length two is

$$\binom{m+1}{2}^n - m^n = \binom{3+1}{2}^2 - 3^2 = 6^2 - 9 = 36 - 9 = 27.$$

These are

$0^{NT} \subset 1^{NT}$, $0^{NT} \subset A_1^{NT}$, $0^{NT} \subset A_2^{NT}$, $0^{NT} \subset A_3^{NT}$, $0^{NT} \subset A_4^{NT}$, $0^{NT} \subset A_5^{NT}$,

$0^{NT} \subset A_6^{NT}$, $0^{NT} \subset A_7^{NT}$, $A_1^{NT} \subset 1^{NT}$, $A_2^{NT} \subset 1^{NT}$, $A_3^{NT} \subset 1^{NT}$, $A_4^{NT} \subset 1^{NT}$,

$A_5^{NT} \subset 1^{NT}$, $A_6^{NT} \subset 1^{NT}$, $A_7^{NT} \subset 1^{NT}$, $A_1^{NT} \subset A_2^{NT}$, $A_1^{NT} \subset A_4^{NT}$, $A_1^{NT} \subset A_5^{NT}$,

$A_1^{NT} \subset A_7^{NT}$, $A_2^{NT} \subset A_5^{NT}$, $A_3^{NT} \subset A_4^{NT}$, $A_3^{NT} \subset A_5^{NT}$, $A_3^{NT} \subset A_6^{NT}$, $A_3^{NT} \subset A_7^{NT}$,

$A_4^{NT} \subset A_5^{NT}$, $A_4^{NT} \subset A_7^{NT}$, $A_6^{NT} \subset A_7^{NT}$.

Proposition 4.3.6 The NCs of length three in $\widehat{P}_{\mathcal{M}}(X)$ is

$$\frac{m^n(m+1)^n(m+2)^n}{6^n} - \frac{4m^n(m+1)^n}{2^n} + 6m^n - 4.$$

Proof: First, let us assume that $n(m-1)+1 \geq 5$. To compute the NCs of length three in $\widehat{P}_{\mathcal{M}}(X)$, we use Algorithm 4.2.1. For each $\left(t_{i_1}, t_{i_2}, \ldots, t_{i_n}\right)$ in $\widehat{P}_{\mathcal{M}}(X)$, let $c_k\left(t_{i_1}, t_{i_2}, \ldots, t_{i_n}\right)$ be the NCs with k elements from $\widehat{P}_{\mathcal{M}}(X)$, and with maximal element $\left(t_{i_1}, t_{i_2}, \ldots, t_{i_n}\right)$. Then,

$$c_1\left(t_{i_1}, t_{i_2}, \ldots, t_{i_n}\right) = 1$$

$$c_2\left(t_{i_1}, t_{i_2}, \ldots, t_{i_n}\right) = (i_1+1)(i_2+1)\ldots(i_n+1) - 2$$ [following the proof of proposition 4.3.4],

and

$$\begin{aligned} c_3\left(t_{i_1}, t_{i_2}, \ldots, t_{i_n}\right) &= \sum\nolimits_{0^{NT} \prec \left(t_{j_1}, t_{j_2}, \ldots, t_{j_n}\right) \prec \left(t_{i_1}, t_{i_2}, \ldots, t_{i_n}\right)} c_2\left(t_{j_1}, t_{j_2}, \ldots, t_{j_n}\right) \\ &= -c_2\left(t_{i_1}, t_{i_2}, \ldots, t_{i_n}\right) \\ &\quad + \sum\nolimits_{0^{NT} \prec \left(t_{j_1}, t_{j_2}, \ldots, t_{j_n}\right) \preceq \left(t_{i_1}, t_{i_2}, \ldots, t_{i_n}\right)} c_2\left(t_{j_1}, t_{j_2}, \ldots, t_{j_n}\right). \end{aligned}$$

Hence, $c_3\left(t_{i_1}, t_{i_2}, \ldots, t_{i_n}\right) + c_2\left(t_{i_1}, t_{i_2}, \ldots, t_{i_n}\right)$

$$\begin{aligned} &= \sum\nolimits_{0^{NT} \prec \left(t_{j_1}, t_{j_2}, \ldots, t_{j_n}\right) \preceq \left(t_{i_1}, t_{i_2}, \ldots, t_{i_n}\right)} c_2\left(t_{j_1}, t_{j_2}, \ldots, t_{j_n}\right) \\ &= \sum\nolimits_{0^{NT} \prec (j_1, j_2, \ldots, j_n) \preceq (i_1, i_2, \ldots, i_n)} \left((j_1+1)(j_2+1)\ldots(j_n+1) - 2\right) \\ &= -\left((0+1)(0+1)\ldots(0+1) - 2\right) \\ &\quad + \sum\nolimits_{0^{NT} \preceq (j_1, j_2, \ldots, j_n) \preceq (i_1, i_2, \ldots, i_n)} \left((j_1+1)(j_2+1)\ldots(j_n+1) - 2\right) \\ &= 1 + \sum\nolimits_{\substack{0 \leq j_k \leq i_k, \\ 1 \leq k \leq n}} \left((j_1+1)(j_2+1)\ldots(j_n+1) - 2\right) \\ &= 1 + \sum\nolimits_{\substack{0 \leq y_k - 1 \leq i_k, \\ 1 \leq k \leq n}} \left(y_1 y_2 \ldots y_n - 2\right) \\ &= 1 + \left(\sum\nolimits_{\substack{0 \leq y_k \leq i_k + 1 \\ 1 \leq k \leq n}} y_1 y_2 \ldots y_n\right) - 2\prod\nolimits_{k=1}^{n} (i_k+1) \\ &= 1 + \prod\nolimits_{k=1}^{n} (1 + 2 + \ldots + (i_k+1)) - 2\prod\nolimits_{k=1}^{n} (i_k+1) \\ &= 1 + \prod\nolimits_{k=1}^{n} \frac{(i_k+2)(i_k+1)}{2} - 2\prod\nolimits_{k=1}^{n} (i_k+1). \end{aligned}$$

Taking the sum of all elements of $\widehat{P}_{\mathcal{M}}(X)$, we have

$$c_3 + c_2 = \sum\nolimits_{(t_{i_1}, t_{i_2}, \ldots, t_{i_n}) \in \widehat{P}_{\mathcal{M}}(X)} c_3\left(t_{i_1}, t_{i_2}, \ldots, t_{i_n}\right) + c_2\left(t_{i_1}, t_{i_2}, \ldots, t_{i_n}\right)$$

$$= \sum\nolimits_{0^{NT} \prec (t_{i_1}, t_{i_2}, \ldots, t_{i_n}) \prec 1^{NT}} \left(1 + \prod\nolimits_{k=1}^{n} \frac{(i_k+2)(i_k+1)}{2} - 2\prod\nolimits_{k=1}^{n} (i_k+1)\right).$$

Therefore, $c_3 + c_2 + 1 + \dfrac{(m+1)^n m^n}{2^n} - 2m^n$

$$= \sum\nolimits_{0^{NT} \preceq (t_{i_1}, t_{i_2}, \ldots, t_{i_n}) \preceq 1^{NT}} \left(1 + \prod\nolimits_{k=1}^{n} \frac{(i_k+2)(i_k+1)}{2} - 2\prod\nolimits_{k=1}^{n} (i_k+1)\right)$$

$$= m^n + \frac{1}{2^n}\left(\sum\nolimits_{i_k=0}^{m-1} (i_k+2)(i_k+1)\right)^n - 2\left(\sum\nolimits_{i_k=0}^{m-1} (i_k+1)\right)^n$$

$$= m^n + \frac{1}{2^n}\left(\sum\nolimits_{i_k=0}^{m-1} (i_k+2)(i_k+1)\right)^n - \frac{2m^n (m+1)^n}{2^n}.$$

Since $c_2 = \left(\dfrac{m(m+1)}{2}\right)^n - 3m^n + 3,$

and $\left(\sum\nolimits_{i_k=0}^{m-1} (i_k+2)(i_k+1)\right)^n = \left(\sum\nolimits_{j=1}^{m} (j+1)j\right)^n$

$$= \left(\sum\nolimits_{j=1}^{m} (j^2 + j)\right)^n$$

$$= \left(\frac{m(m+1)(2m+1)}{6} + \frac{3m(m+1)}{6}\right)^n$$

$$= \left(\frac{m(m+1)(2m+4)}{6}\right)^n$$

$$= \left(\frac{m(m+1)(m+2)}{3}\right)^n.$$

Now we have

$$c_3+\left(\frac{m(m+1)}{2}\right)^n-3m^n+3+1+\frac{m^n(m+1)^n}{2^n}-2m^n$$

$$=m^n+\frac{1}{2^n}\left(\frac{m(m+1)(m+2)}{3}\right)^n-\frac{2m^n(m+1)^n}{2^n}$$

That is, $c_3+2\left(\frac{m(m+1)}{2}\right)^n-5m^n+4$

$$=m^n+\frac{1}{2^n}\left(\frac{m(m+1)(m+2)}{3}\right)^n-\frac{2m^n(m+1)^n}{2^n}.$$

Hence,

$$c_3=-4+6m^n-\frac{4m^n(m+1)^n}{2^n}+\frac{m^n(m+1)^n(m+2)^n}{6^n}.$$

Proposition 4.3.7 The NCs of length three in the neutrosophic power set on X whose neutrosophic values lies in $\mathcal{M}$ is

$$C_N(n,m,3)=\frac{m^n(m+1)^n(m+2)^n}{6^n}-\frac{2m^n(m+1)^n}{2^n}+m^n$$

$$\text{i.e., } C_N(n,m,3)=\binom{m+2}{3}^n-2\binom{m+1}{2}^n+m^n.$$

Proof: Let $|X|=n$ and $|\mathcal{M}|=m$, then $|P_{\mathcal{M}}(X)|=m^n$. To prove the result we have the following cases:

Case 1 Chain of length one with both 0^{NT} and 1^{NT} as subset and supersubset: Every neutrosophic proper subsets of length one together with 0^{NT} and 1^{NT} as subset and super-subset forms a chain of length three. Then, the NCs of length three in the present case is m^n-2.

Case 2 Chain of length two with 0^{NT} as subset: Chain of length two with 0^{NT} as subset forms a chain of length three. Therefore, the NCs of length three in this case is

$$\binom{m+1}{2}^n - 3m^n + 3.$$

Case 3 Chain of length two with 1^{NT} as superset: Similar to case 2, the chain of length two with 1^{NT} as super-subset forms a chain of length three. Therefore, the NCs of length three in this case is

$$\binom{m+1}{2}^n - 3m^n + 3.$$

Case 4 Chain of length three from $\widehat{P}_{\mathcal{M}}(X)$: Following Proposition 4.3.6, the NCs of length three in $\widehat{P}_{\mathcal{M}}(X)$ is $\dfrac{m^n(m+1)^n(m+2)^n}{6^n} - \dfrac{4m^n(m+1)^n}{2^n} + 6m^n - 4.$

Hence, the total NCs of length three is

$$C_N(n,m,3) = m^n - 2 + \binom{m+1}{2}^n - 3m^n + 3 + \binom{m+1}{2}^n - 3m^n + 3$$
$$+ \binom{m+2}{3}^n - 4\binom{m+1}{2}^n + 6m^n - 4.$$

That is, $C_N(n,m,3) = m^n - 2\binom{m+1}{2}^n + \binom{m+2}{3}^n.$

or $$C_N(n,m,3) = \frac{m^n(m+1)^n(m+2)^n}{6^n} - \frac{2m^n(m+1)^n}{2^n} + m^n.$$

Example 4.3.5 If we consider example 4.3.2, then chains of length three are

$0^{NT} \subset A_1^{NT} \subset 1^{NT}$, $0^{NT} \subset A_2^{NT} \subset 1^{NT}$, $0^{NT} \subset A_3^{NT} \subset 1^{NT}$, $0^{NT} \subset A_4^{NT} \subset 1^{NT}$,

$0^{NT} \subset A_5^{NT} \subset 1^{NT}$, $0^{NT} \subset A_6^{NT} \subset 1^{NT}$, $0^{NT} \subset A_7^{NT} \subset 1^{NT}$, $0^{NT} \subset A_1^{NT} \subset A_2^{NT}$,

$0^{NT} \subset A_1^{NT} \subset A_4^{NT}$, $0^{NT} \subset A_1^{NT} \subset A_5^{NT}$, $0^{NT} \subset A_1^{NT} \subset A_7^{NT}$, $0^{NT} \subset A_2^{NT} \subset A_5^{NT}$,

$0^{NT} \subset A_3^{NT} \subset A_4^{NT}$, $0^{NT} \subset A_3^{NT} \subset A_5^{NT}$, $0^{NT} \subset A_3^{NT} \subset A_6^{NT}$, $0^{NT} \subset A_3^{NT} \subset A_7^{NT}$,

$0^{NT} \subset A_4^{NT} \subset A_5^{NT}$, $0^{NT} \subset A_4^{NT} \subset A_7^{NT}$, $0^{NT} \subset A_6^{NT} \subset A_7^{NT}$, $A_1^{NT} \subset A_2^{NT} \subset 1^{NT}$,

$A_1^{NT} \subset A_4^{NT} \subset 1^{NT}$, $A_1^{NT} \subset A_5^{NT} \subset 1^{NT}$, $A_1^{NT} \subset A_7^{NT} \subset 1^{NT}$, $A_2^{NT} \subset A_5^{NT} \subset 1^{NT}$,

$A_3^{NT} \subset A_4^{NT} \subset 1^{NT}$, $A_3^{NT} \subset A_5^{NT} \subset 1^{NT}$, $A_3^{NT} \subset A_6^{NT} \subset 1^{NT}$, $A_3^{NT} \subset A_7^{NT} \subset 1^{NT}$,

$A_4^{NT} \subset A_5^{NT} \subset 1^{NT}$, $A_4^{NT} \subset A_7^{NT} \subset 1^{NT}$, $A_6^{NT} \subset A_7^{NT} \subset 1^{NT}$, $A_1^{NT} \subset A_2^{NT} \subset A_5^{NT}$,

$A_1^{NT} \subset A_4^{NT} \subset A_5^{NT}$, $A_1^{NT} \subset A_4^{NT} \subset A_7^{NT}$, $A_3^{NT} \subset A_4^{NT} \subset A_5^{NT}$,

$A_3^{NT} \subset A_4^{NT} \subset A_7^{NT}$, $A_3^{NT} \subset A_6^{NT} \subset A_7^{NT}$.

Lemma 4.3.1 In $P_{\mathcal{M}}(X)$, the number of antichains of size two (having two elements) with 1^{NT} as union and 0^{NT} as intersection is $2^{n-1}-1$.

Proof: Let $(t_{i_1}, t_{i_2}, \ldots, t_{i_n})$ and $(t_{j_1}, t_{j_2}, \ldots, t_{j_n})$ form such an antichain. These two NSubs are different from 0^{NT} and 1^{NT} and satisfy $t_{i_k} \cap t_{j_k} = (0,1,1)$ and $t_{i_k} \cup t_{j_k} = (1,0,0)$ for each $1 \le k \le n$.

That is, $t_{i_k} = (0,1,1)$ if and only if $t_{j_k} = (1,0,0)$, and $t_{i_k} = (1,0,0)$ if and only if $t_{j_k} = (0,1,1)$. Thus, $(t_{j_1}, t_{j_2}, \ldots, t_{j_n})$ is automatically determined by $(t_{i_1}, t_{i_2}, \ldots, t_{i_n})$. There are exactly $\binom{n}{1}$ such $(t_{i_1}, t_{i_2}, \ldots, t_{i_n})$s containing exactly one $(0,1,1)$, $\binom{n}{2}$ such $(t_{i_1}, t_{i_2}, \ldots, t_{i_n})$s that contain exactly two $(0,1,1)$ s, . . ., and $\binom{n}{n-1}$ such $(t_{i_1}, t_{i_2}, \ldots, t_{i_n})$s containing exactly $n-1$ $(0,1,1)$s. That is, in total we have

$$\binom{n}{1} + \binom{n}{2} + \ldots + \binom{n}{n-1} = 2^n - 2 = 2\left(2^{n-1} - 1\right)$$

different such $(t_{i_1}, t_{i_2}, \ldots, t_{i_n})$s. Since each pair of $(t_{i_1}, t_{i_2}, \ldots, t_{i_n})$ and corresponding $(t_{j_1}, t_{j_2}, \ldots, t_{j_n})$ is repeated twice by this process, we have exactly $2^{n-1}-1$ different antichains.

If n and m are positive integers such that $n(m-1)+1 \le 2$, or equivalently $n(m-1) \le 1$, and since we assumed that $n \ge 1$ and $m \ge 2$, we obtain $n = 1$ and $m = 2$. It is easy to check that there exist no antichains with size two in this case, which is consistent with the obtained formula. This completes the proof of the proposition.

Proposition 4.3.8 For a finite set X whose neutrosophic values lie in $\mathcal{M}$ and $|\mathcal{M}| \geq 2$,

$$|P(X)| \leq |P_{\mathcal{M}}(X)|.$$

Proof: Let $|X| = n$ and $|\mathcal{M}| = m, m \geq 2$. Then $|P(X)| = 2^n$ and $|P_{\mathcal{M}}(X)| = m^n, m \geq 2$. This clearly shows that $|P(X)| \leq |P_{\mathcal{M}}(X)|$.

4.4 CARDINALITIES OF THE NEUTROSOPHIC CRISP SET

Definition 4.4.1 The set of all neutrosophic crisp subsets of a non-empty finite set X is called the neutrosophic crisp power set of X. The notation for the neutrosophic crisp power set of X is $P_{NCr}(X)$, and its cardinality is denoted by $|P_{NCr}(X)|$.

Proposition 4.4.1 A set X with $|X| = n$ has

$$(3.2^n - 4) + 3!\left\{\sum_{i=2}^{n} S(i,2)\binom{n}{i} + \sum_{j=3}^{n} S(j,3)\binom{n}{j}\right\}$$

neutrosophic crisp subsets.

Proof: Let $|X| = n$, then $|P(X)| = 2^n = \sum_{i=0}^{n}\binom{n}{i}$. If $A = \langle A_1, A_2, A_3\rangle$ where A_1, A_2, A_3 are subsets of X such that $A_1 \cap A_2 = A_1 \cap A_3 = A_2 \cap A_3 = \phi$. Then, A is a neutrosophic crisp subset of X. Trivially, ϕ_N and X_N are always in the power set of the neutrosophic crisp subset of X as they are the smallest and the largest subsets. Since, A has three components, A_1, A_2, and A_3, which are chosen in the following three ways.

First, choose two components A as ϕ and another one by any neutrosophic crisp proper subset of X, say A_1. Then A_1 is chosen in $\binom{n}{i}$, $1 \leq i \leq n-1$ ways. Therefore, A_1 is chosen in $\sum_{i=1}^{n-1}\binom{n}{i}$ different ways. We can place A_1 in any of the three places in $3\sum_{i=1}^{n-1}\binom{n}{i} = 3(2^n - 2)$ different ways.

Second, choose one component of A as ϕ and the other two by two neutrosophic crisp proper subsets of X, say A_1 and A_2, such that $A_1 \cap A_2 = \phi$. We can place $A_k, k = 1,2$ in any of the two places in six different ways. For a particular set of A_k, we have $\binom{n}{i}S(i,2)$, $2 \leq i \leq n$ different ways. Therefore, the total number of ways to choose A is $6\sum_{i=2}^{n}\binom{n}{i}S(i,2)$.

Third, choose each of the three components of A as neutrosophic crisp proper subsets of X, say A_1, A_2 and A_3, such that $A_1 \cap A_2 = A_1 \cap A_3 = A_2 \cap A_3 = \phi$. We can place A_k, $k = 1,2,3$ in any of the three places in six different ways. For a particular set of A_k, we have $\binom{n}{j} S(j,3)$, $3 \le j \le n$ different ways. Therefore, the total number of ways to choose A is $6\sum_{j=3}^{n}\binom{n}{j} S(j,3)$.

Hence, the total number of neutrosophic crisp subset on X is

$$2+3\left(2^n-2\right)+3!\left\{\sum_{i=2}^{n} S(i,2)\binom{n}{i}+\sum_{j=3}^{n} S(j,3)\binom{n}{j}\right\}.$$

That is, $(3.2^n-4)+3!\left\{\sum_{i=2}^{n} S(i,2)\binom{n}{i}+\sum_{j=3}^{n} S(j,3)\binom{n}{j}\right\}$.

Proposition 4.4.2 If $|X| = n$, then the cardinality of the power set of NCrS on X is

$$|P_{NCr}(X)| = (3.2^n-4)+3!\left\{\sum_{i=2}^{n} S(i,2)\binom{n}{i}+\sum_{j=3}^{n} S(j,3)\binom{n}{j}\right\}.$$

Proof: By Definition 4.4.1 and Proposition 4.4.1, we can obtain the cardinality of the power set of NCrS on X, which is $|P_{NCr}(X)| = (3.2^n-4)+3!\left\{\sum_{i=2}^{n} S(i,2)\binom{n}{i}+\sum_{j=3}^{n} S(j,3)\binom{n}{j}\right\}$.

Example 4.4.1 Let $X = \{u\}$; then $|X| = n = 1$ the neutrosophic crisp subsets on X are

$$\phi_N = \langle \phi,\phi,X\rangle,\ X_N = \langle X,\phi,\phi\rangle.$$

Therefore,

$$|P_{NCr}(X)| = 2 = (3.2^1-4)+3!\left\{\sum_{i=2}^{1} S(i,2)\binom{1}{i}+\sum_{j=3}^{1} S(j,3)\binom{1}{j}\right\}.$$

Example 4.4.2 Let $X = \{u,v\}$; then $|X| = n = 2$ the neutrosophic crisp subsets on X are

$$\phi_N = \langle \phi,\phi,X\rangle,\ X_N = \langle X,\phi,\phi\rangle,\ A_1 = \langle \phi,\phi,\{u\}\rangle,\ A_2 = \langle \phi,\{u\},\phi\rangle,$$

$$A_3 = \langle \{u\},\phi,\phi\rangle,\ A_4 = \langle \phi,\phi,\{v\}\rangle,\ A_5 = \langle \phi,\{v\},\phi\rangle,\ A_6 = \langle \{v\},\phi,\phi\rangle,$$

$$A_7 = \langle \phi, \{u\}, \{v\} \rangle,\ A_8 = \langle \{u\}, \phi, \{v\} \rangle,\ A_9 = \langle \{u\}, \{v\}, \phi \rangle,$$

$$A_{10} = \langle \phi, \{v\}, \{u\} \rangle,\ A_{11} = \langle \{v\}, \phi, \{u\} \rangle,\ A_{12} = \langle \{v\}, \{u\}, \phi \rangle.$$

Therefore, $|P_{NCr}(X)| = 14 = (3.2^2 - 4) + 3!\left\{\sum_{i=2}^{2} S(i,2)\binom{2}{i} + \sum_{j=3}^{2} S(j,3)\binom{2}{j}\right\}$.

Proposition 4.4.3 For a non-empty finite set X, $|P(X)| \le |P_{NCr}(X)|$.

Proof: Let $|X| = n$. Then $|P(X)| = 2^n$ and

$$|P_{NCr}(X)| = (3.2^n - 4) + 3!\left\{\sum_{i=2}^{n} S(i,2)\binom{n}{i} + \sum_{j=3}^{n} S(j,3)\binom{n}{j}\right\}.$$

Now, let $T = 3!\left\{\sum_{i=2}^{n} S(i,2)\binom{n}{i} + \sum_{j=3}^{n} S(j,3)\binom{n}{j}\right\}$ and clearly $T \ge 0$ for $n \ge 1$.

Then,
$$\begin{aligned} |P_{NCr}(X)| &= (3.2^n - 4) + T \\ &= 2^n + 4(2^{n-1} - 1) + T \\ &\ge 2^n \text{ as } 4(2^{n-1} - 1) \ge 0 \text{ and } T \ge 0 \text{ for } n \ge 1 \\ &\ge |P(X)|. \end{aligned}$$

Hence, $|P(X)| \le |P_{NCr}(X)|$.

4.5 CONCLUSION

In this chapter, we examined formulae for the cardinalities of $P_{\mathcal{M}}(X)$ and $P_{NCr}(X)$ in a finite set X and discussed some results related to the number of chains in $P_{\mathcal{M}}(X)$.

REFERENCES

[1] Smarandache, F. (2005). Neutrosophic set, a generalization of the generalized fuzzy sets. *International Journal of Pure and Applied Mathematics*. 24: 287–297.

[2] Atanassov, K. (1986). Intuitionistic fuzzy sets. *Fuzzy Sets and Systems*. 20(1): 87–96.

[3] Salama, A. A., and Smarandache F. (2015). *Neutrosophic crisp set theory*. Educational Publisher, Columbus.

[4] Salama, A. A. (2013). Neutrosophic crisp points and neutrosophic crisp ideals. *Neutrosophic Sets and Systems*. 1(1): 50–53.

[5] Hanafy, I., Salama, A., and Mahfouz, K. (2013). Neutrosophic classical events and its probability. *International Journal of Mathematics and Computer Applications Research*. 3(1): 171–178.

[6] Smarandache, F. (2016). *Neutrosophic overset, neutrosophic underset, and neutrosophic offset. Similarly for neutrosophic over-/under-/off-logic, probability, and statistics*. Pons Editions, Brussels, p. 171. https://digitalrepository.unm.edu/math_fsp/26
[7] Cantor, G. (1984). *Ueber unendliche, lineare Punktmannichfaltigkeiten*. Springer, Vienna, pp. 45–156.
[8] Yager, R. R. (1987). Cardinality of fuzzy sets via bags. *Mathematical Modelling*. 9(6): 441–446.
[9] Yager, R. R. (1993). Counting the number of classes in a fuzzy set. *IEEE Transactions on Systems, Man, and Cybernetics*. 23(1): 257–264.
[10] Yager, R. R. (2006). On the fuzzy cardinality of a fuzzy set. *International Journal of General Systems*. 35(2): 191–206.
[11] Mohapatra, R. K., and Hong, T. P. (2022). On the number of finite fuzzy subsets with analysis of integer sequences. *Mathematics*. 10(7): 1–19.
[12] Murali, V. (2005). Equivalent finite fuzzy sets and stirling numbers. *Information Sciences*. 174(3–4): 251–263.
[13] Murali, V. (2006). Combinatorics of counting finite fuzzy subsets. *Fuzzy Sets and Systems*. 157(17): 2403–2411.
[14] Chamorro-Martínez, J., Sánchez, D., Soto-Hidalgo, J. M., and Martínez-Jiménez, P. M. (2014). A discussion on fuzzy cardinality and quantification. Some applications in image processing. *Fuzzy Sets and Systems*. 257: 85–101.
[15] Benoumhani, M., and Jaballah, A. (2019). Chains in lattices of mappings and finite fuzzy topological spaces. *Journal of Combinatorial Theory. Series A*. 161: 99–111.
[16] Salama, A., and Alblowi, S. (2012). Neutrosophic set and neutrosophic topological spaces. *IOSR Journal of Mathematics*. 3(4): 31–35.
[17] Benoumhani, M., and Jaballah, A. (2017). Finite fuzzy topological spaces. *Fuzzy Sets and Systems*. 321: 101–114.
[18] Benoumhani, M. (2006). The number of topologies on a finite set. *Journal of Integer Sequences*. 9(2): 1–9.
[19] Cameron, P. J. (1994). *Combinatorics: Topics, Techniques, Algorithms*. Cambridge University Press, Cambridge.

Chapter 5

The Systems of Neutrosophic Cylindrical Coordinates

Prasen Boro, Bhimraj Basumatary, and Said Broumi

5.1 INTRODUCTION

Smarandache [1, 2] put forward the neutrosophic concept to represent a mathematical model of unclear determination, uncertainty, vagueness, unclearness, incompleteness, inconsistency, and redundancy. The neutrosophic measure generalizes the classical measure for cases of indeterminacy. And since the world is full of indeterminacy, neutrosophic sets found their place in contemporary research.

Smarandache [1] also defined complex neutrosophic numbers in standard form and found the root index $n \geq 2$ of real and complex neutrosophic numbers [2, 3], studying the concepts of neutrosophic probability [4, 5] and neutrosophic statistics [3, 6]. Smarandache [7] then incorporated the concepts of differential and integral calculus and introduced the notions of neutrosophic mereo-limit, mereo-continuity, mereo-derivative, and mereo-integral [1, 7]. Madeleine [8] presented results on single-valued neutrosophic (weak) polygroups. Edalatpanah [9] proposed a new direct algorithm to solve neutrosophic linear programming where the variables and right-hand side are represented with triangular neutrosophic numbers. Chakraborty [10, 11] used pentagonal neutrosophic numbers in networking problems and shortest-path problems.

Generally, Cartesian coordinates can also be used to determine a particle in space, but sometimes, we have to analyze problems involving cylindrical shapes like heat transfer through pipes and electrical wires or water flow through pipes, etc. Neutrosophic sets are important in accounting for the uncertainty of such cylindrical coordinates.

Here, we are going to study the gradient of a cylindrical particle in neutrosophic form and find its derivation in neutrosophic form. After the derivation of the gradient, we shall study the divergent and curl in neutrosophic form and derive some results and propositions from them. After these studies, we shall discuss the Laplacian operator and derive the operator in the neutrosophic form.

5.2 PRELIMINARIES

This section contains some basic definitions and results of neutrosophic real numbers.

DOI: 10.1201/9781003487104-5

5.2.1 Neutrosophic Real Numbers [3]

Let w be a neutrosophic real number; then its standard form will be as follows:

$w = a + bI$ where a, b are real coefficients, and I represents indeterminacy, such that $0.I = 0$ and $I^n = I$ for all positive integers n.

5.2.2 Division of Neutrosophic Real Numbers [3]

Let us consider that w_1, w_2 are two neutrosophic real numbers, where

$$w_1 = a_1 + b_1 I, w_2 = a_2 + b_2 I.$$

Now we determine $(a_1 + b_1 I) \div (a_2 + b_2 I)$ considering the following equation:

$$\frac{a_1 + b_1 I}{a_2 + b_2 I} \equiv x + yI$$

where x and y are real unknowns.

$$a_1 + b_1 I \equiv (a_2 + b_2 I)(x + yI)$$

Or $a_1 + b_1 I \equiv a_2 x + (b_2 x + a_2 y + b_2 y) I$

By identifying the coefficients, we get

$$a_1 = a_2 x$$

and $b_1 = b_2 x + (a_2 + b_2) y$

We obtain one unique solution provided that

$$\begin{vmatrix} a_2 & 0 \\ b_2 & a_2 + b_2 \end{vmatrix} \neq 0 \Rightarrow a_2 (a_2 + b_2) \neq 0.$$

Hence, $a_2 \neq 0$ and $a_2 \neq -b_2$ are the conditions so that the division of two neutrosophic real numbers exists.

Therefore, we obtain the following result:

$$\frac{a_1 + b_1 I}{a_2 + b_2 I} = \frac{a_1}{a_2} + \frac{a_2 b_1 - a_1 b_2)}{a_2 (a_2 + b_2)} .I.$$

5.3 NEUTROSOPHIC CYLINDRICAL COORDINATES

Neutrosophic cylindrical coordinates (NCCs) are ordered triplets that describe the location of a point; they are the coordinates of a point in 3D space in the neutrosophic form. These coordinates combine the neutrosophic Z coordinate of the Cartesian system with the neutrosophic polar coordinates in the XY-plane. The neutrosophic radial distance, neutrosophic azimuthal angle, and neutrosophic height from a plane to a point are denoted using NCCs. The NCC system is useful in analyzing systems that exhibit rotational symmetry. These coordinates describe two neutrosophic distances and one neutrosophic angle.

In classical form, the cylindrical coordinates of a point can be expressed as (ρ, φ, z). In the neutrosophic form, they can be expressed as $(\rho + \rho_0 I, \varphi + \varphi_0 I, z + z_0 I)$, where I indicates indeterminacy such that $I^n = I, n \in N$, and $I.0 = 0$ and ρ_0, φ_0, z_0 are some scalars (or real numbers).

Here, $\rho + \rho_0 I$ is the neutrosophic radial distance from the z-axis to point A. $\varphi + \varphi_0 I$ is the neutrosophic azimuthal angle between the x-axis and the line segment that is drawn from the origin to B, where B is the projection of the point A in the XY-plane. The azimuthal angle is measured in radians. $z + z_0 I$ represents the third NCC; it denotes the signed distance of A to the XY-plane. This description is shown in Figure 5.1.

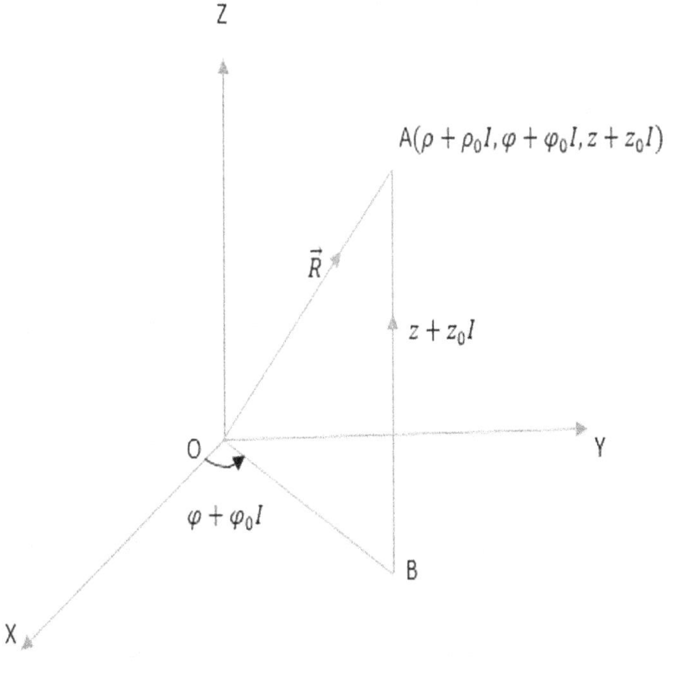

Generally in classical form, the distance between any two points is considered a certain value, but in actuality, this distance cannot be certain; it is more likely to be uncertain. Similarly, velocity is the rate of change of distance with respect to time, so the velocity of a particle between any two points also cannot be certain. Therefore, to minimize the uncertainty in the velocity of a particle, a neutrosophic study in velocity is needed. We discuss some preliminary ideas on neutrosophic velocity and neutrosophic acceleration in Section 5.7.

5.4 CONVERSIONS

There is a relationship between NCCs and neutrosophic Cartesian coordinates. NCCs can be converted to neutrosophic Cartesian coordinates and vice versa.

5.4.1 Conversion of NCCs to Neutrosophic Cartesian Coordinates

Converting NCCs $(\rho+\rho_0 I, \varphi+\varphi_0 I, z+z_0 I)$ to neutrosophic Cartesian coordinates $(x+x_0 I, y+y_0 I, z+z_0 I)$ uses the following equations:

$$x+x_0 I = (\rho+\rho_0 I)cos(\varphi+\varphi_0 I)$$

or $\quad X = Pcos\Phi$

$$y+y_0 I = (\rho+\rho_0 I)sin(\varphi+\varphi_0 I)$$

or $\quad Y = Psin\Phi$

and $\quad z+z_0 I = z+z_0 I$

or $\quad Z = Z.$

Here, $\quad P = \rho+\rho_0 I,\ \Phi = \varphi+\varphi_0 I,\ Z = z+z_0 I$

$$X = x+x_0 I, Y = y+y_0 I\ Z = z+z_0 I.$$

Example 5.1 The NCC $\left(4+2I, \frac{2\pi}{3}, 2+I\right)$ is converted to neutrosophic Cartesian coordinates as follows:

$$P = 4+2I$$

$$\Phi = \frac{2\pi}{3}$$

$$Z = 2+I.$$

Therefore, $X = Pcos\Phi$

$$= (4+2I)cos\frac{2\pi}{3}$$

$$= (4+2I)\left(-\frac{1}{2}\right)$$

$$= -2 - I$$

$$Y = Psin\Phi$$

$$= (4+2I)sin\frac{2\pi}{3}$$

$$= (4+2I)\left(\frac{\sqrt{3}}{2}\right)$$

$$= 2\sqrt{3} + \sqrt{3}I$$

and $Z = Z$

$$= 2 + I.$$

Therefore, the required neutrosophic Cartesian coordinates are $\left(-2-I, 2\sqrt{3}+\sqrt{3}I, 2+I\right)$.

5.4.2 Conversion of Neutrosophic Cartesian Coordinates to NCCs

The equations for converting the neutrosophic Cartesian coordinates to NCCs are as follows:

$$P^2 = X^2 + Y^2$$

$$tan\Phi = \frac{X}{Y}$$

$$Z = Z,$$

where $P = \left(\rho + \rho_0 I\right)$, $\Phi = \varphi + \varphi_0 I$, $Z = z + z_0 I$

$$X = x + x_0 I, Y = y + y_0 I, Z = z + z_0 I.$$

Example 5.2 The neutrosophic Cartesian coordinates $(1+I, 3+2I, 5+I)$ are expressed as NCCs:

$$X = 1+I$$
$$Y = 3+2I$$
$$Z = 5+I.$$

Therefore, $P = \sqrt{X^2 + Y^2}$

$$= \sqrt{(1+I)^2 + (3+2I)^2}$$
$$= \sqrt{1+2I+I+9+12I+4I}$$
$$= \sqrt{10+19I}$$

$$tan\Phi = \frac{Y}{X}$$
$$= \frac{3+2I}{1+I}$$
$$= \frac{3}{1} + \frac{1.2-3.1}{1(1+1)}.I$$
$$= \left(3 - \frac{I}{2}\right)$$
$$= \frac{6-I}{2}.$$

Therefore, $\Phi = \tan^{-1}\left(\frac{6-I}{2}\right)$

and $Z = Z = 5+I.$

5.5 UNIT VECTORS IN AN NCC SYSTEM

The position vector of any point A in NCCs is given by

$$\vec{R} = X\hat{i} + Y\hat{j} + Z\hat{k}$$
$$= Pcos\Phi\hat{i} + Psin\Phi\hat{j} + Z\hat{k}$$

where $P=\left(\rho+\rho_0 I\right)$, $\Phi=\varphi+\varphi_0 I$, $Z=z+z_0 I$

$$X=x+x_0 I,\ Y=y+y_0 I,\ Z=z+z_0 I$$

The tangent vectors in the directions of $P, \Phi\ and\ Z$ are given by

$$\frac{\partial \vec{R}}{\partial P}=cos\Phi\,\hat{i}+sin\Phi\,\hat{j}$$

$$\frac{\partial \vec{R}}{\partial \Phi}=-Psin\Phi\,\hat{i}+Pcos\Phi\,\hat{j}$$

$$\frac{\partial \vec{R}}{\partial Z}=\hat{k}$$

∴ The unit vectors in these directions are given by

$$\hat{e}_P=\frac{\frac{\partial \vec{R}}{\partial P}}{\left|\frac{\partial \vec{R}}{\partial P}\right|}=\frac{cos\Phi\,\hat{i}+sin\Phi\,\hat{j}}{\sqrt{\cos^2\Phi+\sin^2\Phi}}=cos\Phi\,\hat{i}+sin\Phi\,\hat{j} \tag{5.1}$$

$$\hat{e}_\Phi=\frac{\frac{\partial \vec{R}}{\partial \Phi}}{\left|\frac{\partial \vec{R}}{\partial \Phi}\right|}=\frac{-Psin\Phi\,\hat{i}+Pcos\Phi\,\hat{j}}{\sqrt{\mathrm{P}^2\sin^2\Phi+\mathrm{P}^2\cos^2\Phi}}=-sin\Phi\,\hat{i}+cos\Phi\,\hat{j} \tag{5.2}$$

$$\hat{e}_Z=\frac{\frac{\partial \vec{R}}{\partial Z}}{\left|\frac{\partial \vec{R}}{\partial Z}\right|}=\frac{\hat{k}}{\sqrt{1^2}}=\hat{k} \tag{5.3}$$

In matrix form, (5.1), (5.2), and (5.3) are written as

$$\begin{bmatrix}\hat{e}_P\\ \hat{e}_\Phi\\ \hat{e}_Z\end{bmatrix}=\begin{bmatrix}\cos\Phi & \sin\Phi & 0\\ -\sin\Phi & \cos\Phi & 0\\ 0 & 0 & 1\end{bmatrix}\begin{bmatrix}\hat{\mathrm{i}}\\ \hat{\mathrm{j}}\\ \hat{\mathrm{k}}\end{bmatrix}.$$

Again, the unit vector along the direction of $\vec{R}$ is given by

$$\hat{e}_R = \frac{\vec{R}}{|\vec{R}|}$$

$$= \frac{Pcos\Phi\hat{i} + Psin\Phi\hat{j} + Z\hat{k}}{\sqrt{P^2\cos^2\Phi + P^2\sin^2\Phi + Z^2}}$$

$$= \frac{Pcos\Phi\hat{i} + Psin\Phi\hat{j} + Z\hat{k}}{\sqrt{P^2 + Z^2}}.$$

5.6 DOT PRODUCTS AND CROSS PRODUCTS OF NEUTROSOPHIC UNIT VECTORS OF NCCs

The neutrosophic unit vectors of NCCs along the direction of neutrosophic coordinates $P, \Phi, and\, Z$ are given by

$$\hat{e}_P = cos\Phi\hat{i} + sin\Phi\hat{j}$$

$$\hat{e}_\Phi = -sin\Phi\hat{i} + cos\Phi\hat{j}$$

and $\hat{e}_Z = \hat{k}$

Now, the dot products of these unit vectors are

(a) $\hat{e}_P.\hat{e}_P = \left(cos\Phi\hat{i} + sin\Phi\hat{j}\right).\left(cos\Phi\hat{i} + sin\Phi\hat{j}\right)$

$$= \cos^2\Phi\left(\hat{i}.\hat{i}\right) + cos\Phi.sin\Phi\left(\hat{i}.\hat{j}\right) + sin\Phi.cos\Phi\left(\hat{j}.\hat{i}\right) + \sin^2\Phi\left(\hat{j}.\hat{j}\right)$$

$$= \cos^2\Phi + 0 + 0 + \sin^2\Phi$$

$$= 1$$

(b) $\hat{e}_\Phi.\hat{e}_\Phi = \left(-sin\Phi\hat{i} + cos\Phi\hat{j}\right)\left(-sin\Phi\hat{i} + cos\Phi\hat{j}\right)$

$$= \sin^2\Phi + \cos^2\Phi$$

$$= 1$$

(c) $\hat{e}_Z.\hat{e}_Z = \hat{k}.\hat{k} = 1$

(d) $\hat{e}_P.\hat{e}_\Phi = \left(cos\Phi\hat{i} + sin\Phi\hat{j}\right).\left(-sin\Phi\hat{i} + cos\Phi\hat{j}\right)$

$$= -cos\Phi\, sin\Phi + sin\Phi\, cos\Phi$$

$$= 0$$

(e) $\hat{e}_P.\hat{e}_Z = \left(cos\Phi\hat{i} + sin\Phi\hat{j}\right).\hat{k} = 0 + 0 = 0.$

Similarly, we get the following results:

$$\hat{e}_{\Phi}\cdot\hat{e}_Z = 0,\ \hat{e}_{\Phi}\cdot\hat{e}_P = 0,\ \hat{e}_Z\cdot\hat{e}_P = 0, \hat{e}_Z\cdot\hat{e}_{\Phi} = 0.$$

Again, the cross products of the neutrosophic unit vectors are

(f) $\hat{e}_P \times \hat{e}_P = \left(cos\Phi\hat{i} + sin\Phi\hat{j}\right)\times\left(cos\Phi\hat{i} + sin\Phi\hat{j}\right)$

$$= \cos^2\Phi\left(\hat{i}\times\hat{i}\right) + cos\Phi.sin\Phi\left(\hat{i}\times\hat{j}\right) + sin\Phi.cos\Phi\left(\hat{j}\times\hat{i}\right)$$

$$+\sin^2\Phi\left(\hat{j}\times\hat{j}\right)$$

$$= \cos^2\Phi\left(\vec{0}\right) + cos\Phi.sin\Phi\left(\hat{k}\right) + sin\Phi.cos\Phi\left(-\hat{k}\right) + \sin^2\Phi\left(\hat{j}\times\hat{j}\right)$$

$$= \vec{0} + cos\Phi.sin\Phi\hat{k} - sin\Phi.cos\Phi\hat{k} + \vec{0}$$

$$= \vec{0}$$

(g) $\hat{e}_{\Phi} \times \hat{e}_{\Phi} = \left(-sin\Phi\hat{i} + cos\Phi\hat{j}\right)\times\left(-sin\Phi\hat{i} + cos\Phi\hat{j}\right)$

$$= -\sin\Phi cos\Phi\left(\hat{i}\times\hat{j}\right) - \cos\Phi sin\Phi\left(\hat{j}\times\hat{i}\right)$$

$$= -\sin\Phi cos\Phi\hat{k} - \cos\Phi sin\Phi$$

$$= -\sin\Phi cos\Phi\hat{k} + \cos\Phi sin\Phi\hat{k}$$

$$= \vec{0}$$

(h) $\hat{e}_{\Phi} \times \hat{e}_{\Phi} = \hat{k}.\hat{k} = \vec{0}$

(i) $\hat{e}_P \times \hat{e}_{\Phi} = \left(cos\Phi\hat{i} + sin\Phi\hat{j}\right)\times\left(-sin\Phi\hat{i} + cos\Phi\hat{j}\right)$

$$= \cos^2\Phi\left(\hat{i}\times\hat{j}\right) - \sin^2\Phi\left(\hat{j}\times\hat{i}\right)$$

$$= \cos^2\Phi\left(\hat{k}\right) - \sin^2\Phi\left(-\hat{k}\right)$$

$$= \cos^2\Phi\left(\hat{k}\right) + \sin^2\Phi\left(\hat{k}\right)$$

$$= \hat{k} = \hat{e}_Z$$

(j) $\hat{e}_{\Phi} \times \hat{e}_P = \left(-sin\Phi\hat{i} + cos\Phi\hat{j}\right)\times\left(cos\Phi\hat{i} + sin\Phi\hat{j}\right)$

$$= -\sin^2\Phi\left(\hat{i}\times\hat{j}\right) + \cos^2\Phi\left(\hat{j}\times\hat{i}\right)$$

$$= -\sin^2\Phi\left(\hat{k}\right) - \cos^2\Phi\left(\hat{k}\right)$$

$$= -\hat{k} = -\hat{e}_Z$$

Similarly, we get the following results:

$$\hat{e}_P \times \hat{e}_Z = -\hat{e}_\Phi,\ \hat{e}_Z \times \hat{e}_P = -\hat{e}_\Phi,\ \hat{e}_\Phi \times \hat{e}_Z = \hat{e}_P,\ \hat{e}_Z \times \hat{e}_\Phi = -\hat{e}_P$$

Proposition 5.1: We obtain the following results for the NCCs:

$$\frac{\partial \hat{e}_P}{\partial P} = \vec{0} \quad \frac{\partial \hat{e}_P}{\partial \Phi} = \hat{e}_\Phi \quad \frac{\partial \hat{e}_P}{\partial Z} = \vec{0}$$

$$\frac{\partial \hat{e}_\Phi}{\partial P} = \vec{0} \quad \frac{\partial \hat{e}_\Phi}{\partial \Phi} = -\hat{e}_P \quad \frac{\partial \hat{e}_\Phi}{\partial Z} = \vec{0}$$

$$\frac{\partial \hat{e}_Z}{\partial P} = \vec{0} \quad \frac{\partial \hat{e}_Z}{\partial \Phi} = \vec{0} \quad \frac{\partial \hat{e}_Z}{\partial Z} = \vec{0}$$

where $P = \rho + \rho_0 I,\ \Phi = \varphi + \varphi_0 I,\ Z = z + z_0 I.$

Proof: We know that $\hat{e}_P = cos\Phi\hat{i} + sin\Phi\hat{j},\ \hat{e}_\Phi = -sin\Phi\hat{i} + cos\Phi\hat{j},\ \hat{e}_Z = \hat{k}.$
Therefore,

$$\frac{\partial \hat{e}_P}{\partial P} = \vec{0}, \quad \frac{\partial \hat{e}_P}{\partial \Phi} = -sin\Phi\hat{i} + cos\Phi\hat{j} = \hat{e}_\Phi, \quad \frac{\partial \hat{e}_P}{\partial Z} = \vec{0}$$

$$\frac{\partial \hat{e}_\Phi}{\partial P} = \vec{0}, \quad \frac{\partial \hat{e}_\Phi}{\partial \Phi} = -cos\Phi\hat{i} - sin\Phi\hat{j} = -\hat{e}_P, \quad \frac{\partial \hat{e}_\Phi}{\partial Z} = \vec{0},$$

and

$$\frac{\partial \hat{e}_Z}{\partial P} = \vec{0}, \quad \frac{\partial \hat{e}_Z}{\partial \Phi} = \vec{0}, \quad \frac{\partial \hat{e}_Z}{\partial Z} = \vec{0}.$$

Proposition 5.2: We obtain the following results for the NCCs:

$$\frac{d}{dt}(\hat{e}_P) = \dot{\Phi}\hat{e}_\Phi,\ \frac{d}{dt}(\hat{e}_\Phi) = -\dot{\Phi}\hat{e}_P,\ \frac{d}{dt}(\hat{e}_Z) = \vec{0}$$

where dots denote the differentiation with respect to time t.

We have $\hat{e}_P = cos\Phi\hat{i} + sin\Phi\hat{j},\ \hat{e}_\Phi = -sin\Phi\hat{i} + cos\Phi\hat{j},\ \hat{e}_Z = \hat{k}$, and therefore,

$$\frac{d}{dt}(\hat{e}_P) = \frac{d}{dt}\left(cos\Phi\hat{i} + sin\Phi\hat{j}\right)$$

$$= -sin\Phi\frac{d\Phi}{dt}\hat{i} + cos\Phi\frac{d\Phi}{dt}\hat{j}$$

$$= \left(-sin\Phi\hat{i} + cos\Phi\hat{j}\right)\frac{d\Phi}{dt} = \dot{\Phi}\hat{e}_\Phi$$

$$\frac{d}{dt}(\hat{e}_\Phi) = -cos\Phi\frac{d\Phi}{dt}\hat{i} - sin\Phi\frac{d\Phi}{dt}\hat{j} = -\left(cos\Phi\hat{i} + sin\Phi\hat{j}\right)\frac{d\Phi}{dt} = -\dot{\Phi}\hat{e}_P$$

and

$$\frac{d}{dt}(\hat{e}_Z) = \frac{d}{dt}\left(\hat{k}\right) = \vec{0}.$$

5.7 THE VELOCITY AND ACCELERATION OF A NEUTROSOPHIC CYLINDRICAL PARTICLE

The velocity and acceleration of a neutrosophic cylindrical particle (NCP) whose neutrosophic coordinates are (P, Φ, Z) may be expressed by taking into account the associated rates of change in the neutrsphic unit vectors.

Let us consider the velocity of a neutrosophic particle as v: Then,

$$\vec{v} = \dot{\vec{R}} = \frac{d}{dt}\left(\vec{R}\right)$$

$$= \frac{d}{dt}\left(Pcos\Phi\hat{i} + Psin\Phi\hat{j} + Z\hat{k}\right)$$

$$= \frac{dP}{dt}cos\Phi\hat{i} - Psin\Phi\frac{d\Phi}{dt}\hat{i} + \frac{dP}{dt}sin\Phi\hat{j} + Pcos\Phi\frac{d\Phi}{dt}\hat{j} + \frac{dZ}{dt}\hat{k}$$

$$= \dot{P}cos\Phi\hat{i} - \dot{\Phi}Psin\Phi\hat{i} + \dot{P}sin\Phi\hat{j} + \dot{\Phi}Pcos\Phi\hat{j} + \dot{Z}\hat{k}$$

$$= \dot{P}\left(cos\Phi\hat{i} + sin\Phi\hat{j}\right) + P\dot{\Phi}\left(-sin\Phi\hat{i} + cos\Phi\hat{j}\right) + \dot{Z}\hat{k}$$

$$= \dot{P}\hat{e}_P + P\dot{\Phi}\hat{e}_\Phi + \dot{Z}\hat{e}_Z, \qquad \left[since\,\hat{k} = \hat{e}_Z\right],$$

and the acceleration of the NCP will be given by

$$\begin{aligned}\vec{a} &= \ddot{\vec{R}} \\ &= \dot{\vec{v}} \\ &= \frac{d}{dt}\left(\dot{P}\hat{e}_P + P\dot{\Phi}\hat{e}_\Phi + \dot{Z}\hat{e}_Z\right) \\ &= \ddot{P}\hat{e}_P + \dot{P}\dot{\hat{e}}_P + \dot{P}\dot{\Phi}\hat{e}_\Phi + P\ddot{\Phi}\hat{e}_\Phi + P\dot{\Phi}\dot{\hat{e}}_\Phi + \ddot{Z}\hat{e}_Z + Z\dot{\hat{e}}_Z \\ &= \ddot{P}\hat{e}_P + \dot{P}\dot{\Phi}\hat{e}_\Phi + \dot{P}\dot{\Phi}\hat{e}_\Phi + P\ddot{\Phi}\hat{e}_\Phi + P\dot{\Phi}\dot{\Phi}\hat{e}_P + \ddot{Z}\hat{e}_Z + Z(0) \\ &= \left(\ddot{P} - P\dot{\Phi}^2\right)\hat{e}_P + \left(\ddot{\Phi} + 2\dot{P}\dot{\Phi}\right)\hat{e}_\Phi + \ddot{Z}\hat{e}_Z.\end{aligned}$$

5.8 THE MAGNITUDE OF VELOCITY AND ACCELERATION OF A NEUTROSOPHIC CYLINDRICAL PARTICLE AT ANY TIME T

Based on the equations in Section 5.7, we obtained the expressions of velocity and acceleration for an NCP as

$$\vec{v} = \dot{P}\hat{e}_P + P\dot{\Phi}\hat{e}_\Phi + \dot{Z}\hat{e}_Z$$

$$\vec{a} = \left(\ddot{P} - P\dot{\Phi}^2\right)\hat{e}_P + \left(\ddot{\Phi} + 2\dot{P}\dot{\Phi}\right)\hat{e}_\Phi + \ddot{Z}\hat{e}_Z.$$

Thus, the magnitude of velocity of an NCP at any time t is given by

$$|\vec{v}| = \sqrt{\dot{P}^2 + P^2\dot{\Phi}^2 + \dot{Z}^2},$$

and the magnitude of acceleration of an NCP at any time t is given by

$$|\vec{a}| = \sqrt{\left(\ddot{P} - P\dot{\Phi}^2\right)^2 + \left(\ddot{\Phi} + 2\dot{P}\dot{\Phi}\right)^2 + \ddot{Z}^2}.$$

5.9 THE NEUTROSOPHIC DEL OPERATOR FROM THE DEFINITION OF THE GRADIENT

Let $U = u + u_0 I$ be any neutrosophic scalar field and a function of the NCCs P, Φ and Z, where $P = \rho + \rho_0 I$, $\Phi = \varphi + \varphi_0 I$ and $Z = z + z_0 I$. The value of

U changes by the infinitesimal amount dU when the point of observation is changing by $d\vec{R}$. That change is determined from the partial derivatives as

$$dU = \frac{\partial U}{\partial P} dP + \frac{\partial U}{\partial \Phi} d\Phi + \frac{\partial U}{\partial Z} dZ \qquad (5.4),$$

but the gradient is also defined as

$$dU = \vec{\nabla} U . d\vec{R} \qquad (5.5).$$

From (5.4) and (5.5), we get

$$\frac{\partial U}{\partial P} dP + \frac{\partial U}{\partial \Phi} d\Phi + \frac{\partial U}{\partial Z} dZ = \vec{\nabla} U . d\vec{R}$$

$$\frac{\partial U}{\partial P} dP + \frac{\partial U}{\partial \Phi} d\Phi + \frac{\partial U}{\partial Z} dZ = \left(\vec{\nabla} U\right)_P dP + (\vec{\nabla} U)_\Phi P d\Phi + (\vec{\nabla} U)_Z dZ.$$

We demand this relationship hold for any choice of dP, $d\Phi$ *and* dZ. Thus,

$$\left(\vec{\nabla} U\right)_P = \frac{\partial U}{\partial P}, \ \left(\vec{\nabla} U\right)_\Phi = \frac{1}{P}\frac{\partial U}{\partial \Phi}, \ \left(\vec{\nabla} U\right)_Z = \frac{\partial U}{\partial Z},$$

which gives

$$\vec{\nabla} = \hat{e}_P \frac{\partial}{\partial P} + \frac{\hat{e}_\Phi}{P}\frac{\partial}{\partial \Phi} + \hat{e}_Z \frac{\partial}{\partial Z}.$$

This is the required neutrosophic Del operator.

5.10 NEUTROSOPHIC DIVERGENCE

Neutrosophic divergence $\vec{\nabla}.\vec{A}$ is carried out taking into account that the neutrosophic unit vectors themselves are the functions of the NCC. Thus we have

$$\vec{\nabla}.\vec{A} = \left(\hat{e}_P \frac{\partial}{\partial P} + \frac{\hat{e}_\Phi}{P}\frac{\partial}{\partial \Phi} + \hat{e}_P \frac{\partial}{\partial Z} \right).\left(A_P \hat{e}_P + A_\Phi \hat{e}_\Phi + A_Z \hat{e}_P \right),$$

where the derivatives must be taken before the dot product so that

$$\vec{\nabla}.\vec{A}=\left(\hat{e}_P\frac{\partial}{\partial P}+\frac{\hat{e}_\Phi}{P}\frac{\partial}{\partial \Phi}+\hat{e}_Z\frac{\partial}{\partial Z}\right).\vec{A}$$

$$=\hat{e}_P.\frac{\partial \vec{A}}{\partial P}+\frac{\hat{e}_\Phi}{P}.\frac{\partial \vec{A}}{\partial \Phi}+\hat{e}_Z.\frac{\partial \vec{A}}{\partial Z}$$

$$=\hat{e}_P.\left(\frac{\partial A_P}{\partial P}\hat{e}_P+\frac{\partial A_\Phi}{\partial P}\hat{e}_\Phi+\frac{\partial A_Z}{\partial P}\hat{e}_Z+A_P\frac{\partial \hat{e}_P}{\partial P}+A_\Phi\frac{\partial \hat{e}_\Phi}{\partial P}+A_Z\frac{\partial \hat{e}_Z}{\partial P}\right)$$

$$+\frac{\hat{e}_\Phi}{P}.\left(\frac{\partial A_P}{\partial \Phi}\hat{e}_P+\frac{\partial A_\Phi}{\partial \Phi}\hat{e}_\Phi+\frac{\partial A_Z}{\partial \Phi}\hat{e}_Z+A_P\frac{\partial \hat{e}_P}{\partial \Phi}+A_\Phi\frac{\partial \hat{e}_\Phi}{\partial \Phi}+A_Z\frac{\partial \hat{e}_Z}{\partial \Phi}\right)$$

$$+\hat{e}_Z.\left(\frac{\partial A_P}{\partial Z}\hat{e}_P+\frac{\partial A_\Phi}{\partial Z}\hat{e}_\Phi+\frac{\partial A_Z}{\partial Z}\hat{e}_Z+A_P\frac{\partial \hat{e}_P}{\partial Z}+A_\Phi\frac{\partial \hat{e}_\Phi}{\partial Z}+A_Z\frac{\partial \hat{e}_Z}{\partial Z}\right).$$

From these results, we get

$$\vec{\nabla}.\vec{A}=\hat{e}_P.\left(\frac{\partial A_P}{\partial P}\hat{e}_P+\frac{\partial A_\Phi}{\partial P}\hat{e}_\Phi+\frac{\partial A_Z}{\partial P}\hat{e}_Z+0+0+0\right)$$

$$+\frac{\hat{e}_\Phi}{P}.\left(\frac{\partial A_P}{\partial \Phi}\hat{e}_P+\frac{\partial A_\Phi}{\partial \Phi}\hat{e}_\Phi+\frac{\partial A_Z}{\partial \Phi}\hat{e}_Z+A_P\hat{e}_\Phi-A_\Phi\hat{e}_P+0\right)$$

$$+\hat{e}_Z.\left(\frac{\partial A_P}{\partial Z}\hat{e}_P+\frac{\partial A_\Phi}{\partial Z}\hat{e}_\Phi+\frac{\partial A_Z}{\partial Z}\hat{e}_Z+0+0+0\right)$$

$$=\left(\frac{\partial A_P}{\partial P}\right)+\left(\frac{1}{P}\frac{\partial A_\Phi}{\partial \Phi}+\frac{A_P}{P}\right)+\left(\frac{\partial A_Z}{\partial Z}\right)$$

$$=\left(\frac{\partial A_P}{\partial P}+\frac{A_P}{P}\right)+\frac{1}{P}\frac{\partial A_\Phi}{\partial \Phi}+\frac{\partial A_Z}{\partial Z}$$

$$\therefore\ \vec{\nabla}.\vec{A}=\frac{1}{P}\frac{\partial}{\partial P}\left(A_P P\right)+\frac{1}{P}\frac{\partial A_\Phi}{\partial \Phi}+\frac{\partial A_Z}{\partial Z}.$$

5.11 NEUTROSOPHIC CURL

The neutrosophic curl $\vec{\nabla}\times\vec{A}$ is also carried out taking into account that the neutrosophic unit vectors themselves are functions of the neutrosophic cylindrical coordinates.

Thus we have

$$\vec{\nabla}\times\vec{A}=\left(\hat{e}_P\frac{\partial}{\partial P}+\frac{\hat{e}_\Phi}{P}\frac{\partial}{\partial\Phi}+\hat{e}_P\frac{\partial}{\partial Z}\right)\times\left(A_P\hat{e}_P+A_\Phi\hat{e}_\Phi+A_Z\hat{e}_P\right)$$

where the derivatives must be taken before the cross product so that

$$\vec{\nabla}\times\vec{A}=\left(\hat{e}_P\frac{\partial}{\partial P}+\frac{\hat{e}_\Phi}{P}\frac{\partial}{\partial\Phi}+\hat{e}_P\frac{\partial}{\partial Z}\right)\times\vec{A}$$

$$=\hat{e}_P\times\frac{\partial\vec{A}}{\partial P}+\frac{\hat{e}_\Phi}{P}\times\frac{\partial\vec{A}}{\partial\Phi}+\hat{e}_Z\times\frac{\partial\vec{A}}{\partial Z}$$

$$=\hat{e}_P\times\left(\frac{\partial A_P}{\partial P}\hat{e}_P+\frac{\partial A_\Phi}{\partial P}\hat{e}_\Phi+\frac{\partial A_Z}{\partial P}\hat{e}_Z+A_P\frac{\partial\hat{e}_P}{\partial P}+A_\Phi\frac{\partial\hat{e}_\Phi}{\partial P}+A_Z\frac{\partial\hat{e}_Z}{\partial P}\right)$$

$$+\frac{\hat{e}_\Phi}{P}\times\left(\frac{\partial A_P}{\partial\Phi}\hat{e}_P+\frac{\partial A_\Phi}{\partial\Phi}\hat{e}_\Phi+\frac{\partial A_Z}{\partial\Phi}\hat{e}_Z+A_P\frac{\partial\hat{e}_P}{\partial\Phi}+A_\Phi\frac{\partial\hat{e}_\Phi}{\partial\Phi}+A_Z\frac{\partial\hat{e}_Z}{\partial\Phi}\right)$$

$$+\hat{e}_Z\times\left(\frac{\partial A_P}{\partial Z}\hat{e}_P+\frac{\partial A_\Phi}{\partial Z}\hat{e}_\Phi+\frac{\partial A_Z}{\partial Z}\hat{e}_Z+A_P\frac{\partial\hat{e}_P}{\partial Z}+A_\Phi\frac{\partial\hat{e}_\Phi}{\partial Z}+A_Z\frac{\partial\hat{e}_Z}{\partial Z}\right).$$

With the help of the previously obtained results of partial derivatives, we get

$$\vec{\nabla}\times\vec{A}=\hat{e}_P\times\left(\frac{\partial A_P}{\partial P}\hat{e}_P+\frac{\partial A_\Phi}{\partial P}\hat{e}_\Phi+\frac{\partial A_Z}{\partial P}\hat{e}_Z+0+0+0\right)$$

$$+\frac{\hat{e}_\Phi}{P}\times\left(\frac{\partial A_P}{\partial\Phi}\hat{e}_P+\frac{\partial A_\Phi}{\partial\Phi}\hat{e}_\Phi+\frac{\partial A_Z}{\partial\Phi}\hat{e}_Z+A_P\hat{e}_\Phi-A_\Phi\hat{e}_P+0\right)$$

$$+\hat{e}_Z\times\left(\frac{\partial A_P}{\partial Z}\hat{e}_P+\frac{\partial A_\Phi}{\partial Z}\hat{e}_\Phi+\frac{\partial A_Z}{\partial Z}\hat{e}_Z+0+0+0\right)$$

$$=\left(\frac{\partial A_\Phi}{\partial P}\hat{e}_Z-\frac{\partial A_Z}{\partial P}\hat{e}_\Phi\right)+\left(-\frac{1}{P}\frac{\partial A_P}{\partial\Phi}\hat{e}_Z+\frac{1}{P}\frac{\partial A_Z}{\partial\Phi}\hat{e}_P+\frac{A_\Phi}{P}\hat{e}_Z\right)$$

$$+\left(\frac{\partial A_P}{\partial Z}\hat{e}_\Phi-\frac{\partial A_\Phi}{\partial Z}\hat{e}_P\right)$$

$$=\hat{e}_P\left(\frac{1}{P}\frac{\partial A_Z}{\partial\Phi}-\frac{\partial A_\Phi}{\partial Z}\right)+\hat{e}_\Phi\left(\frac{\partial A_P}{\partial Z}-\frac{\partial A_Z}{\partial P}\right)+\hat{e}_Z\left(\frac{\partial A_\Phi}{\partial P}+\frac{A_\Phi}{P}-\frac{1}{P}\frac{\partial A_P}{\partial\Phi}\right).$$

Therefore,

$$\vec{\nabla}\times\vec{A}=\hat{e}_P\left(\frac{1}{P}\frac{\partial A_Z}{\partial \Phi}-\frac{\partial A_\Phi}{\partial Z}\right)+\hat{e}_\Phi\left(\frac{\partial A_P}{\partial Z}-\frac{\partial A_Z}{\partial P}\right)+\hat{e}_Z\left(\frac{1}{P}\frac{\partial}{\partial P}(A_\Phi P)-\frac{1}{P}\frac{\partial A_P}{\partial \Phi}\right).$$

5.12 THE NEUTROSOPHIC LAPLASIAN OPERATOR

The neutrosophic Laplasian operator is a scalar operator that is determined from its definition as

$$\nabla^2 u=\vec{\nabla}.\left(\vec{\nabla}u\right)$$

$$=\left(\hat{e}_P\frac{\partial}{\partial P}+\frac{\hat{e}_\Phi}{P}\frac{\partial}{\partial \Phi}+\hat{e}_Z\frac{\partial}{\partial Z}\right).\left(\hat{e}_P\frac{\partial u}{\partial P}+\frac{\hat{e}_\Phi}{P}\frac{\partial u}{\partial \Phi}+\hat{e}_Z\frac{\partial u}{\partial Z}\right)$$

$$=\hat{e}_P.\frac{\partial}{\partial P}\left(\hat{e}_P\frac{\partial u}{\partial P}+\frac{\hat{e}_\Phi}{P}\frac{\partial u}{\partial \Phi}+\hat{e}_Z\frac{\partial u}{\partial Z}\right)$$

$$+\frac{\hat{e}_\Phi}{P}.\frac{\partial}{\partial \Phi}+\hat{e}_Z.\frac{\partial}{\partial Z}\left(\hat{e}_P\frac{\partial u}{\partial P}+\frac{\hat{e}_\Phi}{P}\frac{\partial u}{\partial \Phi}+\hat{e}_Z\frac{\partial u}{\partial Z}\right)$$

With the help of the results of the previously obtained partial derivatives, we get

$$\nabla^2 u=\hat{e}_P.\left(\hat{e}_P\frac{\partial^2 u}{\partial P^2}-\frac{\hat{e}_\Phi}{P^2}\frac{\partial u}{\partial \Phi}+\frac{\hat{e}_\Phi}{P}\frac{\partial^2 u}{\partial \Phi\partial P}+\hat{e}_Z\frac{\partial^2 u}{\partial Z\partial P}\right)$$

$$+\frac{\hat{e}_\Phi}{P}.\left(\hat{e}_\Phi\frac{\partial u}{\partial P}+\hat{e}_P\frac{\partial^2 u}{\partial P\partial \Phi}-\frac{\hat{e}_P}{P}\frac{\partial u}{\partial \Phi}+\frac{\hat{e}_\Phi}{P}\frac{\partial^2 u}{\partial \Phi^2}+\hat{e}_Z\frac{\partial^2 u}{\partial Z\partial \Phi}\right)$$

$$+\hat{e}_Z.\left(\hat{e}_P\frac{\partial^2 u}{\partial P\partial Z}+\frac{\hat{e}_\Phi}{P}\frac{\partial^2 u}{\partial \Phi\partial Z}+\hat{e}_Z\frac{\partial^2 u}{\partial Z^2}\right)$$

$$=\frac{\partial^2 u}{\partial P^2}+\frac{1}{P}\frac{\partial u}{\partial P}+\frac{1}{P^2}\frac{\partial^2 u}{\partial \Phi^2}+\frac{\partial^2 u}{\partial Z^2}$$

$$=\frac{1}{P}\frac{\partial}{\partial P}\left(P\frac{\partial u}{\partial P}\right)+\frac{1}{P^2}\frac{\partial^2 u}{\partial \Phi^2}+\frac{\partial^2 u}{\partial Z^2}.$$

Thus, the neutrosophic Laplasian operator is written as

$$\nabla^2 = \frac{1}{P}\frac{\partial}{\partial P}\left(P\frac{\partial}{\partial P}\right) + \frac{1}{P^2}\frac{\partial^2}{\partial \Phi^2} + \frac{\partial^2}{\partial Z^2}.$$

(Here, $\mathrm{P} = \rho + \rho_0 I,\ \Phi = \varphi + \varphi_0 I,\ Z = z + z_0 I$.)

5.13 CONCLUSION

In this chapter, we have discussed neutrosophic cylindrical coordinates, conversions from NCCs to neutrosophic Cartesian coordinates, and vice versa. We discussed neutrosophic unit vectors and their results of dot products and also discussed the velocity and acceleration of NCPs. We also discussed the neutrosophic Del operator, neutrosophic divergence, neutrosophic curl, and neutrosophic Laplacian operator in terms of NCCs.

REFERENCES

[1] Smarandache, F., "Introduction to Neutrosophic Measure, Neutrosophic Integral, and Neutrosophic Probability", Sitech-Education Publisher, Craiova – Columbus, 2013.

[2] Smarandache, F., "Finite Neutrosophic Complex Numbers, by W. B. Vasantha Kandasamy", Zip Publisher, Columbus, OH, pp. 1–16, 2011.

[3] Smarandache, F., "Introduction to Neutrosophic Statistics", Sitech-Education Publisher, Columbus, pp. 34–44, 2014.

[4] Smarandache, F., "Neutrosophy./Neutrosophic Probability, Set, and Logic", American Research Press, Rehoboth, 1998.

[5] Smarandache, F., "A Unifying Field in Logics: Neutrosophic Logic", Preface by Charles Le, American Research Press, Rehoboth, 1999, 2000. Second edition of the Proceedings of the First International Conference on Neutrosophy, Neutrosophic Logic, Neutrosophic Set, Neutrosophic Probability and Statistics, University of New Mexico, Gallup, 2001.

[6] Smarandache, F., "Proceedings of the First International Conference on Neutrosophy", Neutrosophic Set, Neutrosophic Probability and Statistics, University of New Mexico, USA, 2001.

[7] Smarandache, F. and Khalid, Huda E. (eds.), "Neutrosophic Precalculus and Neutrosophic Calculus", Second enlarged edition. Pons Editions, Brussels, Belgium, 2018, p. 176. ISBN: 978-1-59973-555-9

[8] Al-Tahan, M., "Some Results on Single Valued Neutrosophic (Weak) Polygroups", International Journal of Neutrosophic Science, Volume 2, Issue 1, pp. 38–46, 2020.

[9] Edalatpanah, S., "A Direct Model for Triangular Neutrosophic Linear Programming", International Journal of Neutrosophic Science, Volume 1, Issue 1, pp. 19–28, 2020.

[10] Chakraborty, A., "A New Score Function of Pentagonal Neutrosophic Number and Its Application in Networking Problem", International Journal of Neutrosophic Science, Volume 1, Issue 1, pp. 40–51, 2020.

[11] Chakraborty, A., "Application of Pentagonal Neutrosophic Number in Shortest Path Problem", International Journal of Neutrosophic Science, Volume 3, Issue 1, pp. 21–28, 2020.

Chapter 6

Applying Interval Quadripartitioned Single-Valued Neutrosophic Sets to Graphs and Climatic Analysis

Satham S. Hussain, Muhammad Aslam, Hossein Rahmonlou, and N. Durga

6.1 INTRODUCTION

The author Zadeh in 1975 invented interval-valued fuzzy sets as the generalization of fuzzy sets [1]. To handle the uncertainty, interval-valued fuzzy sets utilize the intervals of numbers instead of numbers as the membership function. To reflect the grade of membership of fuzzy set $\boldsymbol{A}$, $\left[\boldsymbol{T}_{AL}(\boldsymbol{x}), \boldsymbol{T}_{AU}(\boldsymbol{x})\right]$ with $0 \leq \boldsymbol{T}_{AL} + \boldsymbol{T}_{AU} \leq 1$ is widely employed [2]. It is crucial to use interval-valued fuzzy sets in applications such as fuzzy control.

On the other hand, in order to represent the grade of true and false membership functions, some constraints permit only incomplete data to be held, but the handling of indeterminate data persists. For clear understanding, one can characterize the weather as cold = true, moderate = undetermined and hot = false using a neutrosophic set. The neutrosophic field allows for accounting for the indeterminant data as well. From a philosophical standpoint, it approximates the aforementioned sets.

The single-valued neutrosophic set is a generalization of intuitionistic fuzzy sets that is applied to solve real-world problems, particularly in decision support [3]. The sum of the computations of notion in that element (truth), incredulity in that element (falsehood) and indeterminacy of that element are strictly less than 1 as proposed by Smarandache [4] and references therein. A generalization of the concept of single-valued neutrosophic sets is the interval-valued neutrosophic set, in which three membership $(\boldsymbol{T}, \boldsymbol{I}, \boldsymbol{F})$. functions are independent and their values belong to the unit interval $[\mathbf{0,1}]$.

In the case of neutrosophic sets, indeterminacy is treated as a separate term, and each element x is characterized by a truth function $\mathcal{T}_A(x)$, an indeterminacy function $\mathcal{I}_A(x)$ and a falsity function $\mathcal{F}_A(x)$, each from the non-standard unit interval $\left]0^-, 1^+\right[$. Despite the fact that neutrosophic indeterminacy is independent of truth and falsity membership numbers, it is more extensive than the intuitionistic fuzzy set's hesitation range. As a consequence, some authors prefer to model

 DOI: 10.1201/9781003487104-6

indeterminacy's behaviour similarly to truth membership, while others prefer to model it similarly to falsity membership.

Wang et al. [3] initiated the concept of a single-valued neutrosophic set and provide its various properties. It has been widely applied in various fields, such as information fusion in which data are combined from different sensors [5], control theory [6], image processing [7], medical diagnosis [8], decision making [9] and graph theory [10, 11–17]. When the indeterminacy portion of the neutrosophic set is divided into two parts, we get four components: contradiction (both true and false) and unknown (neither true nor false), that is $\mathcal{T}, \mathcal{C}, \mathcal{U}$ and $\mathcal{F}$, which define a new quadripartitioned single-valued neutrosophic set, as introduced by Chatterjee et al. [18].

The study for this chapter is completely based on Belnap's four-valued logic [19] and Smarandache's four-valued neutrosophic logic [20]. The chapter presents interval quadripartitioned single-valued neutrosophic sets (QSVNSs). Operations on single-valued neutrosophic graphs are discussed in [21], and operations on interval-valued neutrosophic graphs are established in [22]. Further, operations on neutrosophic vague graphs are discussed in [13]. Authors in [23] studied bipolar QSVNSs. The authors in [24] studied interval intuitionist neutrosophic sets with the operations and properties of its graphs. More information can be seen in [25–28].

Based on our review of the literature we have discussed, we conclude that there is no extant research work on the concept of interval QSVNSs and their application to graphs. The main contributions of this chapter are therefore explained as follows: We introduce the concept of interval QSVNSs and their applications to graphs. In addition, we define the complete, strong, complement and operations on interval-valued quadripartitioned neutrosophic sets. We also apply the developed results and discuss the operations as Cartesian product, union, join and the combination of interval quadripartitioned single-valued neutrosophic graphs with their properties. Finally, we analyze climatic characterization in Toronto based on the information related to interval quadripartitioned neutrosophic field.

The remainder of this proceeds as follows: The basic definitions are given in Section 6.2 Preliminaries. We introduce the interval QSVNSs and graphs in Seciton 6.3 and explain their operations. In Section 6.4, we apply QSVNSs to characterize the weather in Toronto.

6.2 PRELIMINARIES

Some basic definitions are provided in this section for proving the main results.

Definition 6.2.1 [29] Let the universe of discourse be γ and an interval-valued neutrosophic set Λ be stated as

$$\Lambda = \left\{ \left\langle p, \left[\mathcal{T}_{\Lambda_L}(p), \mathcal{T}_{\Lambda_U}(p)\right], \left[\mathcal{I}_{\Lambda_L}(p), \mathcal{I}_{\Lambda_U}(p)\right], \left[\mathcal{F}_{\Lambda_L}(p), \mathcal{F}_{\Lambda_U}(p)\right] \right\rangle, p \in \gamma \right\},$$

$\mathcal{T}_{\Lambda_L}(p), \mathcal{T}_{\Lambda_U}(p), \mathcal{I}_{\Lambda_L}(p), \mathcal{I}_{\Lambda_U}(p), \mathcal{F}_{\Lambda_L}(p), \mathcal{F}_{\Lambda_U}(p)$are neutrosophic subsets of γ thatsatisfy$\mathcal{T}_{\Lambda_L}(p) \leq \mathcal{T}_{\Lambda_U}(p), \mathcal{I}_{\Lambda_L}(p) \leq \mathcal{I}_{\Lambda_U}(p), \mathcal{F}_{\Lambda_L}(p) \leq \mathcal{F}_{\Lambda_U}(p), \forall p \in \gamma$.

Definition 6.2.2 [30] Let γ be the universe. A neutrosophic set Λ in γ is characterized by a truth membership function $\mathcal{T}_{\Lambda}$, an indeterminacy membership function $\mathcal{I}_{\Lambda}$ and a falsity membership function $\mathcal{F}_{\Lambda}$ where $\mathcal{T}_{\Lambda}$, $\mathcal{I}_{\Lambda}$ and $\mathcal{F}_{\Lambda}$ are belong $[0,1]$. It follows that for each point p inγ, $\mathcal{T}_{\Lambda}(p), \mathcal{F}_{\Lambda}(p), \mathcal{I}_{\Lambda}(p) \in [0,1]$. Additionally,

$$\Lambda = \langle p, \mathcal{T}_{\Lambda}(p), \mathcal{F}_{\Lambda}(p), \mathcal{I}_{\Lambda}(p) : p \in \Upsilon \rangle \text{ and } 0 \leq \mathcal{T}_{\Lambda}(p) + \mathcal{I}_{\Lambda}(p) + \mathcal{F}_{\Lambda}(p) \leq 3.$$

Definition 6.2.3 [3, 4, 30] Let γ be a space of points (objects) with generic elements in γ denoted by p. A single-valued neutrosophic set Λ in γ is characterized by $\mathcal{T}_{\Lambda}(p)$, $\mathcal{I}_{\Lambda}(p)$ and $\mathcal{F}_{\Lambda}(p)$.

For each point p in γ, $\mathcal{T}_{\Lambda}(p), \mathcal{I}_{\Lambda}(p), \mathcal{F}_{\Lambda}(p) \in [0,1]$,

$$\Lambda = \{p, \mathcal{T}_{\Lambda}(p), \mathcal{I}_{\Lambda}(p), \mathcal{F}_{\Lambda}(p)\} \text{ and } 0 \leq \mathcal{T}_{\Lambda}(p) + \mathcal{I}_{\Lambda}(p) + \mathcal{F}_{\Lambda}(p) \leq 3.$$

Definition 6.2.4 [4, 30] A neutrosophic set $\mathbb{A}$ is contained in another neutrosophic set$\mathbb{B}$, (i.e, $\mathbb{A} \subseteq \mathbb{B}$ if $\forall p \in \gamma$, $\mathcal{T}_{\mathbb{A}}(p) \leq \mathcal{T}_{\mathbb{B}}(p), \mathcal{I}_{\mathbb{A}}(p) \geq \mathcal{I}_{\mathbb{B}}(p)$ and $\mathcal{F}_{\mathbb{A}}(p) \geq \mathcal{F}_{\mathbb{B}}(p)$).

Definition 6.2.5 [21] A single-valued neutrosophic graph is defined as a pair $\mathbb{G}^* = (\mathbb{V}, \mathbb{E})$

where

(i) $\mathbb{V} = \{a_1, a_2, ..., a_n\}$such that$\mathcal{T}_1 : \mathbb{V} \to [0,1]$, $\mathcal{I}_1 : \mathbb{V} \to [0,1]$ and $\mathcal{F}_1 : \mathbb{V} \to [0,1]$ denote $\mathcal{T}_{\Lambda}(p)$, $\mathcal{I}_{\Lambda}(p)$ and $\mathcal{F}_{\Lambda}(p)$, respectively and $0 \leq \mathcal{T}_1(a_1) + \mathcal{I}_1(a_1) + \mathcal{F}_1(a_1) \leq 3$;

(ii) $\mathbb{E} \subseteq \mathbb{V} \times \mathbb{V}$ where $\mathcal{T}_2 : \mathbb{E} \to [0,1]$, $\mathcal{I}_2 : \mathbb{E} \to [0,1]$ and $\mathcal{F}_2 : \mathbb{E} \to [0,1]$ are such that
$\mathcal{T}_2(a_1 a_2) \leq \min\{\mathcal{T}_1(a_1), \mathcal{T}_1(a_2)\}$,
$\mathcal{I}_2(a_1 a_2) \leq \min\{\mathcal{I}_1(a_1), \mathcal{I}_2(a_2)\}$,
$\mathcal{F}_2(a_1 a_2) \leq \max\{\mathcal{F}_1(a), \mathcal{F}_2(a)\}$,
and $0 \leq \mathcal{T}_2(a_1 a_2) + \mathcal{I}_2(a_1 a_2) + \mathcal{F}_2(a_1 a_2) \leq 3, \forall a_1 a_2 \in \mathbb{E}$.

Definition 6.2.6 [22] Consider a non-empty set γ: An interval-valued neutrosophic graph is a pair $\mathbb{G} = (\mathbb{A}, \mathbb{B})$ where $\mathbb{A}$ and $\mathbb{B}$ are the interval-valued neutrosophic set and interval-valued neutrosophic relationship with γ:

$$\mathcal{T}_{\mathbb{B}_L}(a_1 a_2) \leq \mathcal{T}_{\mathbb{A}_L}(a_1) \wedge \mathcal{T}_{\mathbb{A}_L}(a_2), \; \mathcal{T}_{\mathbb{B}_U}(a_1 a_2) \leq \mathcal{T}_{\mathbb{A}_U}(a_1) \wedge \mathcal{T}_{\mathbb{A}_U}(a_2),$$

$$\mathcal{I}_{\mathbb{B}_L}(a_1a_2) \le \mathcal{I}_{\mathbb{A}_L}(a_1) \wedge \mathcal{I}_{\mathbb{A}_L}(a_2),\ \mathcal{I}_{\mathbb{B}_U}(a_1a_2) \le \mathcal{I}_{\mathbb{A}_U}(a_1) \wedge \mathcal{I}_{\mathbb{A}_U}(a_2),$$

$$\mathcal{F}_{\mathbb{B}_L}(a_1a_2) \le \mathcal{F}_{\mathbb{A}_L}(a_1) \wedge \mathcal{F}_{\mathbb{A}_L}(a_2),\ \mathcal{F}_{\mathbb{B}_U}(a_1a_2) \le \mathcal{F}_{\mathbb{A}_U}(a_1) \wedge \mathcal{F}_{\mathbb{A}_U}(a_2).$$

Here, $\mathbb{B}$ is a symmetric relation on $\mathbb{A}$. For more details about the following definitions and results, see [18].

Definition 6.2.7 Let γ be a non-empty set. A QSVNS Λ over γ characterizes each element p in γ by a truth membership function $\mathcal{T}_\Lambda$, a contradiction membership function $\mathcal{C}_\Lambda$, an ignorance membership function $\mathcal{U}_\Lambda$ and a false membership function $\mathcal{F}_\Lambda$ such that for each, $p \in \gamma, \mathcal{T}_\Lambda, \mathcal{C}_\Lambda, \mathcal{U}_\Lambda, \mathcal{F}_\Lambda \in [0,1]$ and $0 \le \mathcal{T}_\Lambda(p) + \mathcal{C}_\Lambda(p) + \mathcal{U}_\Lambda(p) + \mathcal{F}_\Lambda(p) \le 4$.

Remark 6.2.8 A QSVNS Λ can be decomposed to yields two SVNNs, say, Λ_t and Λ_b, where the respective membership functions of the two sets are stated as

$$\mathcal{T}_{\Lambda_t}(p) = \mathcal{T}_\Lambda(p) = \mathcal{T}_{\Lambda_b}(p)$$

$$\mathcal{I}_{\Lambda_t}(p) = \mathcal{C}_\Lambda(p),\ \mathcal{I}_{\Lambda_b}(p) = \mathcal{U}_\Lambda(p)$$

$$\mathcal{F}_{\Lambda_t}(p) = \mathcal{F}_\Lambda(p) = \mathcal{F}_{\Lambda_b}(p),\quad \forall p \in \gamma.$$

According to this, we need to be clarify that while performing set-theoretic operations over these SVNNs, the behavior of $\mathcal{I}_{\Lambda_t}$ is treated similar to that of $\mathcal{T}_{\Lambda_t}$ while the behavior of $\mathcal{I}_{\Lambda_b}$ is modeled in a way similar to that of $\mathcal{F}_{\Lambda_b}$.

Definition 6.2.9 A QSVNS is said to be an absolute QSVNS, denoted by Λ, if its membership values are defined as $\mathcal{T}_\Lambda(p) = 1$, $\mathcal{C}_\Lambda(p) = 1$, $\mathcal{U}_\Lambda(p) = 0$ and $\mathcal{F}_\Lambda(p) = 0$.

Definition 6.2.10 Let the two QSVNSs be $\mathfrak{A}$ and $\mathfrak{B}$, over γ; then $\mathfrak{A}$ is said to be contained in $\mathfrak{B}$, represented by $\mathfrak{A} \subseteq \mathfrak{B}$ if and only if $\mathcal{T}_{\mathfrak{A}}(p) \le \mathcal{T}_{\mathfrak{B}}(p)$, $\mathcal{C}_{\mathfrak{A}}(p) \le \mathcal{C}_{\mathfrak{B}}(p)$, $\mathcal{U}_{\mathfrak{A}}(p) \ge \mathcal{U}_{\mathfrak{B}}(p)$ and $\mathcal{F}_{\mathfrak{A}}(p) \ge \mathcal{F}_{\mathfrak{B}}(p)$.

Definition 6.2.11 The complement of a QSVNS $\mathfrak{A}$ is notated by $\mathfrak{A}^c$ and is represented as

$$\mathfrak{A}^c = \sum_{i=1}^{n} \langle \mathcal{F}_{\mathfrak{A}}(p_i), \mathcal{U}_{\mathfrak{A}}(p_i), \mathcal{C}_{\mathfrak{A}}(p_i), \mathcal{T}_{\mathfrak{A}}(p_i) \rangle,\ \forall p_i \in \gamma.$$

That is,

$$\mathcal{T}^c_{\mathfrak{A}}(p_i) = \mathcal{F}_{\mathfrak{A}}(p_i),\ \mathcal{C}^c_{\mathfrak{A}}(p_i) = \mathcal{U}_{\mathfrak{A}}(p_i)\ \ \mathcal{U}^c_{\mathfrak{A}}(p_i) = \mathcal{C}_{\mathfrak{A}}(p_i),$$
$$\mathcal{F}^c_{\mathfrak{A}}(p_i) = \mathcal{T}_{\mathfrak{A}}(p_i),\ \forall p_i \in \gamma.$$

Definition 6. 2.12 The union of two QSVNSs $\mathfrak{A}$ and $\mathfrak{B}$ is denoted by $\mathfrak{A} \cup \mathfrak{B}$ and is represented as

$$\mathfrak{A} \cup \mathfrak{B} = \sum_{i=1}^{n} \langle \mathcal{T}_{\mathfrak{A}}(p_i) \vee \mathcal{T}_{\mathfrak{B}}(p_i), \mathcal{C}_{\mathfrak{A}}(p_i) \vee \mathcal{C}_{\mathfrak{B}}(p_i)$$

$$\mathcal{U}_{\mathfrak{A}}(p_i) \wedge \mathcal{U}_{\mathfrak{B}}(p_i), \mathcal{F}_{\mathfrak{A}}(p_i) \wedge \mathcal{F}_{\mathfrak{B}}(p_i) \rangle / \gamma.$$

Definition 6.2.13 The intersection of two QSVNSs $\mathfrak{A}$ and $\mathfrak{B}$ is denoted by $\mathfrak{A} \cap \mathfrak{B}$ and is represented as

$$\mathfrak{A} \cap \mathfrak{B} = \sum_{i=1}^{n} \langle \mathcal{T}_{\mathfrak{A}}(p_i) \wedge \mathcal{T}_{\mathfrak{B}}(p_i), \mathcal{C}_{\mathfrak{A}}(p_i) \wedge \mathcal{C}_{\mathfrak{B}}(p_i)$$

$$\mathcal{U}_{\mathfrak{A}}(p_i) \vee \mathcal{U}_{\mathfrak{B}}(p_i), \mathcal{F}_{\mathfrak{A}}(p_i) \vee \mathcal{F}_{\mathfrak{B}}(p_i) \rangle / \gamma.$$

6.3 PRELIMINARIES

The developed definition and operations of the proposed concept are established in this section.

Definition 6.3.1 An interval QSVNS $\mathbb{A}$ on the universe of discourse γ is defined as

$$\mathbb{A} = \left\{ \begin{matrix} \langle p, [\mathcal{T}_{\mathbb{A}_L}(p), \mathcal{T}_{\mathbb{A}_U}(p)], [\mathcal{C}_{\mathbb{A}_L}(p), \mathcal{C}_{\mathbb{A}_U}(p)], \\ [\mathcal{U}_{\mathbb{A}_L}(p), \mathcal{U}_{\mathbb{A}_U}(p)], [\mathcal{F}_{\mathbb{A}_L}(p), \mathcal{F}_{\mathbb{A}_U}(p)] \rangle, p \in \mathfrak{X} \end{matrix} \right\},$$

where $\mathcal{T}_{\mathbb{A}_L}(p), \mathcal{T}_{\mathbb{A}_U}(p), \mathcal{C}_{\mathbb{A}_L}(p), \mathcal{C}_{\mathbb{A}_U}(p), \mathcal{U}_{\mathbb{A}_L}(p), \mathcal{U}_{\mathbb{A}_U}(p), \mathcal{F}_{\mathbb{A}_L}(p), \mathcal{F}_{\mathbb{A}_U}(p)$ are neutrosophic subsets of γ that satisfy $\mathcal{T}_{\mathbb{A}_L}(p) \le \mathcal{T}_{\mathbb{A}_U}(p), \mathcal{C}_{\mathbb{A}_L}(p) \le \mathcal{C}_{\mathbb{A}_U}(p)$, $\mathcal{U}_{\mathbb{A}_L}(p) \le \mathcal{U}_{\mathbb{A}_U}(p), \mathcal{F}_{\mathbb{A}_L}(p) \le \mathcal{F}_{\mathbb{A}_U}(p)$, $\forall p \in \gamma$.

Definition 6.3.2 An interval QSVNS $\mathbb{A}$ is contained in another interval QSVNS $\mathbb{B}$, $\mathbb{A} \subseteq \mathbb{B}$, if and only if

$$\mathcal{T}_{\mathbb{A}_L}(p) \le \mathcal{T}_{\mathbb{B}_L}(p),\ \mathcal{T}_{\mathbb{A}_U}(p) \le \mathcal{T}_{\mathbb{B}_U}(p),$$

$$\mathcal{C}_{\mathbb{A}_L}(p) \le \mathcal{C}_{\mathbb{B}_L}(p),\ \mathcal{C}_{\mathbb{A}_U}(p) \le \mathcal{C}_{\mathbb{B}_U}(p),$$

$$\mathcal{U}_{\mathbb{A}_L}(p) \ge \mathcal{U}_{\mathbb{B}_L}(p),\ \mathcal{U}_{\mathbb{A}_U}(p) \ge \mathcal{U}_{\mathbb{B}_U}(p), \text{ and}$$

$$\mathcal{F}_{\mathbb{A}_L}(p) \ge \mathcal{F}_{\mathbb{B}_L}(p),\ \mathcal{F}_{\mathbb{A}_U}(p) \ge \mathcal{F}_{\mathbb{B}_U}(p),\ \forall p \in \gamma.$$

Definition 6.3.3 Two interval QSVNSs $\mathbb{D}$ and $\mathbb{S}$ are equal, denoted as $\mathbb{D} = \mathbb{S}$, if and only if $\mathbb{D} \subseteq \mathbb{S}$ and $\mathbb{S} \subseteq \mathbb{D}$.

Definition 6.3.4 The intersection of two interval QSVNSs $\mathbb{A}$ and $\mathbb{B}$ is an interval quadripartitioned neutrosophic set $\mathbb{C}$ written as $\mathbb{C} = \mathbb{A} \cap \mathbb{B}$, whose truth, contradiction, ignorance and false membership functions are related to those of $\mathbb{A}$ and $\mathbb{B}$ such that

$$\mathcal{T}_{\mathbb{C}_L}(p) = \min\left\{\mathcal{T}_{\mathbb{A}_L}(p), \mathcal{T}_{\mathbb{B}_L}(p)\right\}, \quad \mathcal{T}_{\mathbb{C}_U}(p) = \min\left\{\mathcal{T}_{\mathbb{A}_U}(p), \mathcal{T}_{\mathbb{B}_U}(p)\right\}$$

$$\mathcal{C}_{\mathbb{C}_L}(p) = \min\left\{\mathcal{C}_{\mathbb{A}_L}(p), \mathcal{C}_{\mathbb{B}_L}(p)\right\}, \quad \mathcal{C}_{\mathbb{C}_U}(p) = \min\left\{\mathcal{C}_{\mathbb{A}_U}(p), \mathcal{C}_{\mathbb{B}_U}(p)\right\}$$

$$\mathcal{U}_{\mathbb{C}_L}(p) = \max\left\{\mathcal{U}_{\mathbb{A}_L}(p), \mathcal{U}_{\mathbb{B}_L}(p)\right\}, \quad \mathcal{U}_{\mathbb{C}_U}(p) = \max\left\{\mathcal{U}_{\mathbb{A}_U}(p), \mathcal{U}_{\mathbb{B}_U}(p)\right\}, \text{ and}$$

$$\mathcal{F}_{\mathbb{C}_L}(p) = \max\left\{\mathcal{F}_{\mathbb{A}_L}(p), \mathcal{F}_{\mathbb{B}_L}(p)\right\}, \quad \mathcal{F}_{\mathbb{C}_U}(p) = \max\left\{\mathcal{F}_{\mathbb{A}_U}(p), \mathcal{F}_{\mathbb{B}_U}(p)\right\}, \quad \forall p \in \gamma.$$

Definition 6.3.5 The union of two interval QSVNSs $\mathbb{A}$ and $\mathbb{B}$ is an interval QSVNSs $\mathfrak{C}$, written as $\mathfrak{C} = \mathbb{A} \cup \mathbb{B}$, whose truth, contradiction, ignorance and false functions are related to those of $\mathbb{A}$ and $\mathbb{B}$ such that

$$\mathcal{T}_{\mathfrak{C}_L}(p) = \max\left\{\mathcal{T}_{\mathbb{A}_L}(p), \mathcal{T}_{\mathbb{B}_L}(p)\right\}, \quad \mathcal{T}_{\mathfrak{C}_U}(p) = \max\left\{\mathcal{T}_{\mathbb{A}_U}(p), \mathcal{T}_{\mathbb{B}_U}(p)\right\},$$

$$\mathcal{C}_{\mathfrak{C}_L}(p) = \max\left\{\mathcal{C}_{\mathbb{A}_L}(p), \mathcal{C}_{\mathbb{B}_L}(p)\right\}, \quad \mathcal{C}_{\mathfrak{C}_U}(p) = \max\left\{\mathcal{C}_{\mathbb{A}_U}(p), \mathcal{C}_{\mathbb{B}_U}(p)\right\},$$

$$\mathcal{U}_{\mathfrak{C}_L}(p) = \min\left\{\mathcal{U}_{\mathbb{A}_L}(p), \mathcal{U}_{\mathbb{B}_L}(p)\right\}, \quad \mathcal{U}_{\mathfrak{C}_U}(p) = \min\left\{\mathcal{U}_{\mathbb{A}_U}(p), \mathcal{U}_{\mathbb{B}_U}(p)\right\}, \text{ and}$$

$$\mathcal{F}_{\mathfrak{C}_L}(p) = \min\left\{\mathcal{F}_{\mathbb{A}_L}(p), \mathcal{F}_{\mathbb{B}_L}(p)\right\}, \quad \mathcal{F}_{\mathfrak{C}_U}(p) = \min\left\{\mathcal{F}_{\mathbb{A}_U}(p), \mathcal{F}_{\mathbb{B}_U}(p)\right\}, \quad \forall p \in \gamma.$$

Example 6. 3.1 Consider the following sets:

$$\begin{aligned}\mathbb{A} = \{&\left(p_1, [0.3,0.5], [0.7,0.8], [0.4,0.6], [0.3,0.5]\right),\\ &\left(p_2, [0.7,0.8], [0.6,0.7], [0.6,0.8], [0.2,0.4]\right),\\ &\left(p_3, [0.2,0.4], [0.4,0.6], [0.7,0.8], [0.4,0.8]\right)\}\end{aligned}$$

and

$$\begin{aligned}\mathbb{B} = \{&\left(p_1, [0.2,0.5], [0.6,0.7], [0.3,0.7], [0.4,0.5]\right),\\ &\left(p_2, [0.6,0.8], [0.7,0.8], [0.7,0.9], [0.3,0.5]\right),\\ &\left(p_3, [0.3,0.4], [0.5,0.6], [0.6,0.8], [0.2,0.3]\right)\}, \text{ then}\end{aligned}$$

$$\mathbb{A}\cap\mathbb{B}=\{(p_1,[0.2,0.5],[0.6,0.7],[0.4,0.7],[0.4,0.5]),$$
$$(p_2,[0.6,0.8],[0.6,0.7],[0.7,0.9],[0.3,0.5]),$$
$$(p_3,[0.2,0.4],[0.4,0.6],[0.7,0.8],[0.4,0.8])\}.$$ Also,

$$\mathbb{A}\cup\mathbb{B}=\{(p_1,[0.3,0.5],[0.7,0.8],[0.3,0.6],[0.3,0.5]),$$
$$(p_2,[0.7,0.8],[0.7,0.8],[0.6,0.8],[0.2,0.4]),$$
$$(p_3,[0.3,0.4],[0.5,0.6],[0.6,0.8],[0.2,0.3])\}.$$

Definition 6.3.6 The complement of interval QSVNS

$$\mathbb{A}=\{(p,\mathcal{T}_{\mathbb{A}_L}(p),\mathcal{C}_{\mathbb{A}_L}(p),\mathcal{U}_{\mathbb{A}_L}(p),\mathcal{F}_{\mathbb{A}_L}(p),\mathcal{T}_{\mathbb{A}_U}(p),\mathcal{C}_{\mathbb{A}_U}(p),\mathcal{U}_{\mathbb{A}_U}(p),\mathcal{F}_{\mathbb{A}_U}(p))\}$$

for all $p\in\gamma$ is defined as

$$\mathbb{A}'=\{(x,\mathcal{F}_{\mathbb{A}_L}(p),\mathcal{U}_{\mathbb{A}_L}(p),\mathcal{C}_{\mathbb{A}_L}(p),\mathcal{T}_{\mathbb{A}_L}(p),\mathcal{F}_{\mathbb{A}_U}(p),\mathcal{U}_{\mathbb{A}_U}(p),\mathcal{C}_{\mathbb{A}_U}(p),\mathcal{T}_{\mathbb{A}_U}(p))\}$$

for all $p\in\gamma$.

Example 6.3.2 The complement of $\mathbb{A}$ in (6.3.1) is

$$\mathbb{A}'=\{(p_1,[0.3,0.5],[0.4,0.6],[0.7,0.8],[0.3,0.5]),$$
$$(p_2,[0.2,0.4],[0.6,0.8],[0.6,0.7],[0.7,0.8]),$$
$$(p_3,[0.4,0.8],[0.7,0.8],[0.6,0.4],[0.2,0.4])\}.$$

6.4 INTERVAL QUADRIPARTITIONED SINGLE-VALUED NEUTROSOPHIC GRAPHS

In this section, we apply interval QSVNSs as interval quadripartitioned single-valued neutrosophic graphs.

Definition 6.4.1 An interval quadripartitioned single-valued neutrosophic graph is a pair $\mathcal{G}=(\mathbb{A},\mathbb{B})$ with underlying set $\mathbb{V}$ where $\mathcal{T}_{\mathbb{A}_L},\mathcal{C}_{\mathbb{A}_L},\mathcal{U}_{\mathbb{A}_L},\mathcal{F}_{\mathbb{A}_L},\mathcal{T}_{\mathbb{A}_U},\mathcal{C}_{\mathbb{A}_U},\mathcal{U}_{\mathbb{A}_U},\mathcal{F}_{\mathbb{A}_U}$ represent the lower and upper values of truth, contradiction, ignorance and false membership values of the vertices in $\mathbb{V}$ and $\mathcal{T}_{\mathbb{B}_L},\mathcal{C}_{\mathbb{B}_L},\mathcal{U}_{\mathbb{B}_L},\mathcal{F}_{\mathbb{B}_L},\mathcal{T}_{\mathbb{B}_U},\mathcal{C}_{\mathbb{B}_U},\mathcal{U}_{\mathbb{B}_U},\mathcal{F}_{\mathbb{B}_U}\subseteq\mathbb{V}\times\mathbb{V}$ are the lower and upper of truth, contradiction, ignorance and false membership values of the edges, such that

$$\mathcal{T}_{\mathbb{B}_L}(kl) \le \min\{\mathcal{T}_{\mathbb{A}_L}(k), \mathcal{T}_{\mathbb{A}_L}(l)\},\ \mathcal{T}_{\mathbb{B}_U}(kl) \le \min\{\mathcal{T}_{\mathbb{A}_U}(k), \mathcal{T}_{\mathbb{A}_U}(l)\},$$

$$\mathcal{C}_{\mathbb{B}_L}(kl) \le \min\{\mathcal{C}_{\mathbb{A}_L}(k), \mathcal{C}_{\mathbb{A}_L}(l)\},\ \mathcal{C}_{\mathbb{B}_U}(kl) \le \min\{\mathcal{C}_{\mathbb{A}_U}(k), \mathcal{C}_{\mathbb{A}_U}(l)\},$$

$$\mathcal{U}_{\mathbb{B}_L}(kl) \le \max\{\mathcal{U}_{\mathbb{A}_L}(k), \mathcal{U}_{\mathbb{A}_L}(l))\},\ \mathcal{U}_{\mathbb{B}_U}(kl) \le \max\{\mathcal{U}_{\mathbb{A}_U}(k), \mathcal{U}_{\mathbb{A}_U}(l)\},$$

$$\mathcal{F}_{\mathbb{B}_L}(kl) \le \max\{\mathcal{F}_{\mathbb{A}_L}(k), \mathcal{F}_{\mathbb{A}_L}(l))\},\ \mathcal{F}_{\mathbb{B}_U}(kl) \le \max\{\mathcal{F}_{\mathbb{A}_U}(k), \mathcal{F}_{\mathbb{A}_U}(l)\},$$

$\forall k, l \in \mathbb{A}$ and $kl \in \mathbb{B}$, where $\mathbb{A}$ and $\mathbb{B}$ are the underlying vertex set and edge set, respectively, on $\mathcal{G}$.

Example 6. 4.1 Let an interval quadripartitioned single-valued neutrosophic graph be $\mathcal{G} = (\mathbb{A}, \mathbb{B})$ such that $\mathbb{A} = \{a, b, c\}, \mathbb{B} = \{(ab), (bc), (ca)\}$. By similar calculation, one can obtain Figure 6.1.

Definition 6.4.2 Graph $\mathcal{G}' = (\mathbb{A}', \mathbb{B}')$ is a subgraph of $\mathcal{G} = (\mathbb{A}, \mathbb{B})$ if

$$\mathcal{T}'_{\mathbb{A}_L}(k) \le \mathcal{T}_{\mathbb{A}_L}(k),\ \mathcal{T}'_{\mathbb{A}_U}(k) \le \mathcal{T}_{\mathbb{A}_U}(k),$$

$$\mathcal{C}'_{\mathbb{A}_L}(k) \le \mathcal{C}_{\mathbb{A}_L}(k),\ \mathcal{C}'_{\mathbb{A}_U}(k) \le \mathcal{C}_{\mathbb{A}_U}(k),$$

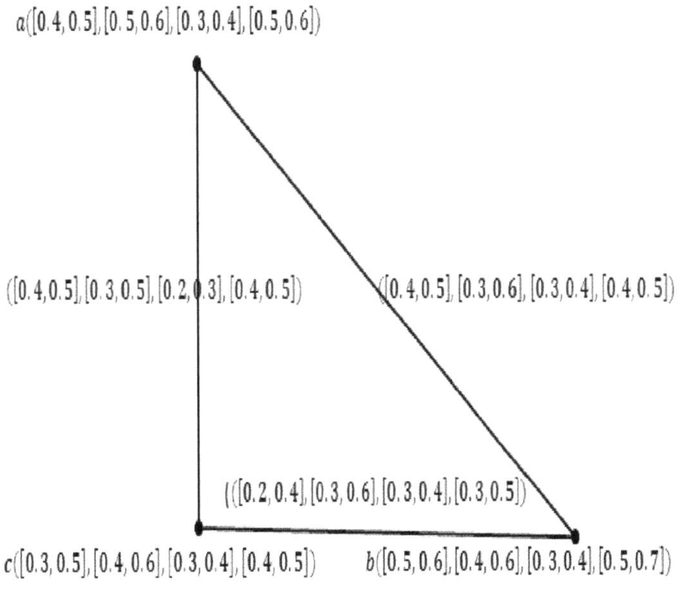

Figure 6.1 $\mathcal{G}$: Interval quadripartitioned neutrosophic graphs.

$$\mathcal{U}'_{\mathbb{A}_L}(k) \geq \mathcal{U}_{\mathbb{A}_L}(k),\ \mathcal{U}'_{\mathbb{A}_U}(k) \geq \mathcal{U}_{\mathbb{A}_U}(k),$$

$$\mathcal{F}'_{\mathbb{A}_L}(k) \geq \mathcal{F}_{\mathbb{A}_L}(k),\ \mathcal{F}'_{\mathbb{A}_U}(k) \geq \mathcal{F}_{\mathbb{A}_U}(k), \forall k \in \mathbb{A}$$

$$\mathcal{T}'_{\mathbb{B}_L}(kl) \leq \mathcal{T}_{\mathbb{B}_L}(kl),\ \mathcal{T}'_{\mathbb{B}_U}(kl) \leq \mathcal{T}_{\mathbb{B}_U}(kl)$$

$$\mathcal{C}'_{\mathbb{B}_L}(kl) \leq \mathcal{C}_{\mathbb{B}_L}(kl),\ \mathcal{C}'_{\mathbb{B}_U}(kl) \leq \mathcal{C}_{\mathbb{B}_U}(kl)$$

$$\mathcal{U}'_{\mathbb{B}_L}(kl) \geq \mathcal{U}_{\mathbb{B}_L}(kl),\ \mathcal{U}'_{\mathbb{B}_U}(k) \geq \mathcal{U}_{\mathbb{B}_U}(kl)$$

$$\mathcal{F}'_{\mathbb{B}_L}(kl) \geq \mathcal{F}_{\mathbb{B}_L}(kl),\ \mathcal{F}'_{\mathbb{B}_U}(k) \geq \mathcal{F}_{\mathbb{B}_U}(kl) \forall\ kl \in \mathbb{B}.$$

Definition 6.4.3 An interval quadripartitioned single-valued neutrosophic graph $\mathcal{G}$ is complete if

$$\mathcal{T}_{\mathbb{B}_L}(kl) = \min\{\mathcal{T}_{\mathbb{A}_L}(k), \mathcal{T}_{\mathbb{A}_L}(l)\},\ \mathcal{T}_{\mathbb{B}_U}(kl) = \min\{\mathcal{T}_{\mathbb{A}_U}(k), \mathcal{T}_{\mathbb{A}_U}(l)\},$$

$$\mathcal{C}_{\mathbb{B}_L}(kl) = \min\{\mathcal{C}_{\mathbb{A}_L}(k), \mathcal{C}_{\mathbb{A}_L}(l)\},\ \mathcal{C}_{\mathbb{B}_U}(kl) = \min\{\mathcal{C}_{\mathbb{A}_U}(k), \mathcal{C}_{\mathbb{A}_U}(l)\},$$

$$\mathcal{U}_{\mathbb{B}_L}(kl) = \max\{\mathcal{U}_{\mathbb{A}_L}(k), \mathcal{U}_{\mathbb{A}_L}(l)\},\ \mathcal{U}_{\mathbb{B}_U}(kl) = \max\{\mathcal{U}_{\mathbb{A}_U}(k), \mathcal{U}_{\mathbb{A}_U}(l)\},$$

$$\mathcal{F}_{\mathbb{B}_L}(kl) = \max\{\mathcal{F}_{\mathbb{A}_L}(k), \mathcal{F}_{\mathbb{A}_L}(l)\},\ \mathcal{F}_{\mathbb{B}_U}(kl) = \max\{\mathcal{F}_{\mathbb{A}_U}(k), \mathcal{F}_{\mathbb{A}_U}(l)\}, \forall k, l \in \mathbb{A}.$$

Definition 6. 4.4 An interval quadripartitioned single-valued neutrosophic graph $\mathcal{G}$ is strong if

$$\mathcal{T}_{\mathbb{B}_L}(kl) = \min\{\mathcal{T}_{\mathbb{A}_L}(k), \mathcal{T}_{\mathbb{A}_L}(l)\},\ \mathcal{T}_{\mathbb{B}_U}(kl) = \min\{\mathcal{T}_{\mathbb{A}_U}(k), \mathcal{T}_{\mathbb{A}_U}(l)\},$$

$$\mathcal{C}_{\mathbb{B}_L}(kl) = \min\{\mathcal{C}_{\mathbb{A}_L}(k), \mathcal{C}_{\mathbb{A}_L}(l)\},\ \mathcal{C}_{\mathbb{B}_U}(kl) = \min\{\mathcal{C}_{\mathbb{A}_U}(k), \mathcal{C}_{\mathbb{A}_U}(l)\},$$

$$\mathcal{U}_{\mathbb{B}_L}(kl) = \max\{\mathcal{U}_{\mathbb{A}_L}(k), \mathcal{U}_{\mathbb{A}_L}(l)\},\ \mathcal{U}_{\mathbb{B}_U}(kl) = \max\{\mathcal{U}_{\mathbb{A}_U}(k), \mathcal{U}_{\mathbb{A}_U}(l)\},$$

$$\mathcal{F}_{\mathbb{B}_L}(kl) = \max\{\mathcal{F}_{\mathbb{A}_L}(k), \mathcal{F}_{\mathbb{A}_L}(l)\}, \mathcal{F}_{\mathbb{B}_U}(kl) = \max\{\mathcal{F}_{\mathbb{A}_U}(k), \mathcal{F}_{\mathbb{A}_U}(l)\}, \forall kl \in \mathbb{B}.$$

Definition 6.4.5 The Cartesian product $\mathcal{G}_1 \times \mathcal{G}_2$ of two interval quadripartitioned single-valued neutrosophic graphs, $\mathcal{G}_1 = (\mathbb{A}_1, \mathbb{B}_1)$ and $\mathcal{G}_2 = (\mathbb{A}_2, \mathbb{B}_2)$, of the crisp graphs $\mathcal{G}_1^* = (\mathbb{V}_1, \mathbb{E}_1)$ and $\mathcal{G}_2^* = (\mathbb{V}_2, \mathbb{E}_2)$ is defined as a pair $\mathcal{G}_1 \times \mathcal{G}_2 = (\mathbb{A}_1 \times \mathbb{A}_2, \mathbb{B}_1 \times \mathbb{B}_2)$. such that

1. $\left(\mathcal{T}_{\mathbb{A}_1L} \times \mathcal{T}_{\mathbb{A}_2L}\right)(k_1, k_2) = \min\{\mathcal{T}_{\mathbb{A}_1L}(k_1), \mathcal{T}_{\mathbb{A}_2L}(k_2)\}$

   $\left(\mathcal{T}_{\mathbb{A}_1U} \times \mathcal{T}_{\mathbb{A}_2U}\right)(k_1, k_2) = \min\{\mathcal{T}_{\mathbb{A}_1U}(k_1), \mathcal{T}_{\mathbb{A}_2U}(k_2)\}$

   $\left(\mathcal{C}_{\mathbb{A}_1L} \times \mathcal{C}_{\mathbb{A}_2L}\right)(k_1, k_2) = \min\{\mathcal{C}_{\mathbb{A}_1L}(k_1), \mathcal{C}_{\mathbb{A}_2L}(k_2)\}$

$$\left(\mathcal{C}_{\mathbb{A}_1U} \times \mathcal{C}_{\mathbb{A}_2U}\right)(k_1,k_2) = \min\left\{\mathcal{C}_{\mathbb{A}_1U}(k_1),\mathcal{C}_{\mathbb{A}_2U}(k_2)\right\}$$

$$\left(\mathcal{U}_{\mathbb{A}_1L} \times \mathcal{U}_{\mathbb{A}_2L}\right)(k_1,k_2) = \max\left\{\mathcal{U}_{\mathbb{A}_1L}(k_1),\mathcal{U}_{\mathbb{A}_2L}(k_2)\right\}$$

$$\left(\mathcal{U}_{\mathbb{A}_1U} \times \mathcal{U}_{\mathbb{A}_2U}\right)(k_1,k_2) = \max\left\{\mathcal{U}_{\mathbb{A}_1U}(k_1),\mathcal{U}_{\mathbb{A}_2U}(k_2)\right\},$$

$$\left(\mathcal{F}_{\mathbb{A}_1L} \times \mathcal{F}_{\mathbb{A}_2L}\right)(k_1,k_2) = \max\left\{\mathcal{F}_{\mathbb{A}_1L}(k_1),\mathcal{F}_{\mathbb{A}_2L}(k_2)\right\}$$

$$\left(\mathcal{F}_{\mathbb{A}_1U} \times \mathcal{U}_{\mathbb{A}_2U}\right)(k_1,k_2) = \max\left\{\mathcal{F}_{\mathbb{A}_1U}(k_1),\mathcal{F}_{\mathbb{A}_2U}(k_2)\right\}, \ \forall k_1,k_2 \in \mathbb{V},$$

2. $$\left(\mathcal{T}_{\mathbb{B}_1L} \times \mathcal{T}_{\mathbb{B}_2L}\right)\left((k,k_2)(k,l_2)\right) = \min\left\{\mathcal{T}_{\mathbb{A}_1L}(k),\mathcal{T}_{\mathbb{B}_2L}(k_2l_2)\right\}$$

$$\left(\mathcal{T}_{\mathbb{B}_1U} \times \mathcal{T}_{\mathbb{B}_2U}\right)\left((k,k_2)(k,l_2)\right) = \min\left\{\mathcal{T}_{\mathbb{A}_1U}(k),\mathcal{T}_{\mathbb{B}_2U}(k_2l_2)\right\}$$

$$\left(\mathcal{C}_{\mathbb{B}_1L} \times \mathcal{C}_{\mathbb{B}_2L}\right)\left((k,k_2)(k,l_2)\right) = \min\left\{\mathcal{C}_{\mathbb{A}_1L}(k),\mathcal{C}_{\mathbb{B}_2L}(k_2l_2)\right\}$$

$$\left(\mathcal{C}_{\mathbb{B}_1U} \times \mathcal{C}_{\mathbb{B}_2U}\right)\left((k,k_2)(k,l_2)\right) = \min\left\{\mathcal{C}_{\mathbb{A}_1U}(k),\mathcal{C}_{\mathbb{B}_2U}(k_2l_2)\right\}$$

$$\left(\mathcal{U}_{\mathbb{B}_1L} \times \mathcal{U}_{\mathbb{B}_2L}\right)\left((k,k_2)(k,l_2)\right) = \max\left\{\mathcal{U}_{\mathbb{A}_1L}(k),\mathcal{U}_{\mathbb{B}_2L}(k_2l_2)\right\}$$

$$\left(\mathcal{U}_{\mathbb{B}_1U} \times \mathcal{U}_{\mathbb{B}_2U}\right)\left((k,k_2)(k,l_2)\right) = \max\left\{\mathcal{U}_{\mathbb{A}_1U}(k),\mathcal{U}_{\mathbb{B}_2U}(k_2l_2)\right\},$$

$$\left(\mathcal{F}_{\mathbb{B}_1L} \times \mathcal{F}_{\mathbb{B}_2L}\right)\left((k,k_2)(k,l_2)\right) = \max\left\{\mathcal{F}_{\mathbb{A}_1L}(k),\mathcal{F}_{\mathbb{B}_2L}(k_2l_2)\right\}$$

$$\left(\mathcal{F}_{\mathbb{B}_1U} \times \mathcal{F}_{\mathbb{B}_2U}\right)\left((k,k_2)(k,l_2)\right) = \max\left\{\mathcal{F}_{\mathbb{A}_1U}(k),\mathcal{F}_{\mathbb{B}_2U}(k_2l_2)\right\},,$$

$$\forall k \in \mathbb{V}_1 \ and \ k_2l_2 \in \mathbb{E}_2,$$

and

3. $$\left(\mathcal{T}_{\mathbb{B}_1L} \times \mathcal{T}_{\mathbb{B}_2L}\right)\left((k_1,z)(l_1,z)\right) = \min\left\{\mathcal{T}_{\mathbb{B}_1L}(k_1l_1),\mathcal{T}_{\mathbb{A}_2L}(z)\right\}$$

$$\left(\mathcal{T}_{\mathbb{B}_1U} \times \mathcal{T}_{\mathbb{B}_2U}\right)\left((k_1,z)(l_1,z)\right) = \min\left\{\mathcal{T}_{\mathbb{B}_1U}(k_1l_1),\mathcal{T}_{\mathbb{A}_2U}(z)\right\}$$

$$\left(\mathcal{C}_{\mathbb{B}_1L} \times \mathcal{C}_{\mathbb{B}_2L}\right)\left((k_1,z)(l_1,z)\right) = \min\left\{\mathcal{C}_{\mathbb{B}_1L}(k_1l_1),\mathcal{C}_{\mathbb{A}_2L}(z)\right\}$$

$$\left(\mathcal{C}_{\mathbb{B}_1U} \times \mathcal{C}_{\mathbb{B}_2U}\right)\left((k_1,z)(l_1,z)\right) = \min\left\{\mathcal{C}_{\mathbb{B}_1U}(k_1l_1),\mathcal{C}_{\mathbb{A}_2U}(z)\right\}$$

$$\left(\mathcal{U}_{\mathbb{B}_1L} \times \mathcal{U}_{\mathbb{B}_2L}\right)\left((k_1,z)(l_1,z)\right) = \max\left\{\mathcal{U}_{\mathbb{B}_1L}(k_1l_1),\mathcal{U}_{\mathbb{A}_2L}(z)\right\}$$

$$\left(\mathcal{U}_{\mathbb{B}_1U} \times \mathcal{U}_{\mathbb{B}_2U}\right)\left((k_1,z)(l_1,z)\right) = \max\left\{\mathcal{U}_{\mathbb{B}_1U}(k_1l_1),\mathcal{U}_{\mathbb{A}_2U}(z)\right\},$$

$$\left(\mathcal{F}_{\mathbb{B}_1L} \times \mathcal{F}_{\mathbb{B}_2L}\right)\left((k_1,z)(l_1,z)\right) = \max\left\{\mathcal{F}_{\mathbb{B}_1L}(k_1l_1),\mathcal{F}_{\mathbb{A}_2L}(z)\right\}$$

$$\left(\mathcal{F}_{\mathbb{B}_1U} \times \mathcal{F}_{\mathbb{B}_2U}\right)\left((k_1,z)(l_1,z)\right) = \max\left\{\mathcal{F}_{\mathbb{B}_1U}(k_1l_1),\mathcal{F}_{\mathbb{A}_2U}(z)\right\},$$

$$\forall z \in \mathbb{V}_2 \ and \ k_1l_1 \in \mathbb{E}_1.$$

Proposition 6.4.6 The Cartesian product $\mathcal{G}_1 \times \mathcal{G}_2 = (\mathbb{A}_1 \times \mathbb{A}_2, \mathbb{B}_1 \times \mathbb{B}_2)$ of two interval quadripartitioned single-valued neutrosophic graphs $\mathcal{G}_1 = (\mathbb{A}_1, \mathbb{B}_1)$ and $\mathcal{G}_2 = (\mathbb{A}_2, \mathbb{B}_2)$ is an interval quadripartitioned single-valued neutrosophic graph.

Proof. The conditions for $\mathbb{A}_1 \times \mathbb{A}_2$ are evident and last are the only conditions for $\mathbb{B}_1 \times \mathbb{B}_2$. Let $k \in \mathbb{V}_1$ and $k_2 l_2 \in \mathbb{E}_2$. Then

$$\left(\mathcal{T}_{\mathbb{B}_1 L} \times \mathcal{T}_{\mathbb{B}_2 L}\right)\left((k,k_2)(k,l_2)\right) = \min\left\{\mathcal{T}_{\mathbb{A}_1 L}(k), \mathcal{T}_{\mathbb{B}_2 L}(k_2 l_2)\right\}$$

$$\leq \min\left\{\mathcal{T}_{\mathbb{A}_1 L}(k), \min\left\{\mathcal{T}_{\mathbb{A}_2 L}(k_2), \mathcal{T}_{\mathbb{A}_2 L}(l_2)\right\}\right\}$$

$$= \min\left\{\min\left\{\mathcal{T}_{\mathbb{A}_1 L}(k), \mathcal{T}_{\mathbb{A}_2 L}(k_2)\right\}, \min\left\{\mathcal{T}_{\mathbb{A}_1 L}(k), \mathcal{T}_{\mathbb{A}_2 L}(l_2)\right\}\right\}$$

$$= \min\left\{\left(\mathcal{T}_{\mathbb{A}_1 L} \times \mathcal{T}_{\mathbb{A}_2 L}\right)(k,k_2), \left(\mathcal{T}_{\mathbb{A}_1 L} \times \mathcal{T}_{\mathbb{A}_2 L}\right)(k,l_2)\right\}$$

$$\left(\mathcal{T}_{\mathbb{B}_1 U} \times \mathcal{T}_{\mathbb{B}_2 U}\right)\left((k,k_2)(k,l_2)\right) = \min\left\{\mathcal{T}_{\mathbb{A}_1 U}(k), \mathcal{T}_{\mathbb{B}_2 U}(k_2 l_2)\right\}$$

$$\leq \min\left\{\mathcal{T}_{\mathbb{A}_1 U}(k), \min\left\{\mathcal{T}_{\mathbb{A}_2 U}(k_2), \mathcal{T}_{\mathbb{A}_2 U}(l_2)\right\}\right\}$$

$$= \min\left\{\min\left\{\mathcal{T}_{\mathbb{A}_1 U}(k), \mathcal{T}_{\mathbb{A}_2 U}(k_2)\right\}, \min\left\{\mathcal{T}_{\mathbb{A}_1 U}(k), \mathcal{T}_{\mathbb{A}_2 U}(l_2)\right\}\right\}$$

$$= \min\left\{\left(\mathcal{T}_{\mathbb{A}_1 U} \times \mathcal{T}_{\mathbb{A}_2 U}\right)(k,k_2), \left(\mathcal{T}_{\mathbb{A}_1 U} \times \mathcal{T}_{\mathbb{A}_2 U}\right)(k,l_2)\right\}.$$

$$\left(\mathcal{C}_{\mathbb{B}_1 L} \times \mathcal{C}_{\mathbb{B}_2 L}\right)\left((k,k_2)(k,l_2)\right) = \min\left\{\mathcal{C}_{\mathbb{A}_1 L}(k), \mathcal{C}_{\mathbb{B}_2 L}(k_2 l_2)\right\}$$

$$\leq \min\left\{\mathcal{C}_{\mathbb{A}_1 L}(k), \min\left\{\mathcal{C}_{\mathbb{A}_2 L}(k_2), \mathcal{C}_{\mathbb{A}_2 L}(l_2)\right\}\right\}$$

$$= \min\left\{\min\left\{\mathcal{C}_{\mathbb{A}_1 L}(k), \mathcal{C}_{\mathbb{A}_2 L}(k_2)\right\}, \min\left\{\mathcal{C}_{\mathbb{A}_1 L}(k), \mathcal{C}_{\mathbb{A}_2 L}(l_2)\right\}\right\}$$

$$= \min\left\{\left(\mathcal{C}_{\mathbb{A}_1 L} \times \mathcal{C}_{\mathbb{A}_2 L}\right)(k,k_2), \left(\mathcal{C}_{\mathbb{A}_1 L} \times \mathcal{C}_{\mathbb{A}_2 L}\right)(k,l_2)\right\}$$

$$\left(\mathcal{C}_{\mathbb{B}_1 U} \times \mathcal{C}_{\mathbb{B}_2 U}\right)\left((k,k_2)(k,l_2)\right) = \min\left\{\mathcal{C}_{\mathbb{A}_1 U}(k), \mathcal{C}_{\mathbb{B}_2 U}(k_2 l_2)\right\}$$

$$\leq \min\left\{\mathcal{C}_{\mathbb{A}_1 U}(k), \min\left\{\mathcal{C}_{\mathbb{A}_2 U}(k_2), \mathcal{C}_{\mathbb{A}_2 U}(l_2)\right\}\right\}$$

$$= \min\left\{\min\left\{\mathcal{I}_{\mathbb{A}_1 U}(k), \mathcal{I}_{\mathbb{A}_2 U}(k_2)\right\}, \left\{\min\mathcal{C}_{\mathbb{A}_1 U}(k), \mathcal{C}_{\mathbb{A}_2 U}(l_2)\right\}\right\}$$

$$= \min\left\{\left(\mathcal{C}_{\mathbb{A}_1 U} \times \mathcal{C}_{\mathbb{A}_2 U}\right)(k,k_2), \left(\mathcal{C}_{\mathbb{A}_1 U} \times \mathcal{C}_{\mathbb{A}_2 U}\right)(k,l_2)\right\}.$$

$$\left(\mathcal{U}_{\mathbb{B}_1 L} \times \mathcal{U}_{\mathbb{B}_2 L}\right)\left((k,k_2)(k,l_2)\right) = \max\left\{\mathcal{U}_{\mathbb{A}_1 L}(k), \mathcal{U}_{\mathbb{B}_2 L}(k_2 l_2)\right\}$$

$$\leq \max\left\{\mathcal{U}_{\mathbb{A}_1 L}(k), \max\left\{\mathcal{U}_{\mathbb{A}_2 L}(k_2), \mathcal{U}_{\mathbb{A}_2 L}(l_2)\right\}\right\}$$

$$= \max\left\{\max\left\{\mathcal{U}_{\mathbb{A}_1 L}(k), \mathcal{U}_{\mathbb{A}_2 L}(k_2)\right\}, \max\left\{\mathcal{U}_{\mathbb{A}_1 L}(k), \mathcal{U}_{\mathbb{A}_2 L}(l_2)\right\}\right\}$$

$$= \max\left\{\left(\mathcal{U}_{\mathbb{A}_1L} \times \mathcal{U}_{\mathbb{A}_2L}\right)(k,k_2),\left(\mathcal{U}_{\mathbb{A}_1L} \times \mathcal{U}_{\mathbb{A}_2L}\right)(k,l_2)\right\}$$

$$\left(\mathcal{U}_{\mathbb{B}_1U} \times \mathcal{U}_{\mathbb{B}_2U}\right)\left((k,k_2)(k,l_2)\right) = \max\left\{\mathcal{U}_{\mathbb{A}_1U}(k),\mathcal{U}_{\mathbb{B}_2U}(k_2l_2)\right\}$$

$$\leq \max\left\{\mathcal{U}_{\mathbb{A}_1U}(k),\max\left\{\mathcal{U}_{\mathbb{A}_2U}(k_2),\mathcal{U}_{\mathbb{A}_2U}(l_2)\right\}\right\}$$

$$= \max\left\{\max\left\{\mathcal{I}_{\mathbb{A}_1U}(k),\mathcal{I}_{\mathbb{A}_2U}(k_2)\right\},\left\{\max\mathcal{U}_{\mathbb{A}_1U}(k),\mathcal{U}_{\mathbb{A}_2U}(l_2)\right\}\right\}$$

$$= \max\left\{\left(\mathcal{U}_{\mathbb{A}_1U} \times \mathcal{U}_{\mathbb{A}_2U}\right)(k,k_2),\left(\mathcal{U}_{\mathbb{A}_1U} \times \mathcal{U}_{\mathbb{A}_2U}\right)(k,l_2)\right\}.$$

$$\left(\mathcal{F}_{\mathbb{B}_1L} \times \mathcal{F}_{\mathbb{B}_2L}\right)\left((k,k_2)(k,l_2)\right) = \max\left\{\mathcal{F}_{\mathbb{A}_1L}(k),\mathcal{F}_{\mathbb{B}_2L}(k_2l_2)\right\}$$

$$\leq \max\left\{\mathcal{F}_{\mathbb{A}_1L}(k),\max\left\{\mathcal{F}_{\mathbb{A}_2L}(k_2),\mathcal{F}_{\mathbb{A}_2L}(l_2)\right\}\right\}$$

$$= \max\left\{\max\left\{\mathcal{F}_{\mathbb{A}_1L}(k),\mathcal{F}_{\mathbb{A}_2L}(k_2)\right\},\max\left\{\mathcal{F}_{\mathbb{A}_1L}(k),\mathcal{F}_{\mathbb{A}_2L}(l_2)\right\}\right\}$$

$$= \max\left\{\left(\mathcal{F}_{\mathbb{A}_1L} \times \mathcal{F}_{\mathbb{A}_2L}\right)(k,k_2),\left(\mathcal{F}_{\mathbb{A}_1L} \times \mathcal{F}_{\mathbb{A}_2L}\right)(k,l_2)\right\}$$

$$\left(\mathcal{F}_{\mathbb{B}_1U} \times \mathcal{F}_{\mathbb{B}_2U}\right)\left((k,k_2)(k,l_2)\right) = \max\left\{\mathcal{F}_{\mathbb{A}_1U}(k),\mathcal{F}_{\mathbb{B}_2U}(k_2l_2)\right\}$$

$$\leq \max\left\{\mathcal{F}_{\mathbb{A}_1U}(k),\max\left\{\mathcal{F}_{\mathbb{A}_2U}(k_2),\mathcal{F}_{\mathbb{A}_2U}(l_2)\right\}\right\}$$

$$= \max\left\{\max\left\{\mathcal{F}_{\mathbb{A}_1U}(k),\mathcal{F}_{\mathbb{A}_2U}(k_2)\right\},\left\{\max\mathcal{F}_{\mathbb{A}_1U}(k),\mathcal{F}_{\mathbb{A}_2U}(l_2)\right\}\right\}$$

$$= \max\left\{\left(\mathcal{F}_{\mathbb{A}_1U} \times \mathcal{F}_{\mathbb{A}_2U}\right)(k,k_2),\left(\mathcal{F}_{\mathbb{A}_1U} \times \mathcal{F}_{\mathbb{A}_2U}\right)(k,l_2)\right\}.$$

Similarly, for $z \in \mathbb{V}_2$ and $k_1, l_1 \in \mathbb{E}_1$, we obtained

$$\left(\mathcal{T}_{\mathbb{B}_1L} \times \mathcal{T}_{\mathbb{B}_2L}\right)\left((k_1,z)(l_1,z)\right) = \min\left\{\mathcal{T}_{\mathbb{B}_1L}(k_1l_1),\mathcal{T}_{\mathbb{A}_2L}(z)\right\}$$

$$\leq \min\left\{\min\left\{\mathcal{T}_{\mathbb{A}_1L}(k_1),\mathcal{T}_{\mathbb{A}_1L}(l_1)\right\},\mathcal{T}_{\mathbb{A}_2L}(z)\right\}$$

$$= \min\left\{\min\left\{\mathcal{T}_{\mathbb{A}_1L}(k_1),\mathcal{T}_{\mathbb{A}_2L}(z)\right\},\min\left\{\mathcal{T}_{\mathbb{A}_1L}(l_1),\mathcal{T}_{\mathbb{A}_2L}(z)\right\}\right\}$$

$$= \min\left\{\left(\mathcal{T}_{\mathbb{A}_1L} \times \mathcal{T}_{\mathbb{A}_2L}\right)(k_1,z),\left(\mathcal{T}_{\mathbb{A}_1L} \times \mathcal{T}_{\mathbb{A}_2L}\right)(l_1,z)\right\}$$

$$\left(\mathcal{T}_{\mathbb{B}_1U} \times \mathcal{T}_{\mathbb{B}_2U}\right)\left((k_1,z)(l_1,z)\right) = \min\left\{\mathcal{T}_{\mathbb{B}_1U}(k_1l_1),\mathcal{T}_{\mathbb{A}_2U}(z)\right\}$$

$$\leq \min\left\{\min\left\{\mathcal{T}_{\mathbb{A}_1U}(k_1),\mathcal{T}_{\mathbb{A}_1U}(l_1)\right\},\mathcal{T}_{\mathbb{A}_2U}(z)\right\}$$

$$= \min\left\{\min\left\{\mathcal{T}_{\mathbb{A}_1U}(k_1),\mathcal{T}_{\mathbb{A}_2U}(z)\right\},\min\left\{\mathcal{T}_{\mathbb{A}_1U}(l_1),\mathcal{T}_{\mathbb{A}_2U}(z)\right\}\right\}$$

$$= \min\left\{\left(\mathcal{T}_{\mathbb{A}_1U} \times \mathcal{T}_{\mathbb{A}_2U}\right)(k_1,z),\left(\mathcal{T}_{\mathbb{A}_1U} \times \mathcal{T}_{\mathbb{A}_2U}\right)(l_1,z)\right\}$$

$$\left(\mathcal{C}_{\mathbb{B}_1L} \times \mathcal{C}_{\mathbb{B}_2L}\right)\left((k_1,z)(l_1,z)\right) = \min\left\{\mathcal{C}_{\mathbb{B}_1L}\left(k_1l_1\right), \mathcal{C}_{\mathbb{A}_2L}\left(z\right)\right\}$$

$$\leq \min\left\{\min\left(\mathcal{C}_{\mathbb{A}_1L}\left(k_1\right), \mathcal{C}_{\mathbb{A}_1L}\left(l_1\right)\right), \mathcal{C}_{\mathbb{A}_2L}\left(z\right)\right\}$$

$$= \min\left\{\min\left\{\mathcal{C}_{\mathbb{A}_1L}\left(k_1\right), \mathcal{C}_{\mathbb{A}_2L}\left(z\right)\right\}, \min\left\{\mathcal{C}_{\mathbb{A}_1L}\left(l_1\right), \mathcal{C}_{\mathbb{A}_2L}\left(z\right)\right\}\right\}$$

$$= \min\left\{\left(\mathcal{C}_{\mathbb{A}_1L} \times \mathcal{C}_{\mathbb{A}_2L}\right)\left(k_1,z\right), \left(\mathcal{C}_{\mathbb{A}_1L} \times \mathcal{C}_{\mathbb{A}_2L}\right)\left(l_1,z\right)\right\}$$

$$\left(\mathcal{C}_{\mathbb{B}_1U} \times \mathcal{C}_{\mathbb{B}_2U}\right)\left((k_1,z)(l_1,z)\right) = \min\left\{\mathcal{C}_{\mathbb{B}_1U}\left(k_1l_1\right), \mathcal{C}_{\mathbb{A}_2U}\left(z\right)\right\}$$

$$\leq \min\left\{\min\left(\mathcal{C}_{\mathbb{A}_1U}\left(k_1\right), \mathcal{C}_{\mathbb{A}_1U}\left(l_1\right)\right), \mathcal{C}_{\mathbb{A}_2U}\left(z\right)\right\}$$

$$= \min\left\{\min\left(\mathcal{C}_{\mathbb{A}_1U}\left(k_1\right), \mathcal{C}_{\mathbb{A}_2U}\left(z\right)\right), \min\left\{\mathcal{C}_{\mathbb{A}_1U}\left(l_1\right), \mathcal{C}_{\mathbb{A}_2U}\left(z\right)\right\}\right\}$$

$$= \min\left\{\left(\mathcal{C}_{\mathbb{A}_1U} \times \mathcal{C}_{\mathbb{A}_2U}\right)\left(k_1,z\right), \left(\mathcal{C}_{\mathbb{A}_1U} \times \mathcal{C}_{\mathbb{A}_2U}\right)\left(l_1,z\right)\right\}$$

$$\left(\mathcal{U}_{\mathbb{B}_1L} \times \mathcal{U}_{\mathbb{B}_2L}\right)\left((k_1,z)(l_1,z)\right) = \max\left\{\mathcal{U}_{\mathbb{B}_1L}\left(k_1l_1\right), \mathcal{U}_{\mathbb{A}_2L}\left(z\right)\right\}$$

$$\leq \max\left\{\max\left(\mathcal{U}_{\mathbb{A}_1L}\left(k_1\right), \mathcal{U}_{\mathbb{A}_1L}\left(l_1\right)\right), \mathcal{U}_{\mathbb{A}_2L}\left(z\right)\right\}$$

$$= \max\left\{\max\left\{\mathcal{U}_{\mathbb{A}_1L}\left(k_1\right), \mathcal{U}_{\mathbb{A}_2L}\left(z\right)\right\}, \max\left\{\mathcal{U}_{\mathbb{A}_1L}\left(l_1\right), \mathcal{U}_{\mathbb{A}_2L}\left(z\right)\right\}\right\}$$

$$= \max\left\{\left(\mathcal{U}_{\mathbb{A}_1L} \times \mathcal{U}_{\mathbb{A}_2L}\right)\left(k_1,z\right), \left(\mathcal{U}_{\mathbb{A}_1L} \times \mathcal{U}_{\mathbb{A}_2L}\right)\left(l_1,z\right)\right\}$$

$$\left(\mathcal{U}_{\mathbb{B}_1U} \times \mathcal{U}_{\mathbb{B}_2U}\right)\left((k_1,z)(l_1,z)\right) = \max\left\{\mathcal{U}_{\mathbb{B}_1U}\left(k_1l_1\right), \mathcal{U}_{\mathbb{A}_2U}\left(z\right)\right\}$$

$$\leq \max\left\{\max\left(\mathcal{U}_{\mathbb{A}_1U}\left(k_1\right), \mathcal{U}_{\mathbb{A}_1U}\left(l_1\right)\right), \mathcal{U}_{\mathbb{A}_2U}\left(z\right)\right\}$$

$$= \max\left\{\max\left(\mathcal{U}_{\mathbb{A}_1U}\left(k_1\right), \mathcal{U}_{\mathbb{A}_2U}\left(z\right)\right), \max\left\{\mathcal{U}_{\mathbb{A}_1U}\left(l_1\right), \mathcal{U}_{\mathbb{A}_2U}\left(z\right)\right\}\right\}$$

$$= \max\left\{\left(\mathcal{U}_{\mathbb{A}_1U} \times \mathcal{U}_{\mathbb{A}_2U}\right)\left(k_1,z\right), \left(\mathcal{U}_{\mathbb{A}_1U} \times \mathcal{U}_{\mathbb{A}_2U}\right)\left(l_1,z\right)\right\}$$

$$\left(\mathcal{F}_{\mathbb{B}_1L} \times \mathcal{F}_{\mathbb{B}_2L}\right)\left((k_1,z)(l_1,z)\right) = \max\left\{\mathcal{F}_{\mathbb{B}_1L}\left(k_1l_1\right), \mathcal{F}_{\mathbb{A}_2L}\left(z\right)\right\}$$

$$\leq \max\left\{\max\left\{\mathcal{F}_{\mathbb{A}_1L}\left(k_1\right), \mathcal{F}_{\mathbb{A}_1L}\left(l_1\right)\right\}, \mathcal{F}_{\mathbb{A}_2L}\left(z\right)\right\}$$

$$= \max\left\{\max\left\{\mathcal{F}_{\mathbb{A}_1L}\left(k_1\right), \mathcal{F}_{\mathbb{A}_2L}\left(z\right)\right\}, \max\left\{\mathcal{F}_{\mathbb{A}_1L}\left(l_1\right), \mathcal{F}_{\mathbb{A}_2L}\left(z\right)\right\}\right\}$$

$$= \max\left\{\left(\mathcal{F}_{\mathbb{A}_1L} \times \mathcal{F}_{\mathbb{A}_2L}\right)\left(k_1,z\right), \left(\mathcal{F}_{\mathbb{A}_1L} \times \mathcal{F}_{\mathbb{A}_2L}\right)\left(l_1,z\right)\right\}$$

$$\left(\mathcal{F}_{\mathbb{B}_1U} \times \mathcal{F}_{\mathbb{B}_2U}\right)\left((k_1,z)(l_1,z)\right) = \max\left\{\mathcal{F}_{\mathbb{B}_1U}\left(k_1l_1\right), \mathcal{F}_{\mathbb{A}_2U}\left(z\right)\right\}$$

$$\leq \max\left\{\max\left\{\mathcal{F}_{\mathbb{A}_1U}\left(k_1\right), \mathcal{F}_{\mathbb{A}_1U}\left(l_1\right)\right\}, \mathcal{F}_{\mathbb{A}_2U}\left(z\right)\right\}$$

$$= \max\left\{\max\left\{\mathcal{F}_{\mathbb{A}_1U}(k_1), \mathcal{F}_{\mathbb{A}_2U}(z)\right\}, \max\left\{\mathcal{F}_{\mathbb{A}_1U}(l_1), \mathcal{F}_{\mathbb{A}_2U}(z)\right\}\right\}$$

$$= \max\left\{\left(\mathcal{F}_{\mathbb{A}_1U} \times \mathcal{F}_{\mathbb{A}_2U}\right)(k_1, z), \left(\mathcal{F}_{\mathbb{A}_1U} \times \mathcal{F}_{\mathbb{A}_2U}\right)(l_1, z)\right\}.$$

Definition 6.4.7 The cross product $\mathcal{G}_1 \star \mathcal{G}_2$ of two interval quadripartitioned single-valued neutrosophic graph $\mathcal{G}_1 = (\mathbb{A}_1, \mathbb{B}_1)$ and $\mathcal{G}_2 = (\mathbb{A}_2, \mathbb{B}_2)$ of the crisp graphs $\mathcal{G}_1^* = (\mathbb{V}_1, \mathbb{E}_1)$ and $\mathcal{G}_2^* = (\mathbb{V}_2, \mathbb{E}_2)$ is defined as a pair $\mathcal{G}_1 \star \mathcal{G}_2 = (\mathbb{A}_1 \star \mathbb{A}_2, \mathbb{B}_1 \star \mathbb{B}_2)$, such that

1. $\left(\mathcal{T}_{\mathbb{A}_1L} \star \mathcal{T}_{\mathbb{A}_2L}\right)(k_1, k_2) = \min\left\{\mathcal{T}_{\mathbb{A}_1L}(k_1), \mathcal{T}_{\mathbb{A}_2L}(k_2)\right\}$

$\left(\mathcal{T}_{\mathbb{A}_1U} \star \mathcal{T}_{\mathbb{A}_2U}\right)(k_1, k_2) = \min\left\{\mathcal{T}_{\mathbb{A}_1U}(k_1), \mathcal{T}_{\mathbb{A}_2U}(k_2)\right\}$

$\left(\mathcal{C}_{\mathbb{A}_1L} \star \mathcal{C}_{\mathbb{A}_2L}\right)(k_1, k_2) = \min\left\{\mathcal{C}_{\mathbb{A}_1L}(k_1), \mathcal{C}_{\mathbb{A}_2L}(k_2)\right\}$

$\left(\mathcal{C}_{\mathbb{A}_1U} \star \mathcal{C}_{\mathbb{A}_2U}\right)(k_1, k_2) = \min\left\{\mathcal{C}_{\mathbb{A}_1U}(k_1), \mathcal{C}_{\mathbb{A}_2U}(k_2)\right\}$

$\left(\mathcal{U}_{\mathbb{A}_1L} \star \mathcal{U}_{\mathbb{A}_2L}\right)(k_1, k_2) = \max\left\{\mathcal{U}_{\mathbb{A}_1L}(k_1), \mathcal{U}_{\mathbb{A}_2L}(k_2)\right\}$

$\left(\mathcal{U}_{\mathbb{A}_1U} \star \mathcal{U}_{\mathbb{A}_2U}\right)(k_1, k_2) = \max\left\{\mathcal{U}_{\mathbb{A}_1U}(k_1), \mathcal{U}_{\mathbb{A}_2U}(k_2)\right\},$

$\left(\mathcal{F}_{\mathbb{A}_1L} \star \mathcal{F}_{\mathbb{A}_2L}\right)(k_1, k_2) = \max\left\{\mathcal{F}_{\mathbb{A}_1L}(k_1), \mathcal{F}_{\mathbb{A}_2L}(k_2)\right\}$

$\left(\mathcal{F}_{\mathbb{A}_1U} \star \mathcal{U}_{\mathbb{A}_2U}\right)(k_1, k_2) = \max\left\{\mathcal{F}_{\mathbb{A}_1U}(k_1), \mathcal{F}_{\mathbb{A}_2U}(k_2)\right\}, \ \forall k_1, k_2 \in \mathbb{V},$

2. $\left(\mathcal{T}_{\mathbb{B}_1L} \star \mathcal{T}_{\mathbb{B}_2L}\right)((k_1, z_1)(k_2, z_2)) = \min\left\{\mathcal{T}_{\mathbb{B}_1L}(k_1k_2), \mathcal{T}_{\mathbb{B}_2L}(z_1z_2)\right\}$

$\left(\mathcal{T}_{\mathbb{B}_1U} \star \mathcal{T}_{\mathbb{B}_2U}\right)((k_1, z_1)(k_2, z_2)) = \min\left\{\mathcal{T}_{\mathbb{B}_1U}(k_1k_2), \mathcal{T}_{\mathbb{B}_2U}(z_1z_2)\right\}$

$\left(\mathcal{C}_{\mathbb{B}_1L} \star \mathcal{C}_{\mathbb{B}_2L}\right)((k_1, z_1)(k_2, z_2)) = \min\left\{\mathcal{C}_{\mathbb{B}_1L}(k_1k_2), \mathcal{C}_{\mathbb{B}_2L}(z_1z_2)\right\}$

$\left(\mathcal{C}_{\mathbb{B}_1U} \star \mathcal{C}_{\mathbb{B}_2U}\right)((k_1, z_1)(k_2, z_2)) = \min\left\{\mathcal{C}_{\mathbb{B}_1U}(k_1k_2), \mathcal{C}_{\mathbb{B}_2U}(z_1z_2)\right\}$

$\left(\mathcal{U}_{\mathbb{B}_1L} \star \mathcal{T}_{\mathbb{B}_2L}\right)((k_1, z_1)(k_2, z_2)) = \max\left\{\mathcal{U}_{\mathbb{B}_1L}(k_1k_2), \mathcal{U}_{\mathbb{B}_2L}(z_1z_2)\right\}$

$\left(\mathcal{U}_{\mathbb{B}_1U} \star \mathcal{U}_{\mathbb{B}_2U}\right)((k_1, z_1)(k_2, z_2)) = \max\left\{\mathcal{U}_{\mathbb{B}_1U}(k_1k_2), \mathcal{U}_{\mathbb{B}_2U}(z_1z_2)\right\}$

$\left(\mathcal{F}_{\mathbb{B}_1L} \star \mathcal{F}_{\mathbb{B}_2L}\right)((k_1, z_1)(k_2, z_2)) = \max\left\{\mathcal{F}_{\mathbb{B}_1L}(k_1k_2), \mathcal{F}_{\mathbb{B}_2L}(z_1z_2)\right\}$

$\left(\mathcal{F}_{\mathbb{B}_1U} \star \mathcal{F}_{\mathbb{B}_2U}\right)((k_1, z_1)(k_2, z_2)) = \max\left\{\mathcal{F}_{\mathbb{B}_1U}(k_1k_2), \mathcal{F}_{\mathbb{B}_2U}(z_1z_2)\right\}$

$\forall k_1k_2 \in \mathbb{E}_1 \ and \ z_1z_2 \in \mathbb{E}_2$

Proposition 6.4.8 The cross product $\mathcal{G}_1 \times \mathcal{G}_2 = (\mathbb{A}_1 \star \mathbb{A}_2, \mathbb{B}_1 \star \mathbb{B}_2)$ of two interval quadripartitioned single-valued neutrosophic graphs $\mathcal{G}_1 = (\mathbb{A}_1, \mathbb{B}_1)$ and $\mathcal{G}_2 = (\mathbb{A}_2, \mathbb{B}_2)$ is an interval quadripartitioned single-valued neutrosophic graph.

Proof. The conditions for $\mathbb{A}_1 \star \mathbb{A}_2$ are obvious and remaining are the only conditions for $\mathbb{B}_1 \star \mathbb{B}_2$. Let $k_1k_2 \in \mathbb{E}_1$ and $z_1z_2 \in \mathbb{E}_2$. Then

$$\left(\mathcal{T}_{\mathbb{B}_1L} \star \mathcal{T}_{\mathbb{B}_2L}\right)\left((k_1,z_1)(k_2,z_2)\right) = \min\left\{\mathcal{T}_{\mathbb{B}_1L}(k_1k_2), \mathcal{T}_{\mathbb{B}_2L}(z_1z_2)\right\}$$

$$\leq \min\left\{\min\left\{\mathcal{T}_{\mathbb{A}_1L}(k_1), \mathcal{T}_{\mathbb{A}_1L}(k_2)\right\}, \min\left\{\mathcal{T}_{\mathbb{A}_2L}(z_1), \mathcal{T}_{\mathbb{A}_2L}(z_2)\right\}\right\}$$

$$= \min\left\{\min\left\{\mathcal{T}_{\mathbb{A}_1L}(k_1), \mathcal{T}_{\mathbb{A}_2L}(z_1)\right\}, \min\left\{\mathcal{T}_{\mathbb{A}_1L}(k_2), \mathcal{T}_{\mathbb{A}_2L}(z_2)\right\}\right\}$$

$$= \min\left\{\left(\mathcal{T}_{\mathbb{A}_1L} \star \mathcal{T}_{\mathbb{A}_2L}\right)(k_1,z_1), \left(\mathcal{T}_{\mathbb{A}_1L} \star \mathcal{T}_{\mathbb{A}_2L}\right)(k_2,z_2)\right\}$$

$$\left(\mathcal{T}_{\mathbb{B}_1U} \star \mathcal{T}_{\mathbb{B}_2U}\right)\left((k_1,z_1)(k_2,z_2)\right) = \min\left\{\mathcal{T}_{\mathbb{B}_1U}(k_1k_2), \mathcal{T}_{\mathbb{B}_2U}(z_1z_2)\right\}$$

$$\leq \min\left\{\min\left\{\mathcal{T}_{\mathbb{A}_1U}(k_1), \mathcal{T}_{\mathbb{A}_1U}(k_2)\right\}, \min\left\{\mathcal{T}_{\mathbb{A}_2U}(z_1), \mathcal{T}_{\mathbb{A}_2U}(z_2)\right\}\right\}$$

$$= \min\left\{\min\left\{\mathcal{T}_{\mathbb{A}_1U}(k_1), \mathcal{T}_{\mathbb{A}_2U}(z_1)\right\}, \min\left\{\mathcal{T}_{\mathbb{A}_1U}(k_2), \mathcal{T}_{\mathbb{A}_2U}(z_2)\right\}\right\}$$

$$= \min\left\{\left(\mathcal{T}_{\mathbb{A}_1U} \star \mathcal{T}_{\mathbb{A}_2U}\right)(k_1,z_1), \left(\mathcal{T}_{\mathbb{A}_1U} \star \mathcal{T}_{\mathbb{A}_2U}\right)(k_2,z_2)\right\}$$

$$\left(\mathcal{C}_{\mathbb{B}_1L} \star \mathcal{C}_{\mathbb{B}_2L}\right)\left((k_1,z_1)(k_2,z_2)\right) = \min\left\{\mathcal{C}_{\mathbb{B}_1L}(k_1k_2), \mathcal{C}_{\mathbb{B}_2L}(z_1z_2)\right\}$$

$$\leq \min\left\{\min\left\{\mathcal{C}_{\mathbb{A}_1L}(k_1), \mathcal{C}_{\mathbb{A}_1L}(k_2)\right\}, \min\left\{\mathcal{C}_{\mathbb{A}_2L}(z_1), \mathcal{C}_{\mathbb{A}_2L}(z_2)\right\}\right\}$$

$$= \min\left\{\min\left\{\mathcal{C}_{\mathbb{A}_1L}(k_1), \mathcal{C}_{\mathbb{A}_2L}(z_1)\right\}, \min\left\{\mathcal{C}_{\mathbb{A}_1L}(k_2), \mathcal{C}_{\mathbb{A}_2L}(z_2)\right\}\right\}$$

$$= \min\left\{\left(\mathcal{C}_{\mathbb{A}_1L} \star \mathcal{C}_{\mathbb{A}_2L}\right)(k_1,z_1), \left(\mathcal{C}_{\mathbb{A}_1L} \star \mathcal{C}_{\mathbb{A}_2L}\right)(k_2,z_2)\right\}$$

$$\left(\mathcal{C}_{\mathbb{B}_1U} \star \mathcal{C}_{\mathbb{B}_2U}\right)\left((k_1,z_1)(k_2,z_2)\right) = \min\left\{\mathcal{C}_{\mathbb{B}_1U}(k_1k_2), \mathcal{C}_{\mathbb{B}_2U}(z_1z_2)\right\}$$

$$\leq \min\left\{\min\left\{\mathcal{C}_{\mathbb{A}_1U}(k_1), \mathcal{C}_{\mathbb{A}_1U}(k_2)\right\}, \min\left\{\mathcal{C}_{\mathbb{A}_2U}(z_1), \mathcal{C}_{\mathbb{A}_2U}(z_2)\right\}\right\}$$

$$= \min\left\{\min\left\{\mathcal{C}_{\mathbb{A}_1U}(k_1), \mathcal{C}_{\mathbb{A}_2U}(z_1)\right\}, \min\left\{\mathcal{C}_{\mathbb{A}_1U}(k_2), \mathcal{C}_{\mathbb{A}_2U}(z_2)\right\}\right\}$$

$$= \min\left\{\left(\mathcal{C}_{\mathbb{A}_1U} \star \mathcal{C}_{\mathbb{A}_2U}\right)(k_1,z_1), \left(\mathcal{C}_{\mathbb{A}_1U} \star \mathcal{C}_{\mathbb{A}_2U}\right)(k_2,z_2)\right\}$$

$$\left(\mathcal{U}_{\mathbb{B}_1L} \star \mathcal{U}_{\mathbb{B}_2L}\right)\left((k_1,z_1)(k_2,z_2)\right) = \max\left\{\mathcal{U}_{\mathbb{B}_1L}(k_1k_2), \mathcal{U}_{\mathbb{B}_2L}(z_1z_2)\right\}$$

$$\leq \max\left\{\max\left\{\mathcal{U}_{\mathbb{A}_1L}(k_1), \mathcal{U}_{\mathbb{A}_1L}(k_2)\right\}, \max\left\{\mathcal{U}_{\mathbb{A}_2L}(z_1), \mathcal{U}_{\mathbb{A}_2L}(z_2)\right\}\right\}$$

$$= \max\left\{\max\left\{\mathcal{U}_{\mathbb{A}_1L}(k_1), \mathcal{U}_{\mathbb{A}_2L}(z_1)\right\}, \max\left\{\mathcal{U}_{\mathbb{A}_1L}(k_2), \mathcal{U}_{\mathbb{A}_2L}(z_2)\right\}\right\}$$

$$= \max\left\{\left(\mathcal{U}_{\mathbb{A}_1L} \star \mathcal{U}_{\mathbb{A}_2L}\right)(k_1,z_1),\left(\mathcal{U}_{\mathbb{A}_1L} \star \mathcal{U}_{\mathbb{A}_2L}\right)(k_2,z_2)\right\}$$

$$\left(\mathcal{U}_{\mathbb{B}_1U} \star \mathcal{U}_{\mathbb{B}_2U}\right)\left((k_1,z_1)(k_2,z_2)\right) = \max\left\{\mathcal{U}_{\mathbb{B}_1U}(k_1k_2),\mathcal{U}_{\mathbb{B}_2U}(z_1z_2)\right\}$$

$$\leq \max\left\{\max\left\{\mathcal{U}_{\mathbb{A}_1U}(k_1),\mathcal{U}_{\mathbb{A}_1U}(k_2)\right\},\max\left\{\mathcal{U}_{\mathbb{A}_2U}(z_1),\mathcal{U}_{\mathbb{A}_2U}(z_2)\right\}\right\}$$

$$= \max\left\{\max\left\{\mathcal{U}_{\mathbb{A}_1U}(k_1),\mathcal{U}_{\mathbb{A}_2U}(z_1)\right\},\max\left\{\mathcal{U}_{\mathbb{A}_1U}(k_2),\mathcal{U}_{\mathbb{A}_2U}(z_2)\right\}\right\}$$

$$= \max\left\{\left(\mathcal{U}_{\mathbb{A}_1U} \star \mathcal{U}_{\mathbb{A}_2U}\right)(k_1,z_1),\left(\mathcal{U}_{\mathbb{A}_1U} \star \mathcal{U}_{\mathbb{A}_2U}\right)(k_2,z_2)\right\}$$

$$\left(\mathcal{F}_{\mathbb{B}_1L} \star \mathcal{F}_{\mathbb{B}_2L}\right)\left((k_1,z_1)(k_2,z_2)\right) = \max\left\{\mathcal{F}_{\mathbb{B}_1L}(k_1k_2),\mathcal{F}_{\mathbb{B}_2L}(z_1z_2)\right\}$$

$$\leq \max\left\{\max\left\{\mathcal{F}_{\mathbb{A}_1L}(k_1),\mathcal{F}_{\mathbb{A}_1L}(k_2)\right\},\max\left\{\mathcal{F}_{\mathbb{A}_2L}(z_1),\mathcal{F}_{\mathbb{A}_2L}(z_2)\right\}\right\}$$

$$= \max\left\{\max\left\{\mathcal{F}_{\mathbb{A}_1L}(k_1),\mathcal{F}_{\mathbb{A}_2L}(z_1)\right\},\max\left\{\mathcal{F}_{\mathbb{A}_1L}(k_2),\mathcal{F}_{\mathbb{A}_2L}(z_2)\right\}\right\}$$

$$= \max\left\{\left(\mathcal{F}_{\mathbb{A}_1L} \star \mathcal{F}_{\mathbb{A}_2L}\right)(k_1,z_1),\left(\mathcal{F}_{\mathbb{A}_1L} \star \mathcal{F}_{\mathbb{A}_2L}\right)(k_2,z_2)\right\}$$

$$\left(\mathcal{F}_{\mathbb{B}_1U} \star \mathcal{F}_{\mathbb{B}_2U}\right)\left((k_1,z_1)(k_2,z_2)\right) = \max\left\{\mathcal{F}_{\mathbb{B}_1U}(k_1k_2),\mathcal{F}_{\mathbb{B}_2U}(z_1z_2)\right\}$$

$$\leq \max\left\{\max\left\{\mathcal{F}_{\mathbb{A}_1U}(k_1),\mathcal{F}_{\mathbb{A}_1U}(k_2)\right\},\max\left\{\mathcal{F}_{\mathbb{A}_2U}(z_1),\mathcal{F}_{\mathbb{A}_2U}(z_2)\right\}\right\}$$

$$= \max\left\{\max\left\{\mathcal{F}_{\mathbb{A}_1U}(k_1),\mathcal{F}_{\mathbb{A}_2U}(z_1)\right\},\max\left\{\mathcal{F}_{\mathbb{A}_1U}(k_2),\mathcal{F}_{\mathbb{A}_2U}(z_2)\right\}\right\}$$

$$= \max\left\{\left(\mathcal{F}_{\mathbb{A}_1U} \star \mathcal{F}_{\mathbb{A}_2U}\right)(k_1,z_1),\left(\mathcal{F}_{\mathbb{A}_1U} \star \mathcal{F}_{\mathbb{A}_2U}\right)(k_2,z_2)\right\}$$

This completes the proof.

Definition 6.4.9 The lexicographic product $\mathcal{G}_1 \times \mathcal{G}_2$ of two interval quadripartitioned single-valued neutrosophic graphs $\mathcal{G}_1 = (\mathbb{A}_1,\mathbb{B}_1)$ and $\mathcal{G}_2 = (\mathbb{A}_2,\mathbb{B}_2)$ of the crisp graphs $\mathcal{G}_1^* = (\mathbb{V}_1,\mathbb{E}_1)$ and $\mathcal{G}_2^* = (\mathbb{V}_2,\mathbb{E}_2)$ is defined as a pair $\mathcal{G}_1 \times \mathcal{G}_2 = (\mathbb{A}_1 \times \mathbb{A}_2, \mathbb{B}_1 \times \mathbb{B}_2)$ such that

1. $\left(\mathcal{T}_{\mathbb{A}_1L} \times \mathcal{T}_{\mathbb{A}_2L}\right)(k_1,k_2) = \min\left\{\mathcal{T}_{\mathbb{A}_1L}(k_1),\mathcal{T}_{\mathbb{A}_2L}(k_2)\right\}$

   $\left(\mathcal{T}_{\mathbb{A}_1U} \times \mathcal{T}_{\mathbb{A}_2U}\right)(k_1,k_2) = \min\left\{\mathcal{T}_{\mathbb{A}_1U}(k_1),\mathcal{T}_{\mathbb{A}_2U}(k_2)\right\}$

   $\left(\mathcal{C}_{\mathbb{A}_1L} \times \mathcal{C}_{\mathbb{A}_2L}\right)(k_1,k_2) = \min\left\{\mathcal{C}_{\mathbb{A}_1L}(k_1),\mathcal{C}_{\mathbb{A}_2L}(k_2)\right\}$

   $\left(\mathcal{C}_{\mathbb{A}_1U} \times \mathcal{C}_{\mathbb{A}_2U}\right)(k_1,k_2) = \min\left\{\mathcal{C}_{\mathbb{A}_1U}(k_1),\mathcal{C}_{\mathbb{A}_2U}(k_2)\right\}$

   $\left(\mathcal{U}_{\mathbb{A}_1L} \times \mathcal{U}_{\mathbb{A}_2L}\right)(k_1,k_2) = \max\left\{\mathcal{U}_{\mathbb{A}_1L}(k_1),\mathcal{U}_{\mathbb{A}_2L}(k_2)\right\}$

$$\left(\mathcal{U}_{\mathbb{A}_1U}\times\mathcal{U}_{\mathbb{A}_2U}\right)(k_1,k_2)=\max\left\{\mathcal{U}_{\mathbb{A}_1U}(k_1),\mathcal{U}_{\mathbb{A}_2U}(k_2)\right\},$$

$$\left(\mathcal{F}_{\mathbb{A}_1L}\times\mathcal{F}_{\mathbb{A}_2L}\right)(k_1,k_2)=\max\left\{\mathcal{F}_{\mathbb{A}_1L}(k_1),\mathcal{F}_{\mathbb{A}_2L}(k_2)\right\}$$

$$\left(\mathcal{F}_{\mathbb{A}_1U}\times\mathcal{U}_{\mathbb{A}_2U}\right)(k_1,k_2)=\max\left\{\mathcal{F}_{\mathbb{A}_1U}(k_1),\mathcal{F}_{\mathbb{A}_2U}(k_2)\right\},\ \forall k_1,k_2\in\mathbb{V},$$

2. $$\left(\mathcal{T}_{\mathbb{B}_1L}\times\mathcal{T}_{\mathbb{B}_2L}\right)\left((k,k_2)(k,l_2)\right)=\min\left\{\mathcal{T}_{\mathbb{A}_1L}(k),\mathcal{T}_{\mathbb{B}_2L}(k_2l_2)\right\}$$

$$\left(\mathcal{T}_{\mathbb{B}_1U}\times\mathcal{T}_{\mathbb{B}_2U}\right)\left((k,k_2)(k,l_2)\right)=\min\left\{\mathcal{T}_{\mathbb{A}_1U}(k),\mathcal{T}_{\mathbb{B}_2U}(k_2l_2)\right\}$$

$$\left(\mathcal{C}_{\mathbb{B}_1L}\times\mathcal{C}_{\mathbb{B}_2L}\right)\left((k,k_2)(k,l_2)\right)=\min\left\{\mathcal{C}_{\mathbb{A}_1L}(k),\mathcal{C}_{\mathbb{B}_2L}(k_2l_2)\right\}$$

$$\left(\mathcal{C}_{\mathbb{B}_1U}\times\mathcal{C}_{\mathbb{B}_2U}\right)\left((k,k_2)(k,l_2)\right)=\min\left\{\mathcal{C}_{\mathbb{A}_1U}(k),\mathcal{C}_{\mathbb{B}_2U}(k_2l_2)\right\}$$

$$\left(\mathcal{U}_{\mathbb{B}_1L}\times\mathcal{U}_{\mathbb{B}_2L}\right)\left((k,k_2)(k,l_2)\right)=\max\left\{\mathcal{U}_{\mathbb{A}_1L}(k),\mathcal{U}_{\mathbb{B}_2L}(k_2l_2)\right\}$$

$$\left(\mathcal{U}_{\mathbb{B}_1U}\times\mathcal{U}_{\mathbb{B}_2U}\right)\left((k,k_2)(k,l_2)\right)=\max\left\{\mathcal{U}_{\mathbb{A}_1U}(k),\mathcal{U}_{\mathbb{B}_2U}(k_2l_2)\right\},$$

$$\left(\mathcal{F}_{\mathbb{B}_1L}\times\mathcal{F}_{\mathbb{B}_2L}\right)\left((k,k_2)(k,l_2)\right)=\max\left\{\mathcal{F}_{\mathbb{A}_1L}(k),\mathcal{F}_{\mathbb{B}_2L}(k_2l_2)\right\}$$

$$\left(\mathcal{F}_{\mathbb{B}_1U}\times\mathcal{F}_{\mathbb{B}_2U}\right)\left((k,k_2)(k,l_2)\right)=\max\left\{\mathcal{F}_{\mathbb{A}_1U}(k),\mathcal{F}_{\mathbb{B}_2U}(k_2l_2)\right\},$$

$$\forall k\in\mathbb{V}_1\ and\ k_2l_2\in\mathbb{E}_2$$

3. $$\left(\mathcal{T}_{\mathbb{B}_1L}\times\mathcal{T}_{\mathbb{B}_2L}\right)\left((k_1,z_1)(k_2,z_2)\right)=\min\left\{\mathcal{T}_{\mathbb{B}_1L}(k_1k_2),\mathcal{T}_{\mathbb{B}_2L}(z_1z_2)\right\}$$

$$\left(\mathcal{T}_{\mathbb{B}_1U}\times\mathcal{T}_{\mathbb{B}_2U}\right)\left((k_1,z_1)(k_2,z_2)\right)=\min\left\{\mathcal{T}_{\mathbb{B}_1U}(k_1k_2),\mathcal{T}_{\mathbb{B}_2U}(z_1z_2)\right\}$$

$$\left(\mathcal{C}_{\mathbb{B}_1L}\times\mathcal{C}_{\mathbb{B}_2L}\right)\left((k_1,z_1)(k_2,z_2)\right)=\min\left\{\mathcal{C}_{\mathbb{B}_1L}(k_1k_2),\mathcal{C}_{\mathbb{B}_2L}(z_1z_2)\right\}$$

$$\left(\mathcal{C}_{\mathbb{B}_1U}\times\mathcal{C}_{\mathbb{B}_2U}\right)\left((k_1,z_1)(k_2,z_2)\right)=\min\left\{\mathcal{C}_{\mathbb{B}_1U}(k_1k_2),\mathcal{C}_{\mathbb{B}_2U}(z_1z_2)\right\}$$

$$\left(\mathcal{U}_{\mathbb{B}_1L}\times\mathcal{T}_{\mathbb{B}_2L}\right)\left((k_1,z_1)(k_2,z_2)\right)=\max\left\{\mathcal{U}_{\mathbb{B}_1L}(k_1k_2),\mathcal{U}_{\mathbb{B}_2L}(z_1z_2)\right\}$$

$$\left(\mathcal{U}_{\mathbb{B}_1U}\times\mathcal{U}_{\mathbb{B}_2U}\right)\left((k_1,z_1)(k_2,z_2)\right)=\max\left\{\mathcal{U}_{\mathbb{B}_1U}(k_1k_2),\mathcal{U}_{\mathbb{B}_2U}(z_1z_2)\right\}$$

$$\left(\mathcal{F}_{\mathbb{B}_1L}\times\mathcal{F}_{\mathbb{B}_2L}\right)\left((k_1,z_1)(k_2,z_2)\right)=\max\left\{\mathcal{F}_{\mathbb{B}_1L}(k_1k_2),\mathcal{F}_{\mathbb{B}_2L}(z_1z_2)\right\}$$

$$\left(\mathcal{F}_{\mathbb{B}_1U}\times\mathcal{F}_{\mathbb{B}_2U}\right)\left((k_1,z_1)(k_2,z_2)\right)=\max\left\{\mathcal{F}_{\mathbb{B}_1U}(k_1k_2),\mathcal{F}_{\mathbb{B}_2U}(z_1z_2)\right\}$$

$\forall k_1k_2\in\mathbb{E}_1\ and\ z_1z_2\in\mathbb{E}_2$

Proposition 6.4.10 The lexicographic product $\mathcal{G}_1 \times \mathcal{G}_2 = (\mathbb{A}_1 \times \mathbb{A}_2, \mathbb{B}_1 \times \mathbb{B}_2)$ of two interval quadripartitioned single-valued neutrosophic graphs $\mathcal{G}_1 = (\mathbb{A}_1, \mathbb{B}_1)$ and $\mathcal{G}_2 = (\mathbb{A}_2, \mathbb{B}_2)$ is an interval quadripartitioned single-valued neutrosophic graph.

Proof. We have two cases:

case 1: Let $k \in \mathbb{V}_1$ and $k_2 l_2 \in \mathbb{E}_2$. Then

$$(\mathcal{T}_{\mathbb{B}_1 L} \times \mathcal{T}_{\mathbb{B}_2 L})((k,k_2)(k,l_2)) = \min\{\mathcal{T}_{\mathbb{A}_1 L}(k), \mathcal{T}_{\mathbb{B}_2 L}(k_2 l_2)\}$$
$$\leq \min\{\mathcal{T}_{\mathbb{A}_1 L}(k), \min\{\mathcal{T}_{\mathbb{A}_2 L}(k_2), \mathcal{T}_{\mathbb{A}_2 L}(l_2)\}\}$$
$$= \min\{\min\{\mathcal{T}_{\mathbb{A}_1 L}(k), \mathcal{T}_{\mathbb{A}_2 L}(k_2)\}, \min\{\mathcal{T}_{\mathbb{A}_1 L}(k), \mathcal{T}_{\mathbb{A}_2 L}(l_2)\}\}$$
$$= \min\{(\mathcal{T}_{\mathbb{A}_1 L} \times \mathcal{T}_{\mathbb{A}_2 L})(k,k_2), (\mathcal{T}_{\mathbb{A}_1 L} \times \mathcal{T}_{\mathbb{A}_2 L})(k,l_2)\}$$
$$(\mathcal{T}_{\mathbb{B}_1 U} \cdot \mathcal{T}_{\mathbb{B}_2 U})((k,k_2)(k,l_2)) = \min\{\mathcal{T}_{\mathbb{A}_1 U}(k), \mathcal{T}_{\mathbb{B}_2 U}(k_2 l_2)\}$$
$$\leq \min\{\mathcal{T}_{\mathbb{A}_1 U}(k), \min\{\mathcal{T}_{\mathbb{A}_2 U}(k_2), \mathcal{T}_{\mathbb{A}_2 U}(l_2)\}\}$$
$$= \min\{\min\{\mathcal{T}_{\mathbb{A}_1 U}(k), \mathcal{T}_{\mathbb{A}_2 U}(k_2)\}, \min\{\mathcal{T}_{\mathbb{A}_1 U}(k), \mathcal{T}_{\mathbb{A}_2 U}(l_2)\}\}$$
$$= \min\{(\mathcal{T}_{\mathbb{A}_1 U} \times \mathcal{T}_{\mathbb{A}_2 U})(k,k_2), (\mathcal{T}_{\mathbb{A}_1 U} \times \mathcal{T}_{\mathbb{A}_2 U})(k,l_2)\}.$$

$$(\mathcal{C}_{\mathbb{B}_1 L} \times \mathcal{C}_{\mathbb{B}_2 L})((k,k_2)(k,l_2)) = \min\{\mathcal{C}_{\mathbb{A}_1 L}(k), \mathcal{C}_{\mathbb{B}_2 L}(k_2 l_2)\}$$
$$\leq \min\{\mathcal{C}_{\mathbb{A}_1 L}(k), \min\{\mathcal{C}_{\mathbb{A}_2 L}(k_2), \mathcal{C}_{\mathbb{A}_2 L}(l_2)\}\}$$
$$= \min\{\min\{\mathcal{C}_{\mathbb{A}_1 L}(k), \mathcal{C}_{\mathbb{A}_2 L}(k_2)\}, \min\{\mathcal{C}_{\mathbb{A}_1 L}(k), \mathcal{C}_{\mathbb{A}_2 L}(l_2)\}\}$$
$$= \min\{(\mathcal{C}_{\mathbb{A}_1 L} \times \mathcal{C}_{\mathbb{A}_2 L})(k,k_2), (\mathcal{C}_{\mathbb{A}_1 L} \times \mathcal{C}_{\mathbb{A}_2 L})(k,l_2)\}$$
$$(\mathcal{C}_{\mathbb{B}_1 U} \times \mathcal{C}_{\mathbb{B}_2 U})((k,k_2)(k,l_2)) = \min\{\mathcal{C}_{\mathbb{A}_1 U}(k), \mathcal{C}_{\mathbb{B}_2 U}(k_2 l_2)\}$$
$$\leq \min\{\mathcal{C}_{\mathbb{A}_1 U}(k), \min\{\mathcal{C}_{\mathbb{A}_2 U}(k_2), \mathcal{C}_{\mathbb{A}_2 U}(l_2)\}\}$$
$$= \min\{\min\{\mathcal{I}_{\mathbb{A}_1 U}(k), \mathcal{I}_{\mathbb{A}_2 U}(k_2)\}, \{\min \mathcal{C}_{\mathbb{A}_1 U}(k), \mathcal{C}_{\mathbb{A}_2 U}(l_2)\}\}$$
$$= \min\{(\mathcal{C}_{\mathbb{A}_1 U} \times \mathcal{C}_{\mathbb{A}_2 U})(k,k_2), (\mathcal{C}_{\mathbb{A}_1 U} \times \mathcal{C}_{\mathbb{A}_2 U})(k,l_2)\}.$$

$$(\mathcal{U}_{\mathbb{B}_1 L} \times \mathcal{U}_{\mathbb{B}_2 L})((k,k_2)(k,l_2)) = \max\{\mathcal{U}_{\mathbb{A}_1 L}(k), \mathcal{U}_{\mathbb{B}_2 L}(k_2 l_2)\}$$
$$\leq \max\{\mathcal{U}_{\mathbb{A}_1 L}(k), \max\{\mathcal{U}_{\mathbb{A}_2 L}(k_2), \mathcal{U}_{\mathbb{A}_2 L}(l_2)\}\}$$

$$= \max\left\{\max\left\{\mathcal{U}_{\mathbb{A}_1L}(k),\mathcal{U}_{\mathbb{A}_2L}(k_2)\right\},\max\left\{\mathcal{U}_{\mathbb{A}_1L}(k),\mathcal{U}_{\mathbb{A}_2L}(l_2)\right\}\right\}$$

$$= \max\left\{\left(\mathcal{U}_{\mathbb{A}_1L}\times\mathcal{U}_{\mathbb{A}_2L}\right)(k,k_2),\left(\mathcal{U}_{\mathbb{A}_1L}\times\mathcal{U}_{\mathbb{A}_2L}\right)(k,l_2)\right\}$$

$$\left(\mathcal{U}_{\mathbb{B}_1U}\times\mathcal{U}_{\mathbb{B}_2U}\right)\left((k,k_2)(k,l_2)\right) = \max\left\{\mathcal{U}_{\mathbb{A}_1U}(k),\mathcal{U}_{\mathbb{B}_2U}(k_2l_2)\right\}$$

$$\leq \max\left\{\mathcal{U}_{\mathbb{A}_1U}(k),\max\left\{\mathcal{U}_{\mathbb{A}_2U}(k_2),\mathcal{U}_{\mathbb{A}_2U}(l_2)\right\}\right\}$$

$$= \max\left\{\max\left\{\mathcal{I}_{\mathbb{A}_1U}(k),\mathcal{I}_{\mathbb{A}_2U}(k_2)\right\},\left\{\max\mathcal{U}_{\mathbb{A}_1U}(k),\mathcal{U}_{\mathbb{A}_2U}(l_2)\right\}\right\}$$

$$= \max\left\{\left(\mathcal{U}_{\mathbb{A}_1U}\times\mathcal{U}_{\mathbb{A}_2U}\right)(k,k_2),\left(\mathcal{U}_{\mathbb{A}_1U}\times\mathcal{U}_{\mathbb{A}_2U}\right)(k,l_2)\right\}.$$

$$\left(\mathcal{F}_{\mathbb{B}_1L}\times\mathcal{F}_{\mathbb{B}_2L}\right)\left((k,k_2)(k,l_2)\right) = \max\left\{\mathcal{F}_{\mathbb{A}_1L}(k),\mathcal{F}_{\mathbb{B}_2L}(k_2l_2)\right\}$$

$$\leq \max\left\{\mathcal{F}_{\mathbb{A}_1L}(k),\max\left\{\mathcal{F}_{\mathbb{A}_2L}(k_2),\mathcal{F}_{\mathbb{A}_2L}(l_2)\right\}\right\}$$

$$= \max\left\{\max\left\{\mathcal{F}_{\mathbb{A}_1L}(k),\mathcal{F}_{\mathbb{A}_2L}(k_2)\right\},\max\left\{\mathcal{F}_{\mathbb{A}_1L}(k),\mathcal{F}_{\mathbb{A}_2L}(l_2)\right\}\right\}$$

$$= \max\left\{\left(\mathcal{F}_{\mathbb{A}_1L}\times\mathcal{F}_{\mathbb{A}_2L}\right)(k,k_2),\left(\mathcal{F}_{\mathbb{A}_1L}\times\mathcal{F}_{\mathbb{A}_2L}\right)(k,l_2)\right\}$$

$$\left(\mathcal{F}_{\mathbb{B}_1U}\times\mathcal{F}_{\mathbb{B}_2U}\right)\left((k,k_2)(k,l_2)\right) = \max\left\{\mathcal{F}_{\mathbb{A}_1U}(k),\mathcal{F}_{\mathbb{B}_2U}(k_2l_2)\right\}$$

$$\leq \max\left\{\mathcal{F}_{\mathbb{A}_1U}(k),\max\left\{\mathcal{F}_{\mathbb{A}_2U}(k_2),\mathcal{F}_{\mathbb{A}_2U}(l_2)\right\}\right\}$$

$$= \max\left\{\max\left\{\mathcal{F}_{\mathbb{A}_1U}(k),\mathcal{F}_{\mathbb{A}_2U}(k_2)\right\},\left\{\max\mathcal{F}_{\mathbb{A}_1U}(k),\mathcal{F}_{\mathbb{A}_2U}(l_2)\right\}\right\}$$

$$= \max\left\{\left(\mathcal{F}_{\mathbb{A}_1U}\times\mathcal{F}_{\mathbb{A}_2U}\right)(k,k_2),\left(\mathcal{F}_{\mathbb{A}_1U}\times\mathcal{F}_{\mathbb{A}_2U}\right)(k,l_2)\right\}.$$

case 2: Let $k_1k_2 \in \mathbb{E}_1$ and$z_1z_2 \in \mathbb{E}_2$. Then

$$\left(\mathcal{T}_{\mathbb{B}_1L}\times\mathcal{T}_{\mathbb{B}_2L}\right)\left((k_1,z_1)(k_2,z_2)\right) = \min\left\{\mathcal{T}_{\mathbb{B}_1L}(k_1k_2),\mathcal{T}_{\mathbb{B}_2L}(z_1z_2)\right\}$$

$$\leq \min\left\{\min\left\{\mathcal{T}_{\mathbb{A}_1L}(k_1),\mathcal{T}_{\mathbb{A}_1L}(k_2)\right\},\min\left\{\mathcal{T}_{\mathbb{A}_2L}(z_1),\mathcal{T}_{\mathbb{A}_2L}(z_2)\right\}\right\}$$

$$= \min\left\{\min\left\{\mathcal{T}_{\mathbb{A}_1L}(k_1),\mathcal{T}_{\mathbb{A}_2L}(z_1)\right\},\min\left\{\mathcal{T}_{\mathbb{A}_1L}(k_2),\mathcal{T}_{\mathbb{A}_2L}(z_2)\right\}\right\}$$

$$= \min\left\{\left(\mathcal{T}_{\mathbb{A}_1L}\times\mathcal{T}_{\mathbb{A}_2L}\right)(k_1,z_1),\left(\mathcal{T}_{\mathbb{A}_1L}\times\mathcal{T}_{\mathbb{A}_2L}\right)(k_2,z_2)\right\}$$

$$\left(\mathcal{T}_{\mathbb{B}_1U}\times\mathcal{T}_{\mathbb{B}_2U}\right)\left((k_1,z_1)(k_2,z_2)\right) = \min\left\{\mathcal{T}_{\mathbb{B}_1U}(k_1k_2),\mathcal{T}_{\mathbb{B}_2U}(z_1z_2)\right\}$$

$$\leq \min\left\{\min\left\{\mathcal{T}_{\mathbb{A}_1U}(k_1),\mathcal{T}_{\mathbb{A}_1U}(k_2)\right\},\min\left\{\mathcal{T}_{\mathbb{A}_2U}(z_1),\mathcal{T}_{\mathbb{A}_2U}(z_2)\right\}\right\}$$

$$= \min\left\{\min\left\{\mathcal{T}_{\mathbb{A}_1U}(k_1), \mathcal{T}_{\mathbb{A}_2U}(z_1)\right\}, \min\left\{\mathcal{T}_{\mathbb{A}_1U}(k_2), \mathcal{T}_{\mathbb{A}_2U}(z_2)\right\}\right\}$$

$$= \min\left\{\left(\mathcal{T}_{\mathbb{A}_1U} \times \mathcal{T}_{\mathbb{A}_2U}\right)(k_1, z_1), \left(\mathcal{T}_{\mathbb{A}_1U} \times \mathcal{T}_{\mathbb{A}_2U}\right)(k_2, z_2)\right\}$$

$$\left(\mathcal{C}_{\mathbb{B}_1L} \times \mathcal{C}_{\mathbb{B}_2L}\right)\left((k_1, z_1)(k_2, z_2)\right) = \min\left\{\mathcal{C}_{\mathbb{B}_1L}(k_1k_2), \mathcal{C}_{\mathbb{B}_2L}(z_1z_2)\right\}$$

$$\leq \min\left\{\min\left\{\mathcal{C}_{\mathbb{A}_1L}(k_1), \mathcal{C}_{\mathbb{A}_1L}(k_2)\right\}, \min\left\{\mathcal{C}_{\mathbb{A}_2L}(z_1), \mathcal{C}_{\mathbb{A}_2L}(z_2)\right\}\right\}$$

$$= \min\left\{\min\left\{\mathcal{C}_{\mathbb{A}_1L}(k_1), \mathcal{C}_{\mathbb{A}_2L}(z_1)\right\}, \min\left\{\mathcal{C}_{\mathbb{A}_1L}(k_2), \mathcal{C}_{\mathbb{A}_2L}(z_2)\right\}\right\}$$

$$= \min\left\{\left(\mathcal{C}_{\mathbb{A}_1L} \times \mathcal{C}_{\mathbb{A}_2L}\right)(k_1, z_1), \left(\mathcal{C}_{\mathbb{A}_1L} \times \mathcal{C}_{\mathbb{A}_2L}\right)(k_2, z_2)\right\}$$

$$\left(\mathcal{C}_{\mathbb{B}_1U} \times \mathcal{C}_{\mathbb{B}_2U}\right)\left((k_1, z_1)(k_2, z_2)\right) = \min\left\{\mathcal{C}_{\mathbb{B}_1U}(k_1k_2), \mathcal{C}_{\mathbb{B}_2U}(z_1z_2)\right\}$$

$$\leq \min\left\{\min\left\{\mathcal{C}_{\mathbb{A}_1U}(k_1), \mathcal{C}_{\mathbb{A}_1U}(k_2)\right\}, \min\left\{\mathcal{C}_{\mathbb{A}_2U}(z_1), \mathcal{C}_{\mathbb{A}_2U}(z_2)\right\}\right\}$$

$$= \min\left\{\min\left\{\mathcal{C}_{\mathbb{A}_1U}(k_1), \mathcal{C}_{\mathbb{A}_2U}(z_1)\right\}, \min\left\{\mathcal{C}_{\mathbb{A}_1U}(k_2), \mathcal{C}_{\mathbb{A}_2U}(z_2)\right\}\right\}$$

$$= \min\left\{\left(\mathcal{C}_{\mathbb{A}_1U} \times \mathcal{C}_{\mathbb{A}_2U}\right)(k_1, z_1), \left(\mathcal{C}_{\mathbb{A}_1U} \times \mathcal{C}_{\mathbb{A}_2U}\right)(k_2, z_2)\right\}$$

$$\left(\mathcal{U}_{\mathbb{B}_1L} \times \mathcal{U}_{\mathbb{B}_2L}\right)\left((k_1, z_1)(k_2, z_2)\right) = \max\left\{\mathcal{U}_{\mathbb{B}_1L}(k_1k_2), \mathcal{U}_{\mathbb{B}_2L}(z_1z_2)\right\}$$

$$\leq \max\left\{\max\left\{\mathcal{U}_{\mathbb{A}_1L}(k_1), \mathcal{U}_{\mathbb{A}_1L}(k_2)\right\}, \max\left\{\mathcal{U}_{\mathbb{A}_2L}(z_1), \mathcal{U}_{\mathbb{A}_2L}(z_2)\right\}\right\}$$

$$= \max\left\{\max\left\{\mathcal{U}_{\mathbb{A}_1L}(k_1), \mathcal{U}_{\mathbb{A}_2L}(z_1)\right\}, \max\left\{\mathcal{U}_{\mathbb{A}_1L}(k_2), \mathcal{U}_{\mathbb{A}_2L}(z_2)\right\}\right\}$$

$$= \max\left\{\left(\mathcal{U}_{\mathbb{A}_1L} \times \mathcal{U}_{\mathbb{A}_2L}\right)(k_1, z_1), \left(\mathcal{U}_{\mathbb{A}_1L} \times \mathcal{U}_{\mathbb{A}_2L}\right)(k_2, z_2)\right\}$$

$$\left(\mathcal{U}_{\mathbb{B}_1U} \times \mathcal{U}_{\mathbb{B}_2U}\right)\left((k_1, z_1)(k_2, z_2)\right) = \max\left\{\mathcal{U}_{\mathbb{B}_1U}(k_1k_2), \mathcal{U}_{\mathbb{B}_2U}(z_1z_2)\right\}$$

$$\leq \max\left\{\max\left\{\mathcal{U}_{\mathbb{A}_1U}(k_1), \mathcal{U}_{\mathbb{A}_1U}(k_2)\right\}, \max\left\{\mathcal{U}_{\mathbb{A}_2U}(z_1), \mathcal{U}_{\mathbb{A}_2U}(z_2)\right\}\right\}$$

$$= \max\left\{\max\left\{\mathcal{U}_{\mathbb{A}_1U}(k_1), \mathcal{U}_{\mathbb{A}_2U}(z_1)\right\}, \max\left\{\mathcal{U}_{\mathbb{A}_1U}(k_2), \mathcal{U}_{\mathbb{A}_2U}(z_2)\right\}\right\}$$

$$= \max\left\{\left(\mathcal{U}_{\mathbb{A}_1U} \times \mathcal{U}_{\mathbb{A}_2U}\right)(k_1, z_1), \left(\mathcal{U}_{\mathbb{A}_1U} \times \mathcal{U}_{\mathbb{A}_2U}\right)(k_2, z_2)\right\}$$

$$\left(\mathcal{F}_{\mathbb{B}_1L} \times \mathcal{F}_{\mathbb{B}_2L}\right)\left((k_1, z_1)(k_2, z_2)\right) = \max\left\{\mathcal{F}_{\mathbb{B}_1L}(k_1k_2), \mathcal{F}_{\mathbb{B}_2L}(z_1z_2)\right\}$$

$$\leq \max\left\{\max\left\{\mathcal{F}_{\mathbb{A}_1L}(k_1), \mathcal{F}_{\mathbb{A}_1L}(k_2)\right\}, \max\left\{\mathcal{F}_{\mathbb{A}_2L}(z_1), \mathcal{F}_{\mathbb{A}_2L}(z_2)\right\}\right\}$$

$$= \max\left\{\max\left\{\mathcal{F}_{\mathbb{A}_1L}(k_1),\mathcal{F}_{\mathbb{A}_2L}(z_1)\right\},\max\left\{\mathcal{F}_{\mathbb{A}_1L}(k_2),\mathcal{F}_{\mathbb{A}_2L}(z_2)\right\}\right\}$$

$$= \max\left\{\left(\mathcal{F}_{\mathbb{A}_1L}\times\mathcal{F}_{\mathbb{A}_2L}\right)(k_1,z_1),\left(\mathcal{F}_{\mathbb{A}_1L}\times\mathcal{F}_{\mathbb{A}_2L}\right)(k_2,z_2)\right\}$$

$$\left(\mathcal{F}_{\mathbb{B}_1U}\times\mathcal{F}_{\mathbb{B}_2U}\right)\left((k_1,z_1)(k_2,z_2)\right) = \max\left\{\mathcal{F}_{\mathbb{B}_1U}(k_1k_2),\mathcal{F}_{\mathbb{B}_2U}(z_1z_2)\right\}$$

$$\leq \max\left\{\max\left\{\mathcal{F}_{\mathbb{A}_1U}(k_1),\mathcal{F}_{\mathbb{A}_1U}(k_2)\right\},\max\left\{\mathcal{F}_{\mathbb{A}_2U}(z_1),\mathcal{F}_{\mathbb{A}_2U}(z_2)\right\}\right\}$$

$$= \max\left\{\max\left\{\mathcal{F}_{\mathbb{A}_1U}(k_1),\mathcal{F}_{\mathbb{A}_2U}(z_1)\right\},\max\left\{\mathcal{F}_{\mathbb{A}_1U}(k_2),\mathcal{F}_{\mathbb{A}_2U}(z_2)\right\}\right\}$$

$$= \max\left\{\left(\mathcal{F}_{\mathbb{A}_1U}\times\mathcal{F}_{\mathbb{A}_2U}\right)(k_1,z_1),\left(\mathcal{F}_{\mathbb{A}_1U}\times\mathcal{F}_{\mathbb{A}_2U}\right)(k_2,z_2)\right\}$$

This completes the proof.

Definition 6.4.11 The strong product $\mathcal{G}_1 \boxtimes \mathcal{G}_2$ of two interval quadripartitioned single-valued neutrosophic graphs $\mathcal{G}_1 = (\mathbb{A}_1,\mathbb{B}_1)$ and $\mathcal{G}_2 = (\mathbb{A}_2,\mathbb{B}_2)$ of the crisp graphs $\mathcal{G}_1^* = (\mathbb{V}_1,\mathbb{E}_1)$ and $\mathcal{G}_2^* = (\mathbb{V}_2,\mathbb{E}_2)$ is defined as a pair $\mathcal{G}_1 \boxtimes \mathcal{G}_2 = (\mathbb{A}_1 \boxtimes \mathbb{A}_2, \mathbb{B}_1 \boxtimes \mathbb{B}_2)$, such that

1. $\left(\mathcal{T}_{\mathbb{A}_1L}\boxtimes\mathcal{T}_{\mathbb{A}_2L}\right)(k_1,k_2) = \min\left\{\mathcal{T}_{\mathbb{A}_1L}(k_1),\mathcal{T}_{\mathbb{A}_2L}(k_2)\right\}$

   $\left(\mathcal{T}_{\mathbb{A}_1U}\boxtimes\mathcal{T}_{\mathbb{A}_2U}\right)(k_1,k_2) = \min\left\{\mathcal{T}_{\mathbb{A}_1U}(k_1),\mathcal{T}_{\mathbb{A}_2U}(k_2)\right\}$

   $\left(\mathcal{C}_{\mathbb{A}_1L}\boxtimes\mathcal{C}_{\mathbb{A}_2L}\right)(k_1,k_2) = \min\left\{\mathcal{C}_{\mathbb{A}_1L}(k_1),\mathcal{C}_{\mathbb{A}_2L}(k_2)\right\}$

   $\left(\mathcal{C}_{\mathbb{A}_1U}\boxtimes\mathcal{C}_{\mathbb{A}_2U}\right)(k_1,k_2) = \min\left\{\mathcal{C}_{\mathbb{A}_1U}(k_1),\mathcal{C}_{\mathbb{A}_2U}(k_2)\right\}$

   $\left(\mathcal{U}_{\mathbb{A}_1L}\boxtimes\mathcal{U}_{\mathbb{A}_2L}\right)(k_1,k_2) = \max\left\{\mathcal{U}_{\mathbb{A}_1L}(k_1),\mathcal{U}_{\mathbb{A}_2L}(k_2)\right\}$

   $\left(\mathcal{U}_{\mathbb{A}_1U}\boxtimes\mathcal{U}_{\mathbb{A}_2U}\right)(k_1,k_2) = \max\left\{\mathcal{U}_{\mathbb{A}_1U}(k_1),\mathcal{U}_{\mathbb{A}_2U}(k_2)\right\},$

   $\left(\mathcal{F}_{\mathbb{A}_1L}\boxtimes\mathcal{F}_{\mathbb{A}_2L}\right)(k_1,k_2) = \max\left\{\mathcal{F}_{\mathbb{A}_1L}(k_1),\mathcal{F}_{\mathbb{A}_2L}(k_2)\right\}$

   $\left(\mathcal{F}_{\mathbb{A}_1U}\boxtimes\mathcal{U}_{\mathbb{A}_2U}\right)(k_1,k_2) = \max\left\{\mathcal{F}_{\mathbb{A}_1U}(k_1),\mathcal{F}_{\mathbb{A}_2U}(k_2)\right\}, \ \forall k_1,k_2 \in \mathbb{V},$

2. $\left(\mathcal{T}_{\mathbb{B}_1L}\boxtimes\mathcal{T}_{\mathbb{B}_2L}\right)\left((k,k_2)(k,l_2)\right) = \min\left\{\mathcal{T}_{\mathbb{A}_1L}(k),\mathcal{T}_{\mathbb{B}_2L}(k_2l_2)\right\}$

   $\left(\mathcal{T}_{\mathbb{B}_1U}\boxtimes\mathcal{T}_{\mathbb{B}_2U}\right)\left((k,k_2)(k,l_2)\right) = \min\left\{\mathcal{T}_{\mathbb{A}_1U}(k),\mathcal{T}_{\mathbb{B}_2U}(k_2l_2)\right\}$

   $\left(\mathcal{C}_{\mathbb{B}_1L}\boxtimes\mathcal{C}_{\mathbb{B}_2L}\right)\left((k,k_2)(k,l_2)\right) = \min\left\{\mathcal{C}_{\mathbb{A}_1L}(k),\mathcal{C}_{\mathbb{B}_2L}(k_2l_2)\right\}$

   $\left(\mathcal{C}_{\mathbb{B}_1U}\boxtimes\mathcal{C}_{\mathbb{B}_2U}\right)\left((k,k_2)(k,l_2)\right) = \min\left\{\mathcal{C}_{\mathbb{A}_1U}(k),\mathcal{C}_{\mathbb{B}_2U}(k_2l_2)\right\}$

   $\left(\mathcal{U}_{\mathbb{B}_1L}\boxtimes\mathcal{U}_{\mathbb{B}_2L}\right)\left((k,k_2)(k,l_2)\right) = \max\left\{\mathcal{U}_{\mathbb{A}_1L}(k),\mathcal{U}_{\mathbb{B}_2L}(k_2l_2)\right\}$

$$\left(\mathcal{U}_{\mathbb{B}_1U} \boxtimes \mathcal{U}_{\mathbb{B}_2U}\right)\left((k,k_2)(k,l_2)\right) = \max\left\{\mathcal{U}_{\mathbb{A}_1U}(k), \mathcal{U}_{\mathbb{B}_2U}(k_2l_2)\right\},$$

$$\left(\mathcal{F}_{\mathbb{B}_1L} \boxtimes \mathcal{F}_{\mathbb{B}_2L}\right)\left((k,k_2)(k,l_2)\right) = \max\left\{\mathcal{F}_{\mathbb{A}_1L}(k), \mathcal{F}_{\mathbb{B}_2L}(k_2l_2)\right\}$$

$$\left(\mathcal{F}_{\mathbb{B}_1U} \boxtimes \mathcal{F}_{\mathbb{B}_2U}\right)\left((k,k_2)(k,l_2)\right) = \max\left\{\mathcal{F}_{\mathbb{A}_1U}(k), \mathcal{F}_{\mathbb{B}_2U}(k_2l_2)\right\},$$

$$\forall k \in \mathbb{V}_1 \text{ and } k_2l_2 \in \mathbb{E}_2$$

and

3. $$\left(\mathcal{T}_{\mathbb{B}_1L} \boxtimes \mathcal{T}_{\mathbb{B}_2L}\right)\left((k_1,z)(l_1,z)\right) = \min\left\{\mathcal{T}_{\mathbb{B}_1L}(k_1l_1), \mathcal{T}_{\mathbb{A}_2L}(z)\right\}$$

$$\left(\mathcal{T}_{\mathbb{B}_1U} \boxtimes \mathcal{T}_{\mathbb{B}_2U}\right)\left((k_1,z)(l_1,z)\right) = \min\left\{\mathcal{T}_{\mathbb{B}_1U}(k_1l_1), \mathcal{T}_{\mathbb{A}_2U}(z)\right\}$$

$$\left(\mathcal{C}_{\mathbb{B}_1L} \boxtimes \mathcal{C}_{\mathbb{B}_2L}\right)\left((k_1,z)(l_1,z)\right) = \min\left\{\mathcal{C}_{\mathbb{B}_1L}(k_1l_1), \mathcal{C}_{\mathbb{A}_2L}(z)\right\}$$

$$\left(\mathcal{C}_{\mathbb{B}_1U} \boxtimes \mathcal{C}_{\mathbb{B}_2U}\right)\left((k_1,z)(l_1,z)\right) = \min\left\{\mathcal{C}_{\mathbb{B}_1U}(k_1l_1), \mathcal{C}_{\mathbb{A}_2U}(z)\right\}$$

$$\left(\mathcal{U}_{\mathbb{B}_1L} \boxtimes \mathcal{U}_{\mathbb{B}_2L}\right)\left((k_1,z)(l_1,z)\right) = \max\left\{\mathcal{U}_{\mathbb{B}_1L}(k_1l_1), \mathcal{U}_{\mathbb{A}_2L}(z)\right\}$$

$$\left(\mathcal{U}_{\mathbb{B}_1U} \boxtimes \mathcal{U}_{\mathbb{B}_2U}\right)\left((k_1,z)(l_1,z)\right) = \max\left\{\mathcal{U}_{\mathbb{B}_1U}(k_1l_1), \mathcal{U}_{\mathbb{A}_2U}(z)\right\},$$

$$\left(\mathcal{F}_{\mathbb{B}_1L} \boxtimes \mathcal{F}_{\mathbb{B}_2L}\right)\left((k_1,z)(l_1,z)\right) = \max\left\{\mathcal{F}_{\mathbb{B}_1L}(k_1l_1), \mathcal{F}_{\mathbb{A}_2L}(z)\right\}$$

$$\left(\mathcal{F}_{\mathbb{B}_1U} \boxtimes \mathcal{F}_{\mathbb{B}_2U}\right)\left((k_1,z)(l_1,z)\right) = \max\left\{\mathcal{F}_{\mathbb{B}_1U}(k_1l_1), \mathcal{F}_{\mathbb{A}_2U}(z)\right\},$$

$$\forall z \in \mathbb{V}_2 \text{ and } k_1l_1 \in \mathbb{E}_1.$$

4. $$\left(\mathcal{T}_{\mathbb{B}_1L} \boxtimes \mathcal{T}_{\mathbb{B}_2L}\right)\left((k_1,z_1)(k_2,z_2)\right) = \min\left\{\mathcal{T}_{\mathbb{B}_1L}(k_1k_2), \mathcal{T}_{\mathbb{B}_2L}(z_1z_2)\right\}$$

$$\left(\mathcal{T}_{\mathbb{B}_1U} \boxtimes \mathcal{T}_{\mathbb{B}_2U}\right)\left((k_1,z_1)(k_2,z_2)\right) = \min\left\{\mathcal{T}_{\mathbb{B}_1U}(k_1k_2), \mathcal{T}_{\mathbb{B}_2U}(z_1z_2)\right\}$$

$$\left(\mathcal{C}_{\mathbb{B}_1L} \boxtimes \mathcal{C}_{\mathbb{B}_2L}\right)\left((k_1,z_1)(k_2,z_2)\right) = \min\left\{\mathcal{C}_{\mathbb{B}_1L}(k_1k_2), \mathcal{C}_{\mathbb{B}_2L}(z_1z_2)\right\}$$

$$\left(\mathcal{C}_{\mathbb{B}_1U} \boxtimes \mathcal{C}_{\mathbb{B}_2U}\right)\left((k_1,z_1)(k_2,z_2)\right) = \min\left\{\mathcal{C}_{\mathbb{B}_1U}(k_1k_2), \mathcal{C}_{\mathbb{B}_2U}(z_1z_2)\right\}$$

$$\left(\mathcal{U}_{\mathbb{B}_1L} \boxtimes \mathcal{T}_{\mathbb{B}_2L}\right)\left((k_1,z_1)(k_2,z_2)\right) = \max\left\{\mathcal{U}_{\mathbb{B}_1L}(k_1k_2), \mathcal{U}_{\mathbb{B}_2L}(z_1z_2)\right\}$$

$$\left(\mathcal{U}_{\mathbb{B}_1U} \boxtimes \mathcal{U}_{\mathbb{B}_2U}\right)\left((k_1,z_1)(k_2,z_2)\right) = \max\left\{\mathcal{U}_{\mathbb{B}_1U}(k_1k_2), \mathcal{U}_{\mathbb{B}_2U}(z_1z_2)\right\}$$

$$\left(\mathcal{F}_{\mathbb{B}_1L} \boxtimes \mathcal{F}_{\mathbb{B}_2L}\right)\left((k_1,z_1)(k_2,z_2)\right) = \max\left\{\mathcal{F}_{\mathbb{B}_1L}(k_1k_2), \mathcal{F}_{\mathbb{B}_2L}(z_1z_2)\right\}$$

$$\left(\mathcal{F}_{\mathbb{B}_1U} \boxtimes \mathcal{F}_{\mathbb{B}_2U}\right)\left((k_1,z_1)(k_2,z_2)\right) = \max\left\{\mathcal{F}_{\mathbb{B}_1U}(k_1k_2), \mathcal{F}_{\mathbb{B}_2U}(z_1z_2)\right\}$$

$$\forall k_1k_2 \in \mathbb{E}_1 \text{ and } z_1z_2 \in \mathbb{E}_2$$

Proposition 6.4.12 The strong product $\mathcal{G}_1 \boxtimes \mathcal{G}_2 = (\mathbb{A}_1 \boxtimes \mathbb{A}_2, \mathbb{B}_1 \boxtimes \mathbb{B}_2)$ of two interval quadripartitioned single-valued neutrosophic graphs $\mathcal{G}_1 = (\mathbb{A}_1, \mathbb{B}_1)$ and $\mathcal{G}_1 = (\mathbb{A}_1, \mathbb{B}_1)$ is an interval quadripartitioned single-valued neutrosophic graph.

Proof. We have three cases:

Case 1: Let $k \in \mathbb{V}_1$ and$k_2 l_2 \in \mathbb{E}_2$. Then

$$(\mathcal{T}_{\mathbb{B}_1 L} \boxtimes \mathcal{T}_{\mathbb{B}_2 L})((k,k_2)(k,l_2)) = \min\{\mathcal{T}_{\mathbb{A}_1 L}(k), \mathcal{T}_{\mathbb{B}_2 L}(k_2 l_2)\}$$

$$\leq \min\{\mathcal{T}_{\mathbb{A}_1 L}(k), \min\{\mathcal{T}_{\mathbb{A}_2 L}(k_2), \mathcal{T}_{\mathbb{A}_2 L}(l_2)\}\}$$

$$= \min\{\min\{\mathcal{T}_{\mathbb{A}_1 L}(k), \mathcal{T}_{\mathbb{A}_2 L}(k_2)\}, \min\{\mathcal{T}_{\mathbb{A}_1 L}(k), \mathcal{T}_{\mathbb{A}_2 L}(l_2)\}\}$$

$$= \min\{(\mathcal{T}_{\mathbb{A}_1 L} \boxtimes \mathcal{T}_{\mathbb{A}_2 L})(k,k_2), (\mathcal{T}_{\mathbb{A}_1 L} \boxtimes \mathcal{T}_{\mathbb{A}_2 L})(k,l_2)\}$$

$$(\mathcal{T}_{\mathbb{B}_1 U} \boxtimes \mathcal{T}_{\mathbb{B}_2 U})((k,k_2)(k,l_2)) = \min\{\mathcal{T}_{\mathbb{A}_1 U}(k), \mathcal{T}_{\mathbb{B}_2 U}(k_2 l_2)\}$$

$$\leq \min\{\mathcal{T}_{\mathbb{A}_1 U}(k), \min\{\mathcal{T}_{\mathbb{A}_2 U}(k_2), \mathcal{T}_{\mathbb{A}_2 U}(l_2)\}\}$$

$$= \min\{\min\{\mathcal{T}_{\mathbb{A}_1 U}(k), \mathcal{T}_{\mathbb{A}_2 U}(k_2)\}, \min\{\mathcal{T}_{\mathbb{A}_1 U}(k), \mathcal{T}_{\mathbb{A}_2 U}(l_2)\}\}$$

$$= \min\{(\mathcal{T}_{\mathbb{A}_1 U} \boxtimes \mathcal{T}_{\mathbb{A}_2 U})(k,k_2), (\mathcal{T}_{\mathbb{A}_1 U} \boxtimes \mathcal{T}_{\mathbb{A}_2 U})(k,l_2)\}.$$

$$(\mathcal{C}_{\mathbb{B}_1 L} \boxtimes \mathcal{C}_{\mathbb{B}_2 L})((k,k_2)(k,l_2)) = \min\{\mathcal{C}_{\mathbb{A}_1 L}(k), \mathcal{C}_{\mathbb{B}_2 L}(k_2 l_2)\}$$

$$\leq \min\{\mathcal{C}_{\mathbb{A}_1 L}(k), \min\{\mathcal{C}_{\mathbb{A}_2 L}(k_2), \mathcal{C}_{\mathbb{A}_2 L}(l_2)\}\}$$

$$= \min\{\min\{\mathcal{C}_{\mathbb{A}_1 L}(k), \mathcal{C}_{\mathbb{A}_2 L}(k_2)\}, \min\{\mathcal{C}_{\mathbb{A}_1 L}(k), \mathcal{C}_{\mathbb{A}_2 L}(l_2)\}\}$$

$$= \min\{(\mathcal{C}_{\mathbb{A}_1 L} \boxtimes \mathcal{C}_{\mathbb{A}_2 L})(k,k_2), (\mathcal{C}_{\mathbb{A}_1 L} \boxtimes \mathcal{C}_{\mathbb{A}_2 L})(k,l_2)\}$$

$$(\mathcal{C}_{\mathbb{B}_1 U} \boxtimes \mathcal{C}_{\mathbb{B}_2 U})((k,k_2)(k,l_2)) = \min\{\mathcal{C}_{\mathbb{A}_1 U}(k), \mathcal{C}_{\mathbb{B}_2 U}(k_2 l_2)\}$$

$$\leq \min\{\mathcal{C}_{\mathbb{A}_1 U}(k), \min\{\mathcal{C}_{\mathbb{A}_2 U}(k_2), \mathcal{C}_{\mathbb{A}_2 U}(l_2)\}\}$$

$$= \min\{\min\{\mathcal{I}_{\mathbb{A}_1 U}(k), \mathcal{I}_{\mathbb{A}_2 U}(k_2)\}, \{\min \mathcal{C}_{\mathbb{A}_1 U}(k), \mathcal{C}_{\mathbb{A}_2 U}(l_2)\}\}$$

$$= \min\{(\mathcal{C}_{\mathbb{A}_1 U} \boxtimes \mathcal{C}_{\mathbb{A}_2 U})(k,k_2), (\mathcal{C}_{\mathbb{A}_1 U} \boxtimes \mathcal{C}_{\mathbb{A}_2 U})(k,l_2)\}.$$

$$(\mathcal{U}_{\mathbb{B}_1 L} \boxtimes \mathcal{U}_{\mathbb{B}_2 L})((k,k_2)(k,l_2)) = \max\{\mathcal{U}_{\mathbb{A}_1 L}(k), \mathcal{U}_{\mathbb{B}_2 L}(k_2 l_2)\}$$

$$\leq \max\{\mathcal{U}_{\mathbb{A}_1 L}(k), \max\{\mathcal{U}_{\mathbb{A}_2 L}(k_2), \mathcal{U}_{\mathbb{A}_2 L}(l_2)\}\}$$

$$= \max\left\{\max\left\{\mathcal{U}_{\mathbb{A}_1L}(k),\mathcal{U}_{\mathbb{A}_2L}(k_2)\right\},\max\left\{\mathcal{U}_{\mathbb{A}_1L}(k),\mathcal{U}_{\mathbb{A}_2L}(l_2)\right\}\right\}$$

$$= \max\left\{\left(\mathcal{U}_{\mathbb{A}_1L}\boxtimes\mathcal{U}_{\mathbb{A}_2L}\right)(k,k_2),\left(\mathcal{U}_{\mathbb{A}_1L}\boxtimes\mathcal{U}_{\mathbb{A}_2L}\right)(k,l_2)\right\}$$

$$\left(\mathcal{U}_{\mathbb{B}_1U}\boxtimes\mathcal{U}_{\mathbb{B}_2U}\right)\left((k,k_2)(k,l_2)\right) = \max\left\{\mathcal{U}_{\mathbb{A}_1U}(k),\mathcal{U}_{\mathbb{B}_2U}(k_2l_2)\right\}$$

$$\leq \max\left\{\mathcal{U}_{\mathbb{A}_1U}(k),\max\left\{\mathcal{U}_{\mathbb{A}_2U}(k_2),\mathcal{U}_{\mathbb{A}_2U}(l_2)\right\}\right\}$$

$$= \max\left\{\max\left\{\mathcal{I}_{\mathbb{A}_1U}(k),\mathcal{I}_{\mathbb{A}_2U}(k_2)\right\},\left\{\max\mathcal{U}_{\mathbb{A}_1U}(k),\mathcal{U}_{\mathbb{A}_2U}(l_2)\right\}\right\}$$

$$= \max\left\{\left(\mathcal{U}_{\mathbb{A}_1U}\boxtimes\mathcal{U}_{\mathbb{A}_2U}\right)(k,k_2),\left(\mathcal{U}_{\mathbb{A}_1U}\boxtimes\mathcal{U}_{\mathbb{A}_2U}\right)(k,l_2)\right\}.$$

$$\left(\mathcal{F}_{\mathbb{B}_1L}\boxtimes\mathcal{F}_{\mathbb{B}_2L}\right)\left((k,k_2)(k,l_2)\right) = \max\left\{\mathcal{F}_{\mathbb{A}_1L}(k),\mathcal{F}_{\mathbb{B}_2L}(k_2l_2)\right\}$$

$$\leq \max\left\{\mathcal{F}_{\mathbb{A}_1L}(k),\max\left\{\mathcal{F}_{\mathbb{A}_2L}(k_2),\mathcal{F}_{\mathbb{A}_2L}(l_2)\right\}\right\}$$

$$= \max\left\{\max\left\{\mathcal{F}_{\mathbb{A}_1L}(k),\mathcal{F}_{\mathbb{A}_2L}(k_2)\right\},\max\left\{\mathcal{F}_{\mathbb{A}_1L}(k),\mathcal{F}_{\mathbb{A}_2L}(l_2)\right\}\right\}$$

$$= \max\left\{\left(\mathcal{F}_{\mathbb{A}_1L}\boxtimes\mathcal{F}_{\mathbb{A}_2L}\right)(k,k_2),\left(\mathcal{F}_{\mathbb{A}_1L}\boxtimes\mathcal{F}_{\mathbb{A}_2L}\right)(k,l_2)\right\}$$

$$\left(\mathcal{F}_{\mathbb{B}_1U}\boxtimes\mathcal{F}_{\mathbb{B}_2U}\right)\left((k,k_2)(k,l_2)\right) = \max\left\{\mathcal{F}_{\mathbb{A}_1U}(k),\mathcal{F}_{\mathbb{B}_2U}(k_2l_2)\right\}$$

$$\leq \max\left\{\mathcal{F}_{\mathbb{A}_1U}(k),\max\left\{\mathcal{F}_{\mathbb{A}_2U}(k_2),\mathcal{F}_{\mathbb{A}_2U}(l_2)\right\}\right\}$$

$$= \max\left\{\max\left\{\mathcal{F}_{\mathbb{A}_1U}(k),\mathcal{F}_{\mathbb{A}_2U}(k_2)\right\},\left\{\max\mathcal{F}_{\mathbb{A}_1U}(k),\mathcal{F}_{\mathbb{A}_2U}(l_2)\right\}\right\}$$

$$= \max\left\{\left(\mathcal{F}_{\mathbb{A}_1U}\boxtimes\mathcal{F}_{\mathbb{A}_2U}\right)(k,k_2),\left(\mathcal{F}_{\mathbb{A}_1U}\boxtimes\mathcal{F}_{\mathbb{A}_2U}\right)(k,l_2)\right\}.$$

Case 2: $z \in \mathbb{V}_2$ and $k_1, l_1 \in \mathbb{E}_1$ give

$$\left(\mathcal{T}_{\mathbb{B}_1L}\boxtimes\mathcal{T}_{\mathbb{B}_2L}\right)\left((k_1,z)(l_1,z)\right) = \min\left\{\mathcal{T}_{\mathbb{B}_1L}(k_1l_1),\mathcal{T}_{\mathbb{A}_2L}(z)\right\}$$

$$\leq \min\left\{\min\left\{\mathcal{T}_{\mathbb{A}_1L}(k_1),\mathcal{T}_{\mathbb{A}_1L}(l_1)\right\},\mathcal{T}_{\mathbb{A}_2L}(z)\right\}$$

$$= \min\left\{\min\left\{\mathcal{T}_{\mathbb{A}_1L}(k_1),\mathcal{T}_{\mathbb{A}_2L}(z)\right\},\min\left\{\mathcal{T}_{\mathbb{A}_1L}(l_1),\mathcal{T}_{\mathbb{A}_2L}(z)\right\}\right\}$$

$$= \min\left\{\left(\mathcal{T}_{\mathbb{A}_1L}\boxtimes\mathcal{T}_{\mathbb{A}_2L}\right)(k_1,z),\left(\mathcal{T}_{\mathbb{A}_1L}\boxtimes\mathcal{T}_{\mathbb{A}_2L}\right)(l_1,z)\right\}$$

$$\left(\mathcal{T}_{\mathbb{B}_1U}\boxtimes\mathcal{T}_{\mathbb{B}_2U}\right)\left((k_1,z)(l_1,z)\right) = \min\left\{\mathcal{T}_{\mathbb{B}_1U}(k_1l_1),\mathcal{T}_{\mathbb{A}_2U}(z)\right\}$$

$$\leq \min\left\{\min\left\{\mathcal{T}_{\mathbb{A}_1U}(k_1),\mathcal{T}_{\mathbb{A}_1U}(l_1)\right\},\mathcal{T}_{\mathbb{A}_2U}(z)\right\}$$

$$= \min\left\{\min\left\{\mathcal{T}_{\mathbb{A}_1U}(k_1), \mathcal{T}_{\mathbb{A}_2U}(z)\right\}, \min\left\{\mathcal{T}_{\mathbb{A}_1U}(l_1), \mathcal{T}_{\mathbb{A}_2U}(z)\right\}\right\}$$

$$= \min\left\{\left(\mathcal{T}_{\mathbb{A}_1U} \boxtimes \mathcal{T}_{\mathbb{A}_2U}\right)(k_1, z), \left(\mathcal{T}_{\mathbb{A}_1U} \boxtimes \mathcal{T}_{\mathbb{A}_2U}\right)(l_1, z)\right\}$$

$$\left(\mathcal{C}_{\mathbb{B}_1L} \boxtimes \mathcal{C}_{\mathbb{B}_2L}\right)\left((k_1, z)(l_1, z)\right) = \min\left\{\mathcal{C}_{\mathbb{B}_1L}(k_1l_1), \mathcal{C}_{\mathbb{A}_2L}(z)\right\}$$

$$\leq \min\left\{\min\left(\mathcal{C}_{\mathbb{A}_1L}(k_1), \mathcal{C}_{\mathbb{A}_1L}(l_1)\right), \mathcal{C}_{\mathbb{A}_2L}(z)\right\}$$

$$= \min\left\{\min\left\{\mathcal{C}_{\mathbb{A}_1L}(k_1), \mathcal{C}_{\mathbb{A}_2L}(z)\right\}, \min\left\{\mathcal{C}_{\mathbb{A}_1L}(l_1), \mathcal{C}_{\mathbb{A}_2L}(z)\right\}\right\}$$

$$= \min\left\{\left(\mathcal{C}_{\mathbb{A}_1L} \boxtimes \mathcal{C}_{\mathbb{A}_2L}\right)(k_1, z), \left(\mathcal{C}_{\mathbb{A}_1L} \boxtimes \mathcal{C}_{\mathbb{A}_2L}\right)(l_1, z)\right\}$$

$$\left(\mathcal{C}_{\mathbb{B}_1U} \boxtimes \mathcal{C}_{\mathbb{B}_2U}\right)\left((k_1, z)(l_1, z)\right) = \min\left\{\mathcal{C}_{\mathbb{B}_1U}(k_1l_1), \mathcal{C}_{\mathbb{A}_2U}(z)\right\}$$

$$\leq \min\left\{\min\left(\mathcal{C}_{\mathbb{A}_1U}(k_1), \mathcal{C}_{\mathbb{A}_1U}(l_1)\right), \mathcal{C}_{\mathbb{A}_2U}(z)\right\}$$

$$= \min\left\{\min\left(\mathcal{C}_{\mathbb{A}_1U}(k_1), \mathcal{C}_{\mathbb{A}_2U}(z)\right), \min\left\{\mathcal{C}_{\mathbb{A}_1U}(l_1), \mathcal{C}_{\mathbb{A}_2U}(z)\right\}\right\}$$

$$= \min\left\{\left(\mathcal{C}_{\mathbb{A}_1U} \boxtimes \mathcal{C}_{\mathbb{A}_2U}\right)(k_1, z), \left(\mathcal{C}_{\mathbb{A}_1U} \boxtimes \mathcal{C}_{\mathbb{A}_2U}\right)(l_1, z)\right\}$$

$$\left(\mathcal{U}_{\mathbb{B}_1L} \boxtimes \mathcal{U}_{\mathbb{B}_2L}\right)\left((k_1, z)(l_1, z)\right) = \max\left\{\mathcal{U}_{\mathbb{B}_1L}(k_1l_1), \mathcal{U}_{\mathbb{A}_2L}(z)\right\}$$

$$\leq \max\left\{\max\left(\mathcal{U}_{\mathbb{A}_1L}(k_1), \mathcal{U}_{\mathbb{A}_1L}(l_1)\right), \mathcal{U}_{\mathbb{A}_2L}(z)\right\}$$

$$= \max\left\{\max\left\{\mathcal{U}_{\mathbb{A}_1L}(k_1), \mathcal{U}_{\mathbb{A}_2L}(z)\right\}, \max\left\{\mathcal{U}_{\mathbb{A}_1L}(l_1), \mathcal{U}_{\mathbb{A}_2L}(z)\right\}\right\}$$

$$= \max\left\{\left(\mathcal{U}_{\mathbb{A}_1L} \boxtimes \mathcal{U}_{\mathbb{A}_2L}\right)(k_1, z), \left(\mathcal{U}_{\mathbb{A}_1L} \boxtimes \mathcal{U}_{\mathbb{A}_2L}\right)(l_1, z)\right\}$$

$$\left(\mathcal{U}_{\mathbb{B}_1U} \boxtimes \mathcal{U}_{\mathbb{B}_2U}\right)\left((k_1, z)(l_1, z)\right) = \max\left\{\mathcal{U}_{\mathbb{B}_1U}(k_1l_1), \mathcal{U}_{\mathbb{A}_2U}(z)\right\}$$

$$\leq \max\left\{\max\left(\mathcal{U}_{\mathbb{A}_1U}(k_1), \mathcal{U}_{\mathbb{A}_1U}(l_1)\right), \mathcal{U}_{\mathbb{A}_2U}(z)\right\}$$

$$= \max\left\{\max\left(\mathcal{U}_{\mathbb{A}_1U}(k_1), \mathcal{U}_{\mathbb{A}_2U}(z)\right), \max\left\{\mathcal{U}_{\mathbb{A}_1U}(l_1), \mathcal{U}_{\mathbb{A}_2U}(z)\right\}\right\}$$

$$= \max\left\{\left(\mathcal{U}_{\mathbb{A}_1U} \boxtimes \mathcal{U}_{\mathbb{A}_2U}\right)(k_1, z), \left(\mathcal{U}_{\mathbb{A}_1U} \boxtimes \mathcal{U}_{\mathbb{A}_2U}\right)(l_1, z)\right\}$$

$$\left(\mathcal{F}_{\mathbb{B}_1L} \boxtimes \mathcal{F}_{\mathbb{B}_2L}\right)\left((k_1, z)(l_1, z)\right) = \max\left\{\mathcal{F}_{\mathbb{B}_1L}(k_1l_1), \mathcal{F}_{\mathbb{A}_2L}(z)\right\}$$

$$\leq \max\left\{\max\left\{\mathcal{F}_{\mathbb{A}_1L}(k_1), \mathcal{F}_{\mathbb{A}_1L}(l_1)\right\}, \mathcal{F}_{\mathbb{A}_2L}(z)\right\}$$

$$= \max\left\{\max\left\{\mathcal{F}_{\mathbb{A}_1L}(k_1), \mathcal{F}_{\mathbb{A}_2L}(z)\right\}, \max\left\{\mathcal{F}_{\mathbb{A}_1L}(l_1), \mathcal{F}_{\mathbb{A}_2L}(z)\right\}\right\}$$

$$= \max\left\{\left(\mathcal{F}_{\mathbb{A}_1L} \boxtimes \mathcal{F}_{\mathbb{A}_2L}\right)(k_1,z),\left(\mathcal{F}_{\mathbb{A}_1L} \boxtimes \mathcal{F}_{\mathbb{A}_2L}\right)(l_1,z)\right\}$$

$$\left(\mathcal{F}_{\mathbb{B}_1U} \boxtimes \mathcal{F}_{\mathbb{B}_2U}\right)\left((k_1,z)(l_1,z)\right) = \max\left\{\mathcal{F}_{\mathbb{B}_1U}(k_1l_1),\mathcal{F}_{\mathbb{A}_2U}(z)\right\}$$

$$\leq \max\left\{\max\left\{\mathcal{F}_{\mathbb{A}_1U}(k_1),\mathcal{F}_{\mathbb{A}_1U}(l_1)\right\},\mathcal{F}_{\mathbb{A}_2U}(z)\right\}$$

$$= \max\left\{\max\left\{\mathcal{F}_{\mathbb{A}_1U}(k_1),\mathcal{F}_{\mathbb{A}_2U}(z)\right\},\max\left\{\mathcal{F}_{\mathbb{A}_1U}(l_1),\mathcal{F}_{\mathbb{A}_2U}(z)\right\}\right\}$$

$$= \max\left\{\left(\mathcal{F}_{\mathbb{A}_1U} \boxtimes \mathcal{F}_{\mathbb{A}_2U}\right)(k_1,z),\left(\mathcal{F}_{\mathbb{A}_1U} \boxtimes \mathcal{F}_{\mathbb{A}_2U}\right)(l_1,z)\right\}.$$

Case 3: $k_1k_2 \in \mathbb{E}_1$ and $z_1z_2 \in \mathbb{E}_2$. Then

$$\left(\mathcal{T}_{\mathbb{B}_1L} \boxtimes \mathcal{T}_{\mathbb{B}_2L}\right)\left((k_1,z_1)(k_2,z_2)\right) = \min\left\{\mathcal{T}_{\mathbb{B}_1L}(k_1k_2),\mathcal{T}_{\mathbb{B}_2L}(z_1z_2)\right\}$$

$$\leq \min\left\{\min\left\{\mathcal{T}_{\mathbb{A}_1L}(k_1),\mathcal{T}_{\mathbb{A}_1L}(k_2)\right\},\min\left\{\mathcal{T}_{\mathbb{A}_2L}(z_1),\mathcal{T}_{\mathbb{A}_2L}(z_2)\right\}\right\}$$

$$= \min\left\{\min\left\{\mathcal{T}_{\mathbb{A}_1L}(k_1),\mathcal{T}_{\mathbb{A}_2L}(z_1)\right\},\min\left\{\mathcal{T}_{\mathbb{A}_1L}(k_2),\mathcal{T}_{\mathbb{A}_2L}(z_2)\right\}\right\}$$

$$= \min\left\{\left(\mathcal{T}_{\mathbb{A}_1L} \boxtimes \mathcal{T}_{\mathbb{A}_2L}\right)(k_1,z_1),\left(\mathcal{T}_{\mathbb{A}_1L} \boxtimes \mathcal{T}_{\mathbb{A}_2L}\right)(k_2,z_2)\right\}$$

$$\left(\mathcal{T}_{\mathbb{B}_1U} \boxtimes \mathcal{T}_{\mathbb{B}_2U}\right)\left((k_1,z_1)(k_2,z_2)\right) = \min\left\{\mathcal{T}_{\mathbb{B}_1U}(k_1k_2),\mathcal{T}_{\mathbb{B}_2U}(z_1z_2)\right\}$$

$$\leq \min\left\{\min\left\{\mathcal{T}_{\mathbb{A}_1U}(k_1),\mathcal{T}_{\mathbb{A}_1U}(k_2)\right\},\min\left\{\mathcal{T}_{\mathbb{A}_2U}(z_1),\mathcal{T}_{\mathbb{A}_2U}(z_2)\right\}\right\}$$

$$= \min\left\{\min\left\{\mathcal{T}_{\mathbb{A}_1U}(k_1),\mathcal{T}_{\mathbb{A}_2U}(z_1)\right\},\min\left\{\mathcal{T}_{\mathbb{A}_1U}(k_2),\mathcal{T}_{\mathbb{A}_2U}(z_2)\right\}\right\}$$

$$= \min\left\{\left(\mathcal{T}_{\mathbb{A}_1U} \boxtimes \mathcal{T}_{\mathbb{A}_2U}\right)(k_1,z_1),\left(\mathcal{T}_{\mathbb{A}_1U} \boxtimes \mathcal{T}_{\mathbb{A}_2U}\right)(k_2,z_2)\right\}$$

$$\left(\mathcal{C}_{\mathbb{B}_1L} \boxtimes \mathcal{C}_{\mathbb{B}_2L}\right)\left((k_1,z_1)(k_2,z_2)\right) = \min\left\{\mathcal{C}_{\mathbb{B}_1L}(k_1k_2),\mathcal{C}_{\mathbb{B}_2L}(z_1z_2)\right\}$$

$$\leq \min\left\{\min\left\{\mathcal{C}_{\mathbb{A}_1L}(k_1),\mathcal{C}_{\mathbb{A}_1L}(k_2)\right\},\min\left\{\mathcal{C}_{\mathbb{A}_2L}(z_1),\mathcal{C}_{\mathbb{A}_2L}(z_2)\right\}\right\}$$

$$= \min\left\{\min\left\{\mathcal{C}_{\mathbb{A}_1L}(k_1),\mathcal{C}_{\mathbb{A}_2L}(z_1)\right\},\min\left\{\mathcal{C}_{\mathbb{A}_1L}(k_2),\mathcal{C}_{\mathbb{A}_2L}(z_2)\right\}\right\}$$

$$= \min\left\{\left(\mathcal{C}_{\mathbb{A}_1L} \boxtimes \mathcal{C}_{\mathbb{A}_2L}\right)(k_1,z_1),\left(\mathcal{C}_{\mathbb{A}_1L} \boxtimes \mathcal{C}_{\mathbb{A}_2L}\right)(k_2,z_2)\right\}$$

$$\left(\mathcal{C}_{\mathbb{B}_1U} \boxtimes \mathcal{C}_{\mathbb{B}_2U}\right)\left((k_1,z_1)(k_2,z_2)\right) = \min\left\{\mathcal{C}_{\mathbb{B}_1U}(k_1k_2),\mathcal{C}_{\mathbb{B}_2U}(z_1z_2)\right\}$$

$$\leq \min\left\{\min\left\{\mathcal{C}_{\mathbb{A}_1U}(k_1),\mathcal{C}_{\mathbb{A}_1U}(k_2)\right\},\min\left\{\mathcal{C}_{\mathbb{A}_2U}(z_1),\mathcal{C}_{\mathbb{A}_2U}(z_2)\right\}\right\}$$

$$= \min\left\{\min\left\{\mathcal{C}_{\mathbb{A}_1U}(k_1),\mathcal{C}_{\mathbb{A}_2U}(z_1)\right\},\min\left\{\mathcal{C}_{\mathbb{A}_1U}(k_2),\mathcal{C}_{\mathbb{A}_2U}(z_2)\right\}\right\}$$

$$= \min\left\{\left(\mathcal{C}_{\mathbb{A}_1U} \boxtimes \mathcal{C}_{\mathbb{A}_2U}\right)(k_1,z_1),\left(\mathcal{C}_{\mathbb{A}_1U} \boxtimes \mathcal{C}_{\mathbb{A}_2U}\right)(k_2,z_2)\right\}$$

$$\left(\mathcal{U}_{\mathbb{B}_1L} \boxtimes \mathcal{U}_{\mathbb{B}_2L}\right)\left((k_1,z_1)(k_2,z_2)\right) = \max\left\{\mathcal{U}_{\mathbb{B}_1L}(k_1k_2),\mathcal{U}_{\mathbb{B}_2L}(z_1z_2)\right\}$$

$$\leq \max\left\{\max\left\{\mathcal{U}_{\mathbb{A}_1L}(k_1),\mathcal{U}_{\mathbb{A}_1L}(k_2)\right\},\max\left\{\mathcal{U}_{\mathbb{A}_2L}(z_1),\mathcal{U}_{\mathbb{A}_2L}(z_2)\right\}\right\}$$

$$= \max\left\{\max\left\{\mathcal{U}_{\mathbb{A}_1L}(k_1),\mathcal{U}_{\mathbb{A}_2L}(z_1)\right\},\max\left\{\mathcal{U}_{\mathbb{A}_1L}(k_2),\mathcal{U}_{\mathbb{A}_2L}(z_2)\right\}\right\}$$

$$= \max\left\{\left(\mathcal{U}_{\mathbb{A}_1L} \boxtimes \mathcal{U}_{\mathbb{A}_2L}\right)(k_1,z_1),\left(\mathcal{U}_{\mathbb{A}_1L} \boxtimes \mathcal{U}_{\mathbb{A}_2L}\right)(k_2,z_2)\right\}$$

$$\left(\mathcal{U}_{\mathbb{B}_1U} \boxtimes \mathcal{U}_{\mathbb{B}_2U}\right)\left((k_1,z_1)(k_2,z_2)\right) = \max\left\{\mathcal{U}_{\mathbb{B}_1U}(k_1k_2),\mathcal{U}_{\mathbb{B}_2U}(z_1z_2)\right\}$$

$$\leq \max\left\{\max\left\{\mathcal{U}_{\mathbb{A}_1U}(k_1),\mathcal{U}_{\mathbb{A}_1U}(k_2)\right\},\max\left\{\mathcal{U}_{\mathbb{A}_2U}(z_1),\mathcal{U}_{\mathbb{A}_2U}(z_2)\right\}\right\}$$

$$= \max\left\{\max\left\{\mathcal{U}_{\mathbb{A}_1U}(k_1),\mathcal{U}_{\mathbb{A}_2U}(z_1)\right\},\max\left\{\mathcal{U}_{\mathbb{A}_1U}(k_2),\mathcal{U}_{\mathbb{A}_2U}(z_2)\right\}\right\}$$

$$= \max\left\{\left(\mathcal{U}_{\mathbb{A}_1U} \boxtimes \mathcal{U}_{\mathbb{A}_2U}\right)(k_1,z_1),\left(\mathcal{U}_{\mathbb{A}_1U} \boxtimes \mathcal{U}_{\mathbb{A}_2U}\right)(k_2,z_2)\right\}$$

$$\left(\mathcal{F}_{\mathbb{B}_1L} \boxtimes \mathcal{F}_{\mathbb{B}_2L}\right)\left((k_1,z_1)(k_2,z_2)\right) = \max\left\{\mathcal{F}_{\mathbb{B}_1L}(k_1k_2),\mathcal{F}_{\mathbb{B}_2L}(z_1z_2)\right\}$$

$$\leq \max\left\{\max\left\{\mathcal{F}_{\mathbb{A}_1L}(k_1),\mathcal{F}_{\mathbb{A}_1L}(k_2)\right\},\max\left\{\mathcal{F}_{\mathbb{A}_2L}(z_1),\mathcal{F}_{\mathbb{A}_2L}(z_2)\right\}\right\}$$

$$= \max\left\{\max\left\{\mathcal{F}_{\mathbb{A}_1L}(k_1),\mathcal{F}_{\mathbb{A}_2L}(z_1)\right\},\max\left\{\mathcal{F}_{\mathbb{A}_1L}(k_2),\mathcal{F}_{\mathbb{A}_2L}(z_2)\right\}\right\}$$

$$= \max\left\{\left(\mathcal{F}_{\mathbb{A}_1L} \boxtimes \mathcal{F}_{\mathbb{A}_2L}\right)(k_1,z_1),\left(\mathcal{F}_{\mathbb{A}_1L} \boxtimes \mathcal{F}_{\mathbb{A}_2L}\right)(k_2,z_2)\right\}$$

$$\left(\mathcal{F}_{\mathbb{B}_1U} \boxtimes \mathcal{F}_{\mathbb{B}_2U}\right)\left((k_1,z_1)(k_2,z_2)\right) = \max\left\{\mathcal{F}_{\mathbb{B}_1U}(k_1k_2),\mathcal{F}_{\mathbb{B}_2U}(z_1z_2)\right\}$$

$$\leq \max\left\{\max\left\{\mathcal{F}_{\mathbb{A}_1U}(k_1),\mathcal{F}_{\mathbb{A}_1U}(k_2)\right\},\max\left\{\mathcal{F}_{\mathbb{A}_2U}(z_1),\mathcal{F}_{\mathbb{A}_2U}(z_2)\right\}\right\}$$

$$= \max\left\{\max\left\{\mathcal{F}_{\mathbb{A}_1U}(k_1),\mathcal{F}_{\mathbb{A}_2U}(z_1)\right\},\max\left\{\mathcal{F}_{\mathbb{A}_1U}(k_2),\mathcal{F}_{\mathbb{A}_2U}(z_2)\right\}\right\}$$

$$= \max\left\{\left(\mathcal{F}_{\mathbb{A}_1U} \boxtimes \mathcal{F}_{\mathbb{A}_2U}\right)(k_1,z_1),\left(\mathcal{F}_{\mathbb{A}_1U} \boxtimes \mathcal{F}_{\mathbb{A}_2U}\right)(k_2,z_2)\right\}$$

This completes the proof.

Definition 6.4.13 The composition $\mathcal{G}_1 \times \mathcal{G}_2$ of two interval quadripartitioned single-valued neutrosophic graphs $\mathcal{G}_1 = (\mathbb{A}_1, \mathbb{B}_1)$ and $\mathcal{G}_2 = (\mathbb{A}_2, \mathbb{B}_2)$ of the graphs $\mathcal{G}_1^* = (\mathbb{V}_1, \mathbb{E}_1)$ and $\mathcal{G}_2^* = (\mathbb{V}_2, \mathbb{E}_2)$ is a pair $\mathcal{G}_1 \times \mathcal{G}_2 = (\mathbb{A}_1 \times \mathbb{A}_2, \mathbb{B}_1 \times \mathbb{B}_2)$, such that

1. $\left(\mathcal{T}_{\mathbb{A}_1L} \times \mathcal{T}_{\mathbb{A}_2L}\right)(k_1,k_2) = \min\left\{\mathcal{T}_{\mathbb{A}_1L}(k_1),\mathcal{T}_{\mathbb{A}_2L}(k_2)\right\}$

   $\left(\mathcal{T}_{\mathbb{A}_1U} \times \mathcal{T}_{\mathbb{A}_2U}\right)(k_1,k_2) = \min\left\{\mathcal{T}_{\mathbb{A}_1U}(k_1),\mathcal{T}_{\mathbb{A}_2U}(k_2)\right\}$

$$\left(\mathcal{C}_{\mathbb{A}_1 L} \times \mathcal{C}_{\mathbb{A}_2 L}\right)(k_1, k_2) = \min\left\{\mathcal{C}_{\mathbb{A}_1 L}(k_1), \mathcal{C}_{\mathbb{A}_2 L}(k_2)\right\}$$

$$\left(\mathcal{C}_{\mathbb{A}_1 U} \times \mathcal{C}_{\mathbb{A}_2 U}\right)(k_1, k_2) = \min\left\{\mathcal{C}_{\mathbb{A}_1 U}(k_1), \mathcal{C}_{\mathbb{A}_2 U}(k_2)\right\}$$

$$\left(\mathcal{U}_{\mathbb{A}_1 L} \times \mathcal{U}_{\mathbb{A}_2 L}\right)(k_1, k_2) = \max\left\{\mathcal{U}_{\mathbb{A}_1 L}(k_1), \mathcal{U}_{\mathbb{A}_2 L}(k_2)\right\}$$

$$\left(\mathcal{U}_{\mathbb{A}_1 U} \times \mathcal{U}_{\mathbb{A}_2 U}\right)(k_1, k_2) = \max\left\{\mathcal{U}_{\mathbb{A}_1 U}(k_1), \mathcal{U}_{\mathbb{A}_2 U}(k_2)\right\}$$

$$\left(\mathcal{F}_{\mathbb{A}_1 L} \times \mathcal{F}_{\mathbb{A}_2 L}\right)(k_1, k_2) = \max\left\{\mathcal{F}_{\mathbb{A}_1 L}(k_1), \mathcal{F}_{\mathbb{A}_2 L}(k_2)\right\}$$

$$\left(\mathcal{F}_{\mathbb{A}_1 U} \times \mathcal{F}_{\mathbb{A}_2 U}\right)(k_1, k_2) = \max\left\{\mathcal{F}_{\mathbb{A}_1 U}(k_1), \mathcal{F}_{\mathbb{A}_2 U}(k_2)\right\}$$

for all $k_1, k_2 \in \mathbb{V}$,

2. $$\left(\mathcal{T}_{\mathbb{B}_1 L} \times \mathcal{T}_{\mathbb{B}_2 L}\right)\left((k, k_2)(k, l_2)\right) = \min\left\{\mathcal{T}_{\mathbb{A}_1 L}(k), \mathcal{T}_{\mathbb{B}_2 L}(k_2 l_2)\right\}$$

$$\left(\mathcal{T}_{\mathbb{B}_1 U} \times \mathcal{T}_{\mathbb{B}_2 U}\right)\left((k, k_2)(k, l_2)\right) = \min\left\{\mathcal{T}_{\mathbb{A}_1 U}(k), \mathcal{T}_{\mathbb{B}_2 U}(k_2 l_2)\right\}$$

$$\left(\mathcal{C}_{\mathbb{B}_1 L} \times \mathcal{C}_{\mathbb{B}_2 L}\right)\left((k, k_2)(k, l_2)\right) = \min\left\{\mathcal{C}_{\mathbb{A}_1 L}(k), \mathcal{C}_{\mathbb{B}_2 L}(k_2 l_2)\right\}$$

$$\left(\mathcal{C}_{\mathbb{B}_1 U} \times \mathcal{C}_{\mathbb{B}_2 U}\right)\left((k, k_2)(k, l_2)\right) = \min\left\{\mathcal{C}_{\mathbb{A}_1 U}(k), \mathcal{C}_{\mathbb{B}_2 U}(k_2 l_2)\right\}$$

$$\left(\mathcal{U}_{\mathbb{B}_1 L} \times \mathcal{U}_{\mathbb{B}_2 L}\right)\left((k, k_2)(k, l_2)\right) = \max\left\{\mathcal{U}_{\mathbb{A}_1 L}(k), \mathcal{U}_{\mathbb{B}_2 L}(k_2 l_2)\right\}$$

$$\left(\mathcal{U}_{\mathbb{B}_1 U} \times \mathcal{U}_{\mathbb{B}_2 U}\right)\left((k, k_2)(k, l_2)\right) = \max\left\{\mathcal{U}_{\mathbb{A}_1 U}(k), \mathcal{U}_{\mathbb{B}_2 U}(k_2 l_2)\right\}$$

$$\left(\mathcal{F}_{\mathbb{B}_1 L} \times \mathcal{F}_{\mathbb{B}_2 L}\right)\left((k, k_2)(k, l_2)\right) = \max\left\{\mathcal{F}_{\mathbb{A}_1 L}(k), \mathcal{F}_{\mathbb{B}_2 L}(k_2 l_2)\right\}$$

$$\left(\mathcal{F}_{\mathbb{B}_1 U} \times \mathcal{F}_{\mathbb{B}_2 U}\right)\left((k, k_2)(k, l_2)\right) = \max\left\{\mathcal{F}_{\mathbb{A}_1 U}(k), \mathcal{F}_{\mathbb{B}_2 U}(k_2 l_2)\right\}$$

for all $k \in \mathbb{V}_1$ *and* $k_2 l_2 \in \mathbb{E}_2$,

3. $$\left(\mathcal{T}_{\mathbb{B}_1 L} \times \mathcal{T}_{\mathbb{B}_2 L}\right)\left((k_1, z)(l_1, z)\right) = \min\left\{\mathcal{T}_{\mathbb{B}_1 L}(k_1 l_1), \mathcal{T}_{\mathbb{A}_2 L}(z)\right\}$$

$$\left(\mathcal{T}_{\mathbb{B}_1 U} \times \mathcal{T}_{\mathbb{B}_2 U}\right)\left((k_1, z)(l_1, z)\right) = \min\left\{\mathcal{T}_{\mathbb{B}_1 U}(k_1 l_1), \mathcal{T}_{\mathbb{A}_2 U}(z)\right\}$$

$$\left(\mathcal{C}_{\mathbb{B}_1 L} \times \mathcal{C}_{\mathbb{B}_2 L}\right)\left((k_1, z)(l_1, z)\right) = \min\left\{\mathcal{C}_{\mathbb{B}_1 L}(k_1 l_1), \mathcal{C}_{\mathbb{A}_2 L}(z)\right\}$$

$$\left(\mathcal{C}_{\mathbb{B}_1 U} \times \mathcal{C}_{\mathbb{B}_2 U}\right)\left((k_1, z)(l_1, z)\right) = \min\left\{\mathcal{C}_{\mathbb{B}_1 U}(k_1 l_1), \mathcal{C}_{\mathbb{A}_2 U}(z)\right\}$$

$$\left(\mathcal{U}_{\mathbb{B}_1 L} \times \mathcal{U}_{\mathbb{B}_2 L}\right)\left((k_1, z)(l_1, z)\right) = \max\left\{\mathcal{U}_{\mathbb{B}_1 L}(k_1 l_1), \mathcal{U}_{\mathbb{A}_2 L}(z)\right\}$$

$$\left(\mathcal{U}_{\mathbb{B}_1 U} \times \mathcal{U}_{\mathbb{B}_2 U}\right)\left((k_1, z)(l_1, z)\right) = \max\left\{\mathcal{U}_{\mathbb{B}_1 U}(k_1 l_1), \mathcal{U}_{\mathbb{A}_2 U}(z)\right\}$$

$$\left(\mathcal{F}_{\mathbb{B}_1L} \times \mathcal{F}_{\mathbb{B}_2L}\right)\left((k_1,z)(l_1,z)\right) = \max\left\{\mathcal{F}_{\mathbb{B}_1L}(k_1l_1), \mathcal{F}_{\mathbb{A}_2L}(z)\right\}$$

$$\left(\mathcal{F}_{\mathbb{B}_1U} \times \mathcal{F}_{\mathbb{B}_2U}\right)\left((k_1,z)(l_1,z)\right) = \max\left\{\mathcal{F}_{\mathbb{B}_1U}(k_1l_1), \mathcal{F}_{\mathbb{A}_2U}(z)\right\}$$

for all $z \in \mathbb{V}_2$ *and* $k_1l_1 \in \mathbb{E}_1$ and

4. $\left(\mathcal{T}_{\mathbb{B}_1L} \times \mathcal{T}_{\mathbb{B}_2L}\right)\left((k_1,k_2)(l_1,l_2)\right) = \min\left\{\mathcal{T}_{\mathbb{A}_2L}(k_2), \mathcal{T}_{\mathbb{A}_2L}(l_2), \mathcal{T}_{\mathbb{B}_1L}(k_1l_1)\right\}$

$$\left(\mathcal{T}_{\mathbb{B}_1U} \times \mathcal{T}_{\mathbb{B}_2U}\right)\left((k_1,k_2)(l_1,l_2)\right) = \min\left\{\mathcal{T}_{\mathbb{A}_2U}(k_2), \mathcal{T}_{\mathbb{A}_2U}(l_2), \mathcal{T}_{\mathbb{B}_1U}(k_1l_1)\right\}$$

$$\left(\mathcal{C}_{\mathbb{B}_1L} \times \mathcal{C}_{\mathbb{B}_2L}\right)\left((k_1,k_2)(l_1,l_2)\right) = \min\left\{\mathcal{C}_{\mathbb{A}_2L}(k_2), \mathcal{C}_{\mathbb{A}_2L}(l_2), \mathcal{C}_{\mathbb{B}_1L}(k_1l_1)\right\}$$

$$\left(\mathcal{C}_{\mathbb{B}_1U} \times \mathcal{C}_{\mathbb{B}_2U}\right)\left((k_1,k_2)(l_1,l_2)\right) = \min\left\{\mathcal{C}_{\mathbb{A}_2U}(k_2), \mathcal{C}_{\mathbb{A}_2U}(l_2), \mathcal{C}_{\mathbb{B}_1U}(k_1l_1)\right\}$$

$$\left(\mathcal{U}_{\mathbb{B}_1L} \times \mathcal{U}_{\mathbb{B}_2L}\right)\left((k_1,k_2)(l_1,l_2)\right) = \max\left\{\mathcal{U}_{\mathbb{A}_2L}(k_2), \mathcal{U}_{\mathbb{A}_2L}(l_2), \mathcal{U}_{\mathbb{B}_1L}(k_1l_1)\right\}$$

$$\left(\mathcal{U}_{\mathbb{B}_1U} \times \mathcal{U}_{\mathbb{B}_2U}\right)\left((k_1,k_2)(l_1,l_2)\right) = \max\left\{\mathcal{U}_{\mathbb{A}_2U}(k_2), \mathcal{U}_{\mathbb{A}_2U}(l_2), \mathcal{U}_{\mathbb{B}_1U}(k_1l_1)\right\}$$

$$\left(\mathcal{F}_{\mathbb{B}_1L} \times \mathcal{F}_{\mathbb{B}_2L}\right)\left((k_1,k_2)(l_1,l_2)\right) = \max\left\{\mathcal{F}_{\mathbb{A}_2L}(k_2), \mathcal{F}_{\mathbb{A}_2L}(l_2), \mathcal{F}_{\mathbb{B}_1L}(k_1l_1)\right\}$$

$$\left(\mathcal{F}_{\mathbb{B}_1U} \times \mathcal{F}_{\mathbb{B}_2U}\right)\left((k_1,k_2)(l_1,l_2)\right) = \max\left\{\mathcal{F}_{\mathbb{A}_2U}(k_2), \mathcal{F}_{\mathbb{A}_2U}(l_2), \mathcal{F}_{\mathbb{B}_1U}(k_1l_1)\right\}$$

for all $k_2, l_2 \in \mathbb{V}_2, k_2 \neq l_2$ *and* $k_1l_1 \in \mathbb{E}_1$.

Proposition 6.4.14 The composition $\mathcal{G}_1 \times \mathcal{G}_2 = (\mathbb{A}_1 \times \mathbb{A}_2, \mathbb{B}_1 \times \mathbb{B}_2)$ of two interval quadripartitionedsingle valued neutrosophic graphs $\mathcal{G}_1 = (\mathbb{A}_1, \mathbb{B}_1)$ and $\mathcal{G}_2 = (\mathbb{A}_2, \mathbb{B}_2)$ is an interval quadripartitioned single valued neutrosophic graph.

Proof. Similar to the previous proof, one can get the conditions for $\mathbb{B}_1 \times \mathbb{B}_2$. Now consider

$$\left(\mathcal{T}_{\mathbb{B}_1L} \times \mathcal{T}_{\mathbb{B}_2L}\right)\left((k,k_2)(k,l_2)\right) = \min\left\{\mathcal{T}_{\mathbb{A}_1L}(k), \mathcal{T}_{\mathbb{B}_2L}(k_2l_2)\right\}$$

$$\leq \min\left\{\mathcal{T}_{\mathbb{A}_1L}(k), \left(\min\mathcal{T}_{\mathbb{A}_2L}(k_2), \mathcal{T}_{\mathbb{A}_2L}(l_2)\right)\right\}$$

$$= \min\left\{\min\left\{\mathcal{T}_{\mathbb{A}_1L}(k), \mathcal{T}_{\mathbb{A}_2L}(k_2)\right\}, \min\left\{\mathcal{T}_{\mathbb{A}_1L}(k), \mathcal{T}_{\mathbb{A}_2L}(l_2)\right\}\right\}$$

$$= \min\left\{\left(\mathcal{T}_{\mathbb{A}_1L} \times \mathcal{T}_{\mathbb{A}_2L}\right)(k,k_2), \left(\mathcal{T}_{\mathbb{A}_1L} \times \mathcal{T}_{\mathbb{A}_2L}\right)(k,l_2)\right\}$$

$$\left(\mathcal{T}_{\mathbb{B}_1U} \times \mathcal{T}_{\mathbb{B}_2U}\right)\left((k,k_2)(k,l_2)\right) = \min\left\{\mathcal{T}_{\mathbb{A}_1U}(k), \mathcal{T}_{\mathbb{B}_2U}(k_2l_2)\right\}$$

$$\leq \min\left\{\mathcal{T}_{\mathbb{A}_1U}(k), \min\left\{\mathcal{T}_{\mathbb{A}_2U}(k_2), \mathcal{T}_{\mathbb{A}_2U}(l_2)\right\}\right\}$$

$$= \min\left\{\min\left\{\mathcal{T}_{\mathbb{A}_1U}(k), \mathcal{T}_{\mathbb{A}_2U}(k_2)\right\}, \min\left\{\mathcal{T}_{\mathbb{A}_1U}(k), \mathcal{T}_{\mathbb{A}_2U}(l_2)\right\}\right\}$$

$$= \min\left\{\left(\mathcal{T}_{\mathbb{A}_1U} \times \mathcal{T}_{\mathbb{A}_2U}\right)(k,k_2), \left(\mathcal{T}_{\mathbb{A}_1U} \times \mathcal{T}_{\mathbb{A}_2U}\right)(k,l_2)\right\}.$$

$$\left(\mathcal{C}_{\mathbb{B}_1L} \times \mathcal{C}_{\mathbb{B}_2L}\right)\left((k,k_2)(k,l_2)\right) = \min\left\{\mathcal{C}_{\mathbb{A}_1L}(k), \mathcal{C}_{\mathbb{B}_2L}(k_2l_2)\right\}$$

$$\leq \min\left\{\mathcal{C}_{\mathbb{A}_1L}(k), \min\left\{\mathcal{C}_{\mathbb{A}_2L}(k_2), \mathcal{C}_{\mathbb{A}_2L}(l_2)\right\}\right\}$$

$$= \min\left\{\min\left\{\mathcal{C}_{\mathbb{A}_1L}(k), \mathcal{C}_{\mathbb{A}_2L}(k_2)\right\}, \min\left\{\mathcal{C}_{\mathbb{A}_1L}(k), \mathcal{C}_{\mathbb{A}_2L}(l_2)\right\}\right\}$$

$$= \min\left\{\left(\mathcal{C}_{\mathbb{A}_1L} \times \mathcal{C}_{\mathbb{A}_2L}\right)(k,k_2), \left(\mathcal{C}_{\mathbb{A}_1L} \times \mathcal{C}_{\mathbb{A}_2L}\right)(k,l_2)\right\}$$

$$\left(\mathcal{C}_{\mathbb{B}_1U} \times \mathcal{C}_{\mathbb{B}_2U}\right)\left((k,k_2)(k,l_2)\right) = \min\left\{\mathcal{C}_{\mathbb{A}_1U}(k), \mathcal{C}_{\mathbb{B}_2U}(k_2l_2)\right\}$$

$$\leq \min\left\{\mathcal{C}_{\mathbb{A}_1U}(k), \min\left\{\mathcal{C}_{\mathbb{A}_2U}(k_2), \mathcal{C}_{\mathbb{A}_2U}(l_2)\right\}\right\}$$

$$= \min\left\{\min\left\{\mathcal{C}_{\mathbb{A}_1U}(k), \mathcal{C}_{\mathbb{A}_2U}(k_2)\right\}, \min\left\{\mathcal{C}_{\mathbb{A}_1U}(k), \mathcal{C}_{\mathbb{A}_2U}(l_2)\right\}\right\}$$

$$= \min\left\{\left(\mathcal{C}_{\mathbb{A}_1U} \times \mathcal{C}_{\mathbb{A}_2U}\right)(k,k_2), \left(\mathcal{C}_{\mathbb{A}_1U} \times \mathcal{C}_{\mathbb{A}_2U}\right)(k,l_2)\right\}.$$

$$\left(\mathcal{U}_{\mathbb{B}_1L} \times \mathcal{U}_{\mathbb{B}_2L}\right)\left((k,k_2)(k,l_2)\right) = \max\left\{\mathcal{U}_{\mathbb{A}_1L}(k), \mathcal{U}_{\mathbb{B}_2L}(k_2l_2)\right\}$$

$$\leq \max\left\{\mathcal{U}_{\mathbb{A}_1L}(k), \max\left\{\mathcal{U}_{\mathbb{A}_2L}(k_2), \mathcal{U}_{\mathbb{A}_2L}(l_2)\right\}\right\}$$

$$= \max\left\{\max\left\{\mathcal{U}_{\mathbb{A}_1L}(k), \mathcal{U}_{\mathbb{A}_2L}(k_2)\right\}, \max\left\{\mathcal{U}_{\mathbb{A}_1L}(k), \mathcal{U}_{\mathbb{A}_2L}(l_2)\right\}\right\}$$

$$= \max\left\{\left(\mathcal{U}_{\mathbb{A}_1L} \times \mathcal{U}_{\mathbb{A}_2L}\right)(k,k_2), \left(\mathcal{U}_{\mathbb{A}_1L} \times \mathcal{U}_{\mathbb{A}_2L}\right)(k,l_2)\right\}$$

$$\left(\mathcal{U}_{\mathbb{B}_1U} \times \mathcal{U}_{\mathbb{B}_2U}\right)\left((k,k_2)(k,l_2)\right) = \max\left\{\mathcal{U}_{\mathbb{A}_1U}(k), \mathcal{U}_{\mathbb{B}_2U}(k_2l_2)\right\}$$

$$\leq \max\left\{\mathcal{U}_{\mathbb{A}_1U}(k), \max\left\{\mathcal{U}_{\mathbb{A}_2U}(k_2), \mathcal{U}_{\mathbb{A}_2U}(l_2)\right\}\right\}$$

$$= \max\left\{\max\left\{\mathcal{U}_{\mathbb{A}_1U}(k), \mathcal{U}_{\mathbb{A}_2U}(k_2)\right\}, \max\left\{\mathcal{U}_{\mathbb{A}_1U}(k), \mathcal{U}_{\mathbb{A}_2U}(l_2)\right\}\right\}$$

$$= \max\left\{\left(\mathcal{U}_{\mathbb{A}_1U} \times \mathcal{U}_{\mathbb{A}_2U}\right)(k,k_2), \left(\mathcal{U}_{\mathbb{A}_1U} \times \mathcal{C}_{\mathbb{A}_2U}\right)(k,l_2)\right\}.$$

$$\left(\mathcal{F}_{\mathbb{B}_1L} \times \mathcal{F}_{\mathbb{B}_2L}\right)\left((k,k_2)(k,l_2)\right) = \max\left\{\mathcal{F}_{\mathbb{A}_1L}(k), \mathcal{F}_{\mathbb{B}_2L}(k_2l_2)\right\}$$

$$\leq \max\left\{\mathcal{F}_{\mathbb{A}_1L}(k), \max\left\{\mathcal{F}_{\mathbb{A}_2L}(k_2), \mathcal{F}_{\mathbb{A}_2L}(l_2)\right\}\right\}$$

$$= \max\left\{\max\left\{\mathcal{F}_{\mathbb{A}_1L}(k), \mathcal{F}_{\mathbb{A}_2L}(k_2)\right\}, \max\left\{\mathcal{F}_{\mathbb{A}_1L}(k), \mathcal{F}_{\mathbb{A}_2L}(l_2)\right\}\right\}$$

$$= \max\left\{\left(\mathcal{F}_{\mathbb{A}_1L} \times \mathcal{F}_{\mathbb{A}_2L}\right)(k,k_2), \left(\mathcal{F}_{\mathbb{A}_1L} \times \mathcal{F}_{\mathbb{A}_2L}\right)(k,l_2)\right\}$$

$$\left(\mathcal{F}_{\mathbb{B}_1U} \times \mathcal{F}_{\mathbb{B}_2U}\right)\left((k,k_2)(k,l_2)\right) = \max\left\{\mathcal{F}_{\mathbb{A}_1U}(k), \mathcal{F}_{\mathbb{B}_2U}(k_2l_2)\right\}$$

$$\leq \max\left\{\mathcal{F}_{\mathbb{A}_1U}(k), \max\left\{\mathcal{F}_{\mathbb{A}_2U}(k_2), \mathcal{F}_{\mathbb{A}_2U}(l_2)\right\}\right\}$$

$$= \max\left\{\max\left\{\mathcal{F}_{\mathbb{A}_1U}(k), \mathcal{F}_{\mathbb{A}_2U}(k_2)\right\}, \max\left\{\mathcal{F}_{\mathbb{A}_1U}(k), \mathcal{F}_{\mathbb{A}_2U}(l_2)\right\}\right\}$$

$$= \max\left\{\left(\mathcal{F}_{\mathbb{A}_1U} \times \mathcal{F}_{\mathbb{A}_2U}\right)(k, k_2), \left(\mathcal{F}_{\mathbb{A}_1U} \times \mathcal{F}_{\mathbb{A}_2U}\right)(k, l_2)\right\}.$$

The similar result is obtained for the case of $z \in \mathbb{V}_2$, and $k_1 l_1 \in \mathbb{E}_1$. Now the case for $k_2, l_2 \in \mathbb{V}_2, k_2 \neq l_2$ and $k_1 l_1 \in \mathbb{E}_1$, yields that,

$$\left(\mathcal{T}_{\mathbb{B}_1L} \times \mathcal{T}_{\mathbb{B}_2L}\right)\left((k_1, k_2)(l_1, l_2)\right) = \min\left\{\mathcal{T}_{\mathbb{A}_2L}(k_2), \mathcal{T}_{\mathbb{A}_2L}(l_2), \mathcal{T}_{\mathbb{B}_1L}(k_1 l_1)\right\}$$

$$\leq \min\left\{\mathcal{T}_{\mathbb{A}_2L}(k_2), \mathcal{T}_{\mathbb{A}_2L}(l_2), \min\left\{\mathcal{T}_{\mathbb{A}_1L}(k_1), \mathcal{T}_{\mathbb{A}_1L}(l_1)\right\}\right\}$$

$$= \min\left\{\min\left\{\mathcal{T}_{\mathbb{A}_1L}(k_1), \mathcal{T}_{\mathbb{A}_2L}(k_2)\right\}, \min\left\{\mathcal{T}_{\mathbb{A}_1L}(l_1), \mathcal{T}_{\mathbb{A}_2L}(l_2)\right\}\right\}$$

$$= \min\left\{\left(\mathcal{T}_{\mathbb{A}_1L} \times \mathcal{T}_{\mathbb{A}_2L}\right)(k_1, k_2), \left(\mathcal{T}_{\mathbb{A}_1L} \times \mathcal{T}_{\mathbb{A}_2L}\right)(l_1, l_2)\right\}$$

$$\left(\mathcal{T}_{\mathbb{B}_1U} \times \mathcal{T}_{\mathbb{B}_2U}\right)\left((k_1, k_2)(l_1, l_2)\right) = \min\left\{\mathcal{T}_{\mathbb{A}_2U}(k_2), \mathcal{T}_{\mathbb{A}_2U}(l_2), \mathcal{T}_{\mathbb{B}_1U}(k_1 l_1)\right\}$$

$$\leq \min\left\{\mathcal{T}_{\mathbb{A}_2U}(k_2), \mathcal{T}_{\mathbb{A}_2U}(l_2), \min\left\{\mathcal{T}_{\mathbb{A}_1U}(k_1), \mathcal{T}_{\mathbb{A}_1U}(l_1)\right\}\right\}$$

$$= \min\left\{\min\left\{\mathcal{T}_{\mathbb{A}_1U}(k_1), \mathcal{T}_{\mathbb{A}_2U}(k_2)\right\}, \min\left\{\mathcal{T}_{\mathbb{A}_1U}(l_1), \mathcal{T}_{\mathbb{A}_2U}(l_2)\right\}\right\}$$

$$= \min\left\{\left(\mathcal{T}_{\mathbb{A}_1U} \times \mathcal{T}_{\mathbb{A}_2U}\right)(k_1, k_2), \left(\mathcal{T}_{\mathbb{A}_1U} \times \mathcal{T}_{\mathbb{A}_2U}\right)(l_1, l_2)\right\}.$$

$$\left(\mathcal{C}_{\mathbb{B}_1L} \times \mathcal{C}_{\mathbb{B}_2L}\right)\left((k_1, k_2)(l_1, l_2)\right) = \min\left\{\mathcal{C}_{\mathbb{A}_2L}(k_2), \mathcal{C}_{\mathbb{A}_2L}(l_2), \mathcal{C}_{\mathbb{B}_1L}(k_1 l_1)\right\}$$

$$\leq \min\left\{\mathcal{C}_{\mathbb{A}_2L}(k_2), \mathcal{C}_{\mathbb{A}_2L}(l_2), \min\left\{\mathcal{C}_{\mathbb{A}_1L}(k_1), \mathcal{C}_{\mathbb{A}_1L}(l_1)\right\}\right\}$$

$$= \min\left\{\min\left\{\mathcal{C}_{\mathbb{A}_1L}(k_1), \mathcal{C}_{\mathbb{A}_2L}(k_2)\right\}, \min\left\{\mathcal{C}_{\mathbb{A}_1L}(l_1), \mathcal{C}_{\mathbb{A}_2L}(l_2)\right\}\right\}$$

$$= \min\left\{\left(\mathcal{C}_{\mathbb{A}_1L} \times \mathcal{C}_{\mathbb{A}_2L}\right)(k_1, k_2), \left(\mathcal{C}_{\mathbb{A}_1L} \times \mathcal{C}_{\mathbb{A}_2L}\right)(l_1, l_2)\right\}$$

$$\left(\mathcal{C}_{\mathbb{B}_1U} \times \mathcal{C}_{\mathbb{B}_2U}\right)\left((k_1, k_2)(l_1, l_2)\right) = \min\left\{\mathcal{C}_{\mathbb{A}_2U}(k_2), \mathcal{C}_{\mathbb{A}_2U}(l_2), \mathcal{C}_{\mathbb{B}_1U}(k_1 l_1)\right\}$$

$$\leq \min\left\{\mathcal{C}_{\mathbb{A}_2U}(k_2), \mathcal{C}_{\mathbb{A}_2U}(l_2), \min\left\{\mathcal{C}_{\mathbb{A}_1U}(k_1), \mathcal{C}_{\mathbb{A}_1U}(l_1)\right\}\right\}$$

$$= \min\left\{\min\left\{\mathcal{C}_{\mathbb{A}_1U}(k_1), \mathcal{C}_{\mathbb{A}_2U}(k_2)\right\}, \min\left\{\mathcal{C}_{\mathbb{A}_1U}(l_1), \mathcal{C}_{\mathbb{A}_2U}(l_2)\right\}\right\}$$

$$= \min\left\{\left(\mathcal{C}_{\mathbb{A}_1U} \times \mathcal{C}_{\mathbb{A}_2U}\right)(k_1, k_2), \left(\mathcal{C}_{\mathbb{A}_1U} \times \mathcal{C}_{\mathbb{A}_2U}\right)(l_1, l_2)\right\}.$$

$$\left(\mathcal{U}_{\mathbb{B}_1L} \times \mathcal{U}_{\mathbb{B}_2L}\right)\left((k_1, k_2)(l_1, l_2)\right) = \max\left\{\mathcal{U}_{\mathbb{A}_2L}(k_2), \mathcal{U}_{\mathbb{A}_2L}(l_2), \mathcal{U}_{\mathbb{B}_1L}(k_1 l_1)\right\}$$

$$\leq \max\left\{\mathcal{U}_{\mathbb{A}_2L}(k_2),\mathcal{U}_{\mathbb{A}_2L}(l_2),\max\left\{\mathcal{U}_{\mathbb{A}_1L}(k_1),\mathcal{U}_{\mathbb{A}_1L}(l_1)\right\}\right\}$$

$$=\max\left\{\max\left\{\mathcal{U}_{\mathbb{A}_1L}(k_1),\mathcal{U}_{\mathbb{A}_2L}(k_2)\right\},\max\left\{\mathcal{U}_{\mathbb{A}_1L}(l_1),\mathcal{U}_{\mathbb{A}_2L}(l_2)\right\}\right\}$$

$$=\max\left\{\left(\mathcal{U}_{\mathbb{A}_1L}\times\mathcal{U}_{\mathbb{A}_2L}\right)(k_1,k_2),\left(\mathcal{U}_{\mathbb{A}_1L}\times\mathcal{U}_{\mathbb{A}_2L}\right)(l_1,l_2)\right\}$$

$$\left(\mathcal{U}_{\mathbb{B}_1U}\times\mathcal{U}_{\mathbb{B}_2U}\right)\left((k_1,k_2)(l_1,l_2)\right)=\max\left\{\mathcal{U}_{\mathbb{A}_2U}(k_2),\mathcal{U}_{\mathbb{A}_2U}(l_2),\mathcal{U}_{\mathbb{B}_1U}(k_1l_1)\right\}$$

$$\leq \max\left\{\mathcal{U}_{\mathbb{A}_2U}(k_2),\mathcal{U}_{\mathbb{A}_2U}(l_2),\max\left\{\mathcal{U}_{\mathbb{A}_1U}(k_1),\mathcal{U}_{\mathbb{A}_1U}(l_1)\right\}\right\}$$

$$=\max\left\{\max\left\{\mathcal{U}_{\mathbb{A}_1U}(k_1),\mathcal{U}_{\mathbb{A}_2U}(k_2)\right\},\max\left\{\mathcal{U}_{\mathbb{A}_1U}(l_1),\mathcal{U}_{\mathbb{A}_2U}(l_2)\right\}\right\}$$

$$=\max\left\{\left(\mathcal{U}_{\mathbb{A}_1U}\times\mathcal{U}_{\mathbb{A}_2U}\right)(k_1,k_2),\left(\mathcal{U}_{\mathbb{A}_1U}\times\mathcal{U}_{\mathbb{A}_2U}\right)(l_1,l_2)\right\}.$$

$$\left(\mathcal{F}_{\mathbb{B}_1L}\times\mathcal{F}_{\mathbb{B}_2L}\right)\left((k_1,k_2)(l_1,l_2)\right)=\max\left\{\mathcal{F}_{\mathbb{A}_2L}(k_2),\mathcal{F}_{\mathbb{A}_2L}(l_2),\mathcal{F}_{\mathbb{B}_1L}(k_1l_1)\right\}$$

$$\leq \max\left\{\mathcal{F}_{\mathbb{A}_2L}(k_2),\mathcal{F}_{\mathbb{A}_2L}(l_2),\max\left\{\mathcal{F}_{\mathbb{A}_1L}(k_1),\mathcal{F}_{\mathbb{A}_1L}(l_1)\right\}\right\}$$

$$=\max\left\{\max\left\{\mathcal{F}_{\mathbb{A}_1L}(k_1),\mathcal{F}_{\mathbb{A}_2L}(k_2)\right\},\max\left\{\mathcal{F}_{\mathbb{A}_1L}(l_1),\mathcal{F}_{\mathbb{A}_2L}(l_2)\right\}\right\}$$

$$=\max\left\{\left(\mathcal{F}_{\mathbb{A}_1L}\times\mathcal{F}_{\mathbb{A}_2L}\right)(k_1,k_2),\left(\mathcal{F}_{\mathbb{A}_1L}\times\mathcal{F}_{\mathbb{A}_2L}\right)(l_1,l_2)\right\}$$

$$\left(\mathcal{F}_{\mathbb{B}_1U}\times\mathcal{F}_{\mathbb{B}_2U}\right)\left((k_1,k_2)(l_1,l_2)\right)=\max\left\{\mathcal{F}_{\mathbb{A}_2U}(k_2),\mathcal{F}_{\mathbb{A}_2U}(l_2),\mathcal{F}_{\mathbb{B}_1U}(k_1l_1)\right\}$$

$$\leq \max\left\{\mathcal{F}_{\mathbb{A}_2U}(k_2),\mathcal{F}_{\mathbb{A}_2U}(l_2),\max\left\{\mathcal{F}_{\mathbb{A}_1U}(k_1),\mathcal{F}_{\mathbb{A}_1U}(l_1)\right\}\right\}$$

$$=\max\left\{\max\left\{\mathcal{F}_{\mathbb{A}_1U}(k_1),\mathcal{F}_{\mathbb{A}_2U}(k_2)\right\},\max\left\{\mathcal{F}_{\mathbb{A}_1U}(l_1),\mathcal{F}_{\mathbb{A}_2U}(l_2)\right\}\right\}$$

$$=\max\left\{\left(\mathcal{F}_{\mathbb{A}_1U}\times\mathcal{F}_{\mathbb{A}_2U}\right)(k_1,k_2),\left(\mathcal{F}_{\mathbb{A}_1U}\times\mathcal{F}_{\mathbb{A}_2U}\right)(l_1,l_2)\right\}.$$

Definition 6.4.15 The union $\mathcal{G}_1\cup\mathcal{G}_2=\left(\mathbb{A}_1\cup\mathbb{A}_2,\mathbb{B}_1\cup\mathbb{B}_2\right)$ of two interval quadripartitioned single-valued neutrosophic graphs $\mathcal{G}_1$ and $\mathcal{G}_2$ of $\mathcal{G}_1^*$ and $\mathcal{G}_2^*$ is defined as follows:

$$\left(\mathcal{T}_{\mathbb{A}_1L}\cup\mathcal{T}_{\mathbb{A}_2L}\right)(k)=\mathcal{T}_{\mathbb{A}_1L}(k)\text{ if } k\in\mathbb{V}_1 \text{ and } k\notin\mathbb{V}_2$$

$$\left(\mathcal{C}_{\mathbb{A}_1L}\cup\mathcal{C}_{A_2L}\right)(k)=\mathcal{C}_{\mathbb{A}_1L}(k)\text{if } k\in\mathbb{V}_1 \text{ and} k\notin\mathbb{V}_2$$

$$\left(\mathcal{U}_{\mathbb{A}_1L}\cup\mathcal{U}_{\mathbb{A}_2L}\right)(k)=\mathcal{U}_{\mathbb{A}_1L}(k)\text{if } k\in\mathbb{V}_1 \text{ and} k\notin\mathbb{V}_2$$

$$\left(\mathcal{F}_{\mathbb{A}_1L}\cup\mathcal{F}_{\mathbb{A}_2L}\right)(k)=\mathcal{F}_{\mathbb{A}_1L}(k)\text{ if } k\in\mathbb{V}_1 \text{ and } k\notin\mathbb{V}_2$$

$$\left(\mathcal{T}_{\mathbb{A}_1L}\cup\mathcal{T}_{\mathbb{A}_2L}\right)(k)=\mathcal{T}_{\mathbb{A}_2L}(k)\text{ if } k\in\mathbb{V}_2 \text{ and } k\notin\mathbb{V}_1$$

$$\left(\mathcal{C}_{\mathbb{A}_1L} \cup \mathcal{C}_{\mathbb{A}_2L}\right)(k) = \mathcal{C}_{\mathbb{A}_2L}(k) \text{ if } k \in \mathbb{V}_2 \text{ and } k \notin \mathbb{V}_1$$

$$\left(\mathcal{U}_{\mathbb{A}_1L} \cup \mathcal{U}_{\mathbb{A}_2L}\right)(k) = \mathcal{U}_{\mathbb{A}_2L}(k) \text{ if } k \in \mathbb{V}_2 \text{ and } k \notin \mathbb{V}_1$$

$$\left(\mathcal{F}_{\mathbb{A}_1L} \cup \mathcal{F}_{\mathbb{A}_2L}\right)(k) = \mathcal{F}_{\mathbb{A}_2L}(k) \text{ if } k \in \mathbb{V}_2 \text{ and } k \notin \mathbb{V}_1$$

$$\left(\mathcal{T}_{\mathbb{A}_1L} \cup \mathcal{T}_{\mathbb{A}_2L}\right)(k) = \max\left\{\mathcal{T}_{\mathbb{A}_1L}(k), \mathcal{T}_{\mathbb{A}_2L}(k)\right\} \text{ if } k \in \mathbb{V}_1 \cap \mathbb{V}_2$$

$$\left(\mathcal{C}_{\mathbb{A}_1L} \cup \mathcal{C}_{\mathbb{A}_2L}\right)(k) = \max\left\{\mathcal{C}_{\mathbb{A}_1L}(k), \mathcal{C}_{\mathbb{A}_2L}(k)\right\} \text{ if } k \in \mathbb{V}_1 \cap \mathbb{V}_2$$

$$\left(\mathcal{U}_{\mathbb{A}_1L} \cup \mathcal{U}_{\mathbb{A}_2L}\right)(k) = \min\left\{\mathcal{U}_{\mathbb{A}_1L}(k), \mathcal{U}_{\mathbb{A}_2L}(k)\right\} \text{ if } k \in \mathbb{V}_1 \cap \mathbb{V}_2$$

$$\left(\mathcal{F}_{\mathbb{A}_1L} \cup \mathcal{F}_{\mathbb{A}_2L}\right)(k) = \min\left\{\mathcal{F}_{\mathbb{A}_1L}(k), \mathcal{F}_{\mathbb{A}_2L}(k)\right\} \text{ if } k \in \mathbb{V}_1 \cap \mathbb{V}_2$$

$$\left(\mathcal{T}_{\mathbb{A}_1U} \cup \mathcal{T}_{\mathbb{A}_2U}\right)(k) = \mathcal{T}_{\mathbb{A}_1U}(k) \text{ if } k \in \mathbb{V}_1 \text{ and } k \notin \mathbb{V}_2$$

$$\left(\mathcal{C}_{\mathbb{A}_1U} \cup \mathcal{C}_{\mathbb{A}_2U}\right)(k) = \mathcal{C}_{\mathbb{A}_1U}(k) \text{ if } k \in \mathbb{V}_1 \text{ and } k \notin \mathbb{V}_2$$

$$\left(\mathcal{U}_{\mathbb{A}_1U} \cup \mathcal{U}_{\mathbb{A}_2U}\right)(k) = \mathcal{U}_{\mathbb{A}_1U}(k) \text{ if } k \in \mathbb{V}_1 \text{ and } k \notin \mathbb{V}_2$$

$$\left(\mathcal{F}_{\mathbb{A}_1U} \cup \mathcal{F}_{\mathbb{A}_2U}\right)(k) = \mathcal{F}_{\mathbb{A}_1U}(k) \text{ if } k \in \mathbb{V}_1 \text{ and } k \notin \mathbb{V}_2$$

$$\left(\mathcal{T}_{\mathbb{A}_1U} \cup \mathcal{T}_{\mathbb{A}_2U}\right)(k) = \mathcal{T}_{\mathbb{A}_2U}(k) \text{ if } k \in \mathbb{V}_2 \text{ and } k \notin \mathbb{V}_1$$

$$\left(\mathcal{C}_{\mathbb{A}_1U} \cup \mathcal{C}_{\mathbb{A}_2U}\right)(k) = \mathcal{C}_{\mathbb{A}_2U}(k) \text{ if } k \in \mathbb{V}_2 \text{ and } k \notin \mathbb{V}_1$$

$$\left(\mathcal{U}_{\mathbb{A}_1U} \cup \mathcal{U}_{\mathbb{A}_2U}\right)(k) = \mathcal{U}_{\mathbb{A}_2U}(k) \text{ if } k \in \mathbb{V}_2 \text{ and } k \notin \mathbb{V}_1$$

$$\left(\mathcal{F}_{\mathbb{A}_1U} \cup \mathcal{F}_{\mathbb{A}_2U}\right)(k) = \mathcal{F}_{\mathbb{A}_2U}(k) \text{ if } k \in \mathbb{V}_2 \text{ and } k \notin \mathbb{V}_1$$

$$\left(\mathcal{T}_{\mathbb{A}_1U} \cup \mathcal{T}_{\mathbb{A}_2U}\right)(k) = \max\left\{\mathcal{T}_{\mathbb{A}_1U}(k), \mathcal{T}_{\mathbb{A}_2U}(k)\right\} \text{ if } k \in \mathbb{V}_1 \cap \mathbb{V}_2$$

$$\left(\mathcal{C}_{\mathbb{A}_1U} \cup \mathcal{C}_{\mathbb{A}_2U}\right)(k) = \max\left\{\mathcal{C}_{\mathbb{A}_1U}(k), \mathcal{C}_{\mathbb{A}_2U}(k)\right\} \text{ if } k \in \mathbb{V}_1 \cap \mathbb{V}_2$$

$$\left(\mathcal{U}_{\mathbb{A}_1U} \cup \mathcal{U}_{\mathbb{A}_2U}\right)(k) = \min\left\{\mathcal{U}_{\mathbb{A}_1U}(k), \mathcal{U}_{\mathbb{A}_2U}(k)\right\} \text{ if } k \in \mathbb{V}_1 \cap \mathbb{V}_2$$

$$\left(\mathcal{F}_{\mathbb{A}_1U} \cup \mathcal{F}_{\mathbb{A}_2U}\right)(k) = \min\left\{\mathcal{F}_{\mathbb{A}_1U}(k), \mathcal{F}_{\mathbb{A}_2U}(k)\right\} \text{ if } k \in \mathbb{V}_1 \cap \mathbb{V}_2$$

$$\left(\mathcal{T}_{\mathbb{B}_1L} \cup \mathcal{T}_{\mathbb{B}_2L}\right)(kl) = \mathcal{T}_{\mathbb{B}_1L}(kl) \text{ if } kl \in \mathbb{E}_1 \text{ and } kl \notin \mathbb{E}_2$$

$$\left(\mathcal{C}_{\mathbb{B}_1L} \cup \mathcal{C}_{\mathbb{B}_2L}\right)(kl) = \mathcal{C}_{\mathbb{B}_1L}(kl) \text{ if } kl \in \mathbb{E}_1 \text{ and } kl \notin \mathbb{E}_2$$

$$\left(\mathcal{U}_{\mathbb{B}_1L} \cup \mathcal{U}_{\mathbb{B}_2L}\right)(kl) = \mathcal{U}_{\mathbb{B}_1L}(kl) \text{ if } kl \in \mathbb{E}_1 \text{ and } kl \notin \mathbb{E}_2$$

$\left(\mathcal{F}_{\mathbb{B}_1L} \cup \mathcal{F}_{\mathbb{B}_2L}\right)(kl) = \mathcal{F}_{\mathbb{B}_1L}(kl)$ if $kl \in \mathbb{E}_1$ and $kl \notin \mathbb{E}_2$

$\left(\mathcal{T}_{\mathbb{B}_1L} \cup \mathcal{T}_{\mathbb{B}_2L}\right)(kl) = \mathcal{T}_{\mathbb{B}_2L}(kl)$ if $kl \in \mathbb{E}_2$ and $kl \notin \mathbb{E}_1$

$\left(\mathcal{C}_{\mathbb{B}_1L} \cup \mathcal{C}_{\mathbb{B}_2L}\right)(kl) = \mathcal{C}_{\mathbb{B}_2L}(kl)$ if $kl \in \mathbb{E}_2$ and $kl \notin \mathbb{E}_1$

$\left(\mathcal{U}_{\mathbb{B}_1L} \cup \mathcal{U}_{\mathbb{B}_2L}\right)(kl) = \mathcal{U}_{\mathbb{B}_2L}(kl)$ if $kl \in \mathbb{E}_2$ and $kl \notin \mathbb{E}_1$

$\left(\mathcal{F}_{\mathbb{B}_1L} \cup \mathcal{F}_{\mathbb{B}_2L}\right)(kl) = \mathcal{F}_{\mathbb{B}_2L}(kl)$ if $kl \in \mathbb{E}_2$ and $kl \notin \mathbb{E}_1$

$\left(\mathcal{T}_{\mathbb{B}_1L} \cup \mathcal{T}_{\mathbb{B}_2L}\right)(kl) = \max\left\{\mathcal{T}_{\mathbb{B}_1L}(kl), \mathcal{T}_{\mathbb{B}_2L}(kl)\right\}$ if $kl \in \mathbb{E}_1 \cap \mathbb{E}_2$

$\left(\mathcal{C}_{\mathbb{B}_1L} \cup \mathcal{C}_{\mathbb{B}_2L}\right)(kl) = \max\left\{\mathcal{C}_{\mathbb{B}_1L}(kl), \mathcal{C}_{\mathbb{B}_2L}(kl)\right\}$ if $kl \in \mathbb{E}_1 \cap \mathbb{E}_2$

$\left(\mathcal{U}_{\mathbb{B}_1L} \cup \mathcal{U}_{\mathbb{B}_2L}\right)(kl) = \min\left\{\mathcal{U}_{\mathbb{B}_1L}(kl), \mathcal{U}_{\mathbb{B}_2L}(kl)\right\}$ if $kl \in \mathbb{E}_1 \cap \mathbb{E}_2$

$\left(\mathcal{F}_{\mathbb{B}_1L} \cup \mathcal{F}_{\mathbb{B}_2L}\right)(kl) = \min\left\{\mathcal{F}_{\mathbb{B}_1L}(kl), \mathcal{F}_{\mathbb{B}_2L}(kl)\right\}$ if $kl \in \mathbb{E}_1 \cap \mathbb{E}_2$

- $\left(\mathcal{T}_{\mathbb{B}_1U} \cup \mathcal{T}_{\mathbb{B}_2U}\right)(kl) = \mathcal{T}_{\mathbb{B}_1U}(kl)$ if $kl \in \mathbb{E}_1$ and $kl \notin \mathbb{E}_2$

$\left(\mathcal{C}_{\mathbb{B}_1U} \cup \mathcal{C}_{\mathbb{B}_2U}\right)(kl) = \mathcal{C}_{\mathbb{B}_1U}(kl)$ if $kl \in \mathbb{E}_1$ and $kl \notin \mathbb{E}_2$

$\left(\mathcal{U}_{\mathbb{B}_1U} \cup \mathcal{U}_{\mathbb{B}_2U}\right)(kl) = \mathcal{U}_{\mathbb{B}_1U}(kl)$ if $kl \in \mathbb{E}_1$ and $kl \notin \mathbb{E}_2$

$\left(\mathcal{F}_{\mathbb{B}_1U} \cup \mathcal{F}_{\mathbb{B}_2U}\right)(kl) = \mathcal{F}_{\mathbb{B}_1U}(kl)$ if $kl \in \mathbb{E}_1$ and $kl \notin \mathbb{E}_2$

$\left(\mathcal{T}_{\mathbb{B}_1U} \cup \mathcal{T}_{\mathbb{B}_2U}\right)(kl) = \mathcal{T}_{\mathbb{B}_2U}(kl)$ if $kl \in \mathbb{E}_2$ and $kl \notin \mathbb{E}_1$

$\left(\mathcal{C}_{\mathbb{B}_1U} \cup \mathcal{C}_{\mathbb{B}_2U}\right)(kl) = \mathcal{C}_{\mathbb{B}_2U}(kl)$ if $kl \in \mathbb{E}_2$ and $kl \notin \mathbb{E}_1$

$\left(\mathcal{U}_{\mathbb{B}_1U} \cup \mathcal{U}_{\mathbb{B}_2U}\right)(kl) = \mathcal{U}_{\mathbb{B}_2U}(kl)$ if $kl \in \mathbb{E}_2$ and $kl \notin \mathbb{E}_1$

$\left(\mathcal{F}_{\mathbb{B}_1U} \cup \mathcal{F}_{\mathbb{B}_2U}\right)(kl) = \mathcal{F}_{\mathbb{B}_2U}(kl)$ if $kl \in \mathbb{E}_2$ and $kl \notin \mathbb{E}_1$

$\left(\mathcal{T}_{\mathbb{B}_1U} \cup \mathcal{T}_{\mathbb{B}_2U}\right)(kl) = \max\left\{\mathcal{T}_{\mathbb{B}_1U}(kl), \mathcal{T}_{\mathbb{B}_2U}(kl)\right\}$ if $kl \in \mathbb{E}_1 \cap \mathbb{E}_2$

$\left(\mathcal{C}_{\mathbb{B}_1U} \cup \mathcal{C}_{\mathbb{B}_2U}\right)(kl) = \max\left\{\mathcal{C}_{\mathbb{B}_1U}(kl), \mathcal{C}_{\mathbb{B}_2U}(kl)\right\}$ if $kl \in \mathbb{E}_1 \cap \mathbb{E}_2$

$\left(\mathcal{U}_{\mathbb{B}_1U} \cup \mathcal{U}_{\mathbb{B}_2U}\right)(kl) = \min\left\{\mathcal{U}_{\mathbb{B}_1U}(kl), \mathcal{U}_{\mathbb{B}_2U}(kl)\right\}$ if $kl \in \mathbb{E}_1 \cap \mathbb{E}_2$

$\left(\mathcal{F}_{\mathbb{B}_1U} \cup \mathcal{F}_{\mathbb{B}_2U}\right)(kl) = \min\left\{\mathcal{F}_{\mathbb{B}_1U}(kl), \mathcal{F}_{\mathbb{B}_2U}(kl)\right\}$ if $kl \in \mathbb{E}_1 \cap \mathbb{E}_2$

Example 6.4.2 Consider the two interval-valued neutrosphic graphs $\mathcal{G}_1 = (\mathbb{A}_1, \mathbb{B}_1)$ and $\mathcal{G}_2 = (\mathbb{A}_2, \mathbb{B}_2)$ of two interval quadripartitioned single-valued neutrosophic graphs is shown as Figure 6.2.

Then, their corresponding union $\mathcal{G}_1 \cup \mathcal{G}_2$ is shown in Figure 6.3.

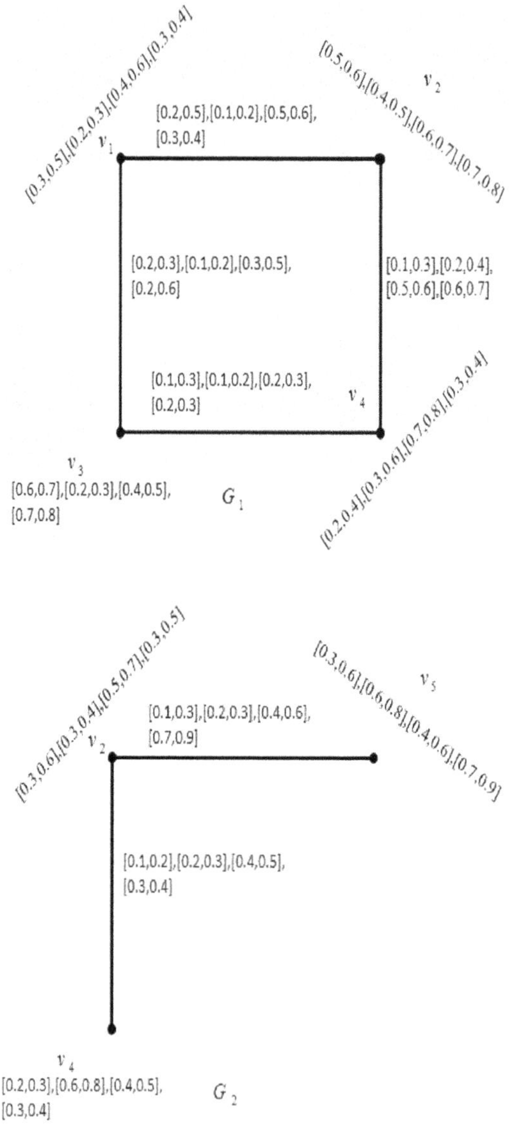

Figure 6.2 $\mathcal{G}_1$ and $\mathcal{G}_2$: Two interval quadripartitioned neutrosophic graphs.

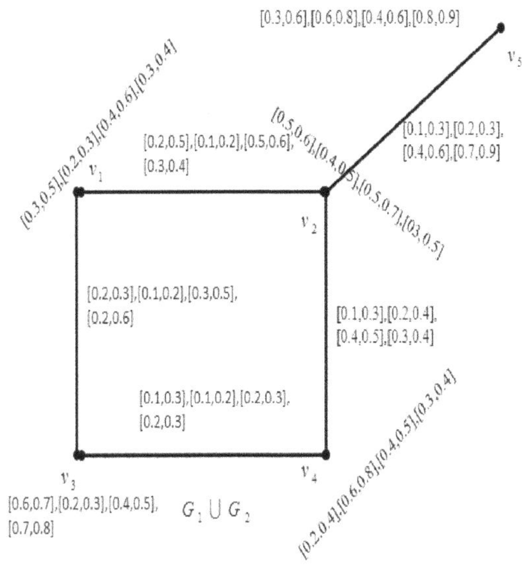

Figure 6.3 $\mathcal{G}$: Interval quadripartitioned neutrosophic graphs.

Proposition 6.4.16 The union $\mathcal{G}_1 \cup \mathcal{G}_2 = (\mathbb{A}_1 \cup \mathbb{A}_2, \mathbb{B}_1 \cup \mathbb{B}_2)$ of two interval quadripartitioned single-valued neutrosophic graphs $\mathcal{G}_1 = (\mathbb{A}_1, \mathbb{B}_1)$ and $\mathcal{G}_2 = (\mathbb{A}_2, \mathbb{B}_2)$ is an interval quadripartitioned single-valued neutrosophic graph.

Proof. Let $\mathcal{G}_1 = (\mathbb{A}_1, \mathbb{B}_1)$ and $\mathcal{G}_2 = (\mathbb{A}_2, \mathbb{B}_2)$ be interval-valued quadripartitioned neutrosophic graphs of the graphs $G_1^* = (V_1, E_1)$ and$G_2^* = (V_2, E_2)$, respectively. We prove that $\mathcal{G}_1 \cup \mathcal{G}_2 = (\mathbb{A}_1 \cup \mathbb{A}_2, \mathbb{B}_1 \cup \mathbb{B}_2)$ is an interval-valued quadripartitioned neutrosophic graph of the graph$G_1^* \cup G_2^*$. Since all the conditions for $\mathbb{A}_1 \cup \mathbb{A}_2$ are automatically satisfied, we verify only conditions for$\mathbb{B}_1 \cup \mathbb{B}_2$.

In the case, when$kl \in E_1 \cap E_2$, then

$$\left(\mathcal{T}_{\mathbb{B}_1L} \cup \mathcal{T}_{\mathbb{B}_2L}\right)(kl) = \max\left\{\mathcal{T}_{\mathbb{B}_1L}(kl), \mathcal{T}_{\mathbb{B}_2L}(kl)\right\}$$

$$\leq \max\left(\min\left(\mathbb{A}_{1L}\right)(k), \left(\mathbb{A}_{1L}\right)(l)\right), \min\left(\mathbb{A}_{2L}\right)(k), \left(\mathbb{A}_{2L}\right)(l)\right)\right)$$

$$= \min\left(\max\left(\mathbb{A}_{1L}\right)(k), \left(\mathbb{A}_{2L}\right)(k)\right), \max\left(\mathbb{A}_{1L}\right)(l), \left(\mathbb{A}_{2L}\right)(l)\right)\right)$$

$$= \min\left(\mathcal{T}_{\mathbb{A}_1L} \cup \mathcal{T}_{\mathbb{A}_1L}\right)(k)\right), \left(\mathcal{T}_{\mathbb{A}_2L} \cup \mathcal{T}_{\mathbb{A}_2L}\right)(l)\right)$$

$$\left(\mathcal{T}_{\mathbb{B}_1U} \cup \mathcal{T}_{\mathbb{B}_2U}\right)(kl) = \max\left\{\mathcal{T}_{\mathbb{B}_1U}(kl), \mathcal{T}_{\mathbb{B}_2U}(kl)\right\}$$

$$\leq \max\left(\min\left(\mathbb{A}_{1U}\right)(k), \left(\mathbb{A}_{1U}\right)(l)\right), \min\left(\mathbb{A}_{2U}\right)(k), \left(\mathbb{A}_{2U}\right)(l)\right)\right)$$

$$= \min\left(\max\left(\mathbb{A}_{1U}\right)(k), \left(\mathbb{A}_{2U}\right)(k)\right), \max\left(\mathbb{A}_{1U}\right)(l), \left(\mathbb{A}_{2U}\right)(l)\right)\right)$$

$$= \min\left(\mathcal{T}_{\mathbb{A}_1 U} \cup \mathcal{T}_{\mathbb{A}_1 U}\right)(k)\right), \left(\mathcal{T}_{\mathbb{A}_2 U} \cup \mathcal{T}_{\mathbb{A}_2 U}\right)(l)\right)$$

$$\left(\mathcal{C}_{\mathbb{B}_1 L} \cup \mathcal{C}_{\mathbb{B}_2 L}\right)(kl) = \max\left\{\mathcal{C}_{\mathbb{B}_1 L}(kl), \mathcal{C}_{\mathbb{B}_2 L}(kl)\right\}$$

$$\leq \max\left(\min\left(\mathbb{A}_{1L}\right)(k), \left(\mathbb{A}_{1L}\right)(l)\right), \min\left(\mathbb{A}_{2L}\right)(k), \left(\mathbb{A}_{2L}\right)(l)\right)\right)$$

$$= \min\left(\max\left(\mathbb{A}_{1L}\right)(k), \left(\mathbb{A}_{2L}\right)(k)\right), \max\left(\mathbb{A}_{1L}\right)(l), \left(\mathbb{A}_{2L}\right)(l)\right)\right)$$

$$= \min\left(\mathcal{C}_{\mathbb{A}_1 L} \cup \mathcal{C}_{\mathbb{A}_1 L}\right)(k)\right), \left(\mathcal{C}_{\mathbb{A}_2 L} \cup \mathcal{C}_{\mathbb{A}_2 L}\right)(l)\right)$$

$$\left(\mathcal{C}_{\mathbb{B}_1 U} \cup \mathcal{C}_{\mathbb{B}_2 U}\right)(kl) = \max\left\{\mathcal{C}_{\mathbb{B}_1 U}(kl), \mathcal{C}_{\mathbb{B}_2 U}(kl)\right\}$$

$$\leq \max\left(\min\left(\mathbb{A}_{1U}\right)(k), \left(\mathbb{A}_{1U}\right)(l)\right), \min\left(\mathbb{A}_{2U}\right)(k), \left(\mathbb{A}_{2U}\right)(l)\right)\right)$$

$$= \min\left(\max\left(\mathbb{A}_{1U}\right)(k), \left(\mathbb{A}_{2U}\right)(k)\right), \max\left(\mathbb{A}_{1U}\right)(l), \left(\mathbb{A}_{2U}\right)(l)\right)\right)$$

$$= \min\left(\mathcal{C}_{\mathbb{A}_1 U} \cup \mathcal{C}_{\mathbb{A}_1 U}\right)(k)\right), \left(\mathcal{C}_{\mathbb{A}_2 U} \cup \mathcal{C}_{\mathbb{A}_2 U}\right)(l)\right)$$

$$\left(\mathcal{U}_{\mathbb{B}_1 L} \cup \mathcal{U}_{\mathbb{B}_2 L}\right)(kl) = \min\left\{\mathcal{U}_{\mathbb{B}_1 L}(kl), \mathcal{U}_{\mathbb{B}_2 L}(kl)\right\}$$

$$\leq \min\left(\min\left(\mathbb{A}_{1L}\right)(k), \left(\mathbb{A}_{1L}\right)(l)\right), \min\left(\mathbb{A}_{2L}\right)(k), \left(\mathbb{A}_{2L}\right)(l)\right)\right)$$

$$= \min\left(\min\left(\mathbb{A}_{1L}\right)(k), \left(\mathbb{A}_{2L}\right)(k)\right), \max\left(\mathbb{A}_{1L}\right)(l), \left(\mathbb{A}_{2L}\right)(l)\right)\right)$$

$$= \min\left(\mathcal{U}_{\mathbb{A}_1 L} \cup \mathcal{U}_{\mathbb{A}_1 L}\right)(k)\right), \left(\mathcal{U}_{\mathbb{A}_2 L} \cup \mathcal{U}_{\mathbb{A}_2 L}\right)(l)\right)$$

$$\left(\mathcal{U}_{\mathbb{B}_1 U} \cup \mathcal{U}_{\mathbb{B}_2 U}\right)(kl) = \min\left\{\mathcal{U}_{\mathbb{B}_1 U}(kl), \mathcal{U}_{\mathbb{B}_2 U}(kl)\right\}$$

$$\leq \min\left(\min\left(\mathbb{A}_{1U}\right)(k), \left(\mathbb{A}_{1U}\right)(l)\right), \min\left(\mathbb{A}_{2U}\right)(k), \left(\mathbb{A}_{2U}\right)(l)\right)\right)$$

$$= \min\left(\min\left(\mathbb{A}_{1U}\right)(k), \left(\mathbb{A}_{2U}\right)(k)\right), \min\left(\mathbb{A}_{1U}\right)(l), \left(\mathbb{A}_{2U}\right)(l)\right)\right)$$

$$= \min\left(\mathcal{U}_{\mathbb{A}_1 U} \cup \mathcal{U}_{\mathbb{A}_1 U}\right)(k)\right), \left(\mathcal{U}_{\mathbb{A}_2 U} \cup \mathcal{U}_{\mathbb{A}_2 U}\right)(l)\right)$$

$$\left(\mathcal{F}_{\mathbb{B}_1 L} \cup \mathcal{F}_{\mathbb{B}_2 L}\right)(kl) = \min\left\{\mathcal{F}_{\mathbb{B}_1 L}(kl), \mathcal{F}_{\mathbb{B}_2 L}(kl)\right\}$$

$$\leq \min\left(\min\left(\mathbb{A}_{1L}\right)(k), \left(\mathbb{A}_{1L}\right)(l)\right), \min\left(\mathbb{A}_{2L}\right)(k), \left(\mathbb{A}_{2L}\right)(l)\right)\right)$$

$$= \min\left(\min\left(\mathbb{A}_{1L}\right)(k), \left(\mathbb{A}_{2L}\right)(k)\right), \max\left(\mathbb{A}_{1L}\right)(l), \left(\mathbb{A}_{2L}\right)(l)\right)\right)$$

$$= \min\left(\mathcal{F}_{\mathbb{A}_1 L} \cup \mathcal{F}_{\mathbb{A}_1 L}\right)(k)\right), \left(\mathcal{F}_{\mathbb{A}_2 L} \cup \mathcal{F}_{\mathbb{A}_2 L}\right)(l)\right)$$

$$\left(\mathcal{F}_{\mathbb{B}_1 U} \cup \mathcal{F}_{\mathbb{B}_2 U}\right)(kl) = \min\left\{\mathcal{F}_{\mathbb{B}_1 U}(kl), \mathcal{F}_{\mathbb{B}_2 U}(kl)\right\}$$

$$\leq \min\left(\min\left(\mathbb{A}_{1U}\right)(k),\left(\mathbb{A}_{1U}\right)(l)\right),\min\left(\mathbb{A}_{2U}\right)(k),\left(\mathbb{A}_{2U}\right)(l)\right)\right)$$

$$= \min\left(\min\left(\mathbb{A}_{1U}\right)(k),\left(\mathbb{A}_{2U}\right)(k)\right),\min\left(\mathbb{A}_{1U}\right)(l),\left(\mathbb{A}_{2U}\right)(l)\right)\right)$$

$$= \min\left(\mathcal{F}_{\mathbb{A}_1U} \cup \mathcal{F}_{\mathbb{A}_1U}\right)(k),\left(\mathcal{F}_{\mathbb{A}_2U} \cup \mathcal{F}_{\mathbb{A}_2U}\right)(l)\right)$$

If $kl \in E_1$ and $kl \notin E_2$, then

$$\left(\mathcal{T}_{\mathbb{B}_1L} \cup \mathcal{T}_{\mathbb{B}_2L}\right)(kl) \leq \min\left(\left(\mathcal{T}_{\mathbb{A}_1L} \cup \mathcal{T}_{\mathbb{A}_2L}\right)(k),\left(\mathcal{T}_{\mathbb{A}_1L} \cup \mathcal{T}_{\mathbb{A}_2L}\right)(l)\right)$$

$$\left(\mathcal{T}_{\mathbb{B}_1U} \cup \mathcal{T}_{\mathbb{B}_2U}\right)(kl) \leq \min\left(\left(\mathcal{T}_{\mathbb{A}_1U} \cup \mathcal{T}_{\mathbb{A}_2U}\right)(k),\left(\mathcal{T}_{\mathbb{A}_1U} \cup \mathcal{T}_{\mathbb{A}_2U}\right)(l)\right)$$

$$\left(\mathcal{C}_{\mathbb{B}_1L} \cup \mathcal{C}_{\mathbb{B}_2L}\right)(kl) \leq \min\left(\left(\mathcal{C}_{\mathbb{A}_1L} \cup \mathcal{C}_{\mathbb{A}_2L}\right)(k),\left(\mathcal{C}_{\mathbb{A}_1L} \cup \mathcal{C}_{\mathbb{A}_2L}\right)(l)\right)$$

$$\left(\mathcal{C}_{\mathbb{B}_1U} \cup \mathcal{C}_{\mathbb{B}_2U}\right)(kl) \leq \min\left(\left(\mathcal{C}_{\mathbb{A}_1U} \cup \mathcal{C}_{\mathbb{A}_2U}\right)(k),\left(\mathcal{C}_{\mathbb{A}_1U} \cup \mathcal{C}_{\mathbb{A}_2U}\right)(l)\right)$$

$$\left(\mathcal{U}_{\mathbb{B}_1L} \cup \mathcal{U}_{\mathbb{B}_2L}\right)(kl) \leq \max\left(\left(\mathcal{U}_{\mathbb{A}_1L} \cup \mathcal{U}_{\mathbb{A}_2L}\right)(k),\left(\mathcal{U}_{\mathbb{A}_1L} \cup \mathcal{U}_{\mathbb{A}_2L}\right)(l)\right)$$

$$\left(\mathcal{U}_{\mathbb{B}_1U} \cup \mathcal{U}_{\mathbb{B}_2U}\right)(kl) \leq \max\left(\left(\mathcal{U}_{\mathbb{A}_1U} \cup \mathcal{U}_{\mathbb{A}_2U}\right)(k),\left(\mathcal{U}_{\mathbb{A}_1U} \cup \mathcal{U}_{\mathbb{A}_2U}\right)(l)\right)$$

$$\left(\mathcal{F}_{\mathbb{B}_1L} \cup \mathcal{F}_{\mathbb{B}_2L}\right)(kl) \leq \max\left(\left(\mathcal{F}_{\mathbb{A}_1L} \cup \mathcal{F}_{\mathbb{A}_2L}\right)(k),\left(\mathcal{F}_{\mathbb{A}_1L} \cup \mathcal{F}_{\mathbb{A}_2L}\right)(l)\right)$$

$$\left(\mathcal{F}_{\mathbb{B}_1U} \cup \mathcal{F}_{\mathbb{B}_2U}\right)(kl) \leq \max\left(\left(\mathcal{F}_{\mathbb{A}_1U} \cup \mathcal{F}_{\mathbb{A}_2U}\right)(k),\left(\mathcal{F}_{\mathbb{A}_1U} \cup \mathcal{F}_{\mathbb{A}_2U}\right)(l)\right)$$

If $kl \notin E_1$ and $kl \in E_2$, then

$$\left(\mathcal{T}_{\mathbb{B}_1L} \cup \mathcal{T}_{\mathbb{B}_2L}\right)(kl) \leq \min\left(\left(\mathcal{T}_{\mathbb{A}_1L} \cup \mathcal{T}_{\mathbb{A}_2L}\right)(k),\left(\mathcal{T}_{\mathbb{A}_1L} \cup \mathcal{T}_{\mathbb{A}_2L}\right)(l)\right)$$

$$\left(\mathcal{T}_{\mathbb{B}_1U} \cup \mathcal{T}_{\mathbb{B}_2U}\right)(kl) \leq \min\left(\left(\mathcal{T}_{\mathbb{A}_1U} \cup \mathcal{T}_{\mathbb{A}_2U}\right)(k),\left(\mathcal{T}_{\mathbb{A}_1U} \cup \mathcal{T}_{\mathbb{A}_2U}\right)(l)\right)$$

$$\left(\mathcal{C}_{\mathbb{B}_1L} \cup \mathcal{C}_{\mathbb{B}_2L}\right)(kl) \leq \min\left(\left(\mathcal{C}_{\mathbb{A}_1L} \cup \mathcal{C}_{\mathbb{A}_2L}\right)(k),\left(\mathcal{C}_{\mathbb{A}_1L} \cup \mathcal{C}_{\mathbb{A}_2L}\right)(l)\right)$$

$$\left(\mathcal{C}_{\mathbb{B}_1U} \cup \mathcal{C}_{\mathbb{B}_2U}\right)(kl) \leq \min\left(\left(\mathcal{C}_{\mathbb{A}_1U} \cup \mathcal{C}_{\mathbb{A}_2U}\right)(k),\left(\mathcal{C}_{\mathbb{A}_1U} \cup \mathcal{C}_{\mathbb{A}_2U}\right)(l)\right)$$

$$\left(\mathcal{U}_{\mathbb{B}_1L} \cup \mathcal{U}_{\mathbb{B}_2L}\right)(kl) \leq \max\left(\left(\mathcal{U}_{\mathbb{A}_1L} \cup \mathcal{U}_{\mathbb{A}_2L}\right)(k),\left(\mathcal{U}_{\mathbb{A}_1L} \cup \mathcal{U}_{\mathbb{A}_2L}\right)(l)\right)$$

$$\left(\mathcal{U}_{\mathbb{B}_1U} \cup \mathcal{U}_{\mathbb{B}_2U}\right)(kl) \leq \max\left(\left(\mathcal{U}_{\mathbb{A}_1U} \cup \mathcal{U}_{\mathbb{A}_2U}\right)(k),\left(\mathcal{U}_{\mathbb{A}_1U} \cup \mathcal{U}_{\mathbb{A}_2U}\right)(l)\right)$$

$$\left(\mathcal{F}_{\mathbb{B}_1L} \cup \mathcal{F}_{\mathbb{B}_2L}\right)(kl) \leq \max\left(\left(\mathcal{F}_{\mathbb{A}_1L} \cup \mathcal{F}_{\mathbb{A}_2L}\right)(k),\left(\mathcal{F}_{\mathbb{A}_1L} \cup \mathcal{F}_{\mathbb{A}_2L}\right)(l)\right)$$

$$\left(\mathcal{F}_{\mathbb{B}_1U} \cup \mathcal{F}_{\mathbb{B}_2U}\right)(kl) \leq \max\left(\left(\mathcal{F}_{\mathbb{A}_1U} \cup \mathcal{F}_{\mathbb{A}_2U}\right)(k),\left(\mathcal{F}_{\mathbb{A}_1U} \cup \mathcal{F}_{\mathbb{A}_2U}\right)(l)\right)$$

This completes the proof.

Definition 6.4.17 The joined $\mathcal{G}_1 + \mathcal{G}_2 = (\mathbb{A}_1 + \mathbb{A}_2, \mathbb{B}_1 + \mathbb{B}_2)$ of two interval quadripartitioned single-valued neutrosophic graphs $\mathcal{G}_1$ and $\mathcal{G}_2$ of $\mathcal{G}_1^*$ and $\mathcal{G}_2^*$ where $V_1 \cap V_2 \neq \phi$, is defined as follows:

[1] $\left(\mathcal{T}_{\mathbb{A}_1L} + \mathcal{T}_{\mathbb{A}_2L}\right)(k) = \left(\mathcal{T}_{\mathbb{A}_1L} \cup \mathcal{T}_{\mathbb{A}_2L}\right)(k)$ if $k \in \mathbb{V}_1 \cup \mathbb{V}_2$

$\left(\mathcal{C}_{\mathbb{A}_1L} + \mathcal{C}_{\mathbb{A}_2L}\right)(k) = \left(\mathcal{C}_{\mathbb{A}_1L} \cup \mathcal{C}_{\mathbb{A}_2L}\right)(k)$ if $k \in \mathbb{V}_1 \cup \mathbb{V}_2$

$\left(\mathcal{U}_{\mathbb{A}_1L} + \mathcal{U}_{\mathbb{A}_2L}\right)(k) = \left(\mathcal{U}_{\mathbb{A}_1L} \cup \mathcal{U}_{\mathbb{A}_2L}\right)(k)$ if $k \in \mathbb{V}_1 \cup \mathbb{V}_2$

$\left(\mathcal{F}_{\mathbb{A}_1L} + \mathcal{F}_{\mathbb{A}_2L}\right)(k) = \left(\mathcal{F}_{\mathbb{A}_1L} \cup \mathcal{F}_{\mathbb{A}_2L}\right)(k)$ if $k \in \mathbb{V}_1 \cup \mathbb{V}_2$.

$\left(\mathcal{T}_{\mathbb{A}_1U} + \mathcal{T}_{\mathbb{A}_2U}\right)(k) = \left(\mathcal{T}_{\mathbb{A}_1U} \cup \mathcal{T}_{\mathbb{A}_2U}\right)(k)$ if $k \in \mathbb{V}_1 \cup \mathbb{V}_2$

$\left(\mathcal{C}_{\mathbb{A}_1U} + \mathcal{C}_{\mathbb{A}_2U}\right)(k) = \left(\mathcal{C}_{\mathbb{A}_1U} \cup \mathcal{C}_{\mathbb{A}_2U}\right)(k)$ if $k \in \mathbb{V}_1 \cup \mathbb{V}_2$

$\left(\mathcal{U}_{\mathbb{A}_1U} + \mathcal{U}_{\mathbb{A}_2U}\right)(k) = \left(\mathcal{U}_{\mathbb{A}_1U} \cup \mathcal{U}_{\mathbb{A}_2U}\right)(k)$ if $k \in \mathbb{V}_1 \cup \mathbb{V}_2$

$\left(\mathcal{F}_{\mathbb{A}_1U} + \mathcal{F}_{\mathbb{A}_2U}\right)(k) = \left(\mathcal{F}_{\mathbb{A}_1U} \cup \mathcal{F}_{\mathbb{A}_2U}\right)(k)$ if $k \in \mathbb{V}_1 \cup \mathbb{V}_2$.

[2] $\left(\mathcal{T}_{\mathbb{B}_1L} + \mathcal{T}_{\mathbb{B}_2L}\right)(kl) = \left(\mathcal{T}_{\mathbb{B}_1L} \cup \mathcal{T}_{\mathbb{B}_2L}\right)(kl)$ if $kl \in \mathbb{E}_1 \cap \mathbb{E}_2$

$\left(\mathcal{C}_{\mathbb{B}_1L} + \mathcal{C}_{\mathbb{B}_2L}\right)(kl) = \left(\mathcal{C}_{\mathbb{B}_1L} \cup \mathcal{C}_{\mathbb{B}_2L}\right)(kl)$ if $kl \in \mathbb{E}_1 \cap \mathbb{E}_2$

$\left(\mathcal{U}_{\mathbb{B}_1L} + \mathcal{U}_{\mathbb{B}_2L}\right)(kl) = \left(\mathcal{U}_{\mathbb{B}_1L} \cup \mathcal{U}_{\mathbb{B}_2L}\right)(kl)$ if $kl \in \mathbb{E}_1 \cap \mathbb{E}_2$

$\left(\mathcal{F}_{\mathbb{B}_1L} + \mathcal{F}_{\mathbb{B}_2L}\right)(kl) = \left(\mathcal{F}_{\mathbb{B}_1L} \cup \mathcal{F}_{\mathbb{B}_2L}\right)(kl)$ if $kl \in \mathbb{E}_1 \cap \mathbb{E}_2$

$\left(\mathcal{T}_{\mathbb{B}_1U} + \mathcal{T}_{\mathbb{B}_2U}\right)(kl) = \left(\mathcal{T}_{\mathbb{B}_1U} \cup \mathcal{T}_{\mathbb{B}_2U}\right)(kl)$ if $kl \in \mathbb{E}_1 \cap \mathbb{E}_2$

$\left(\mathcal{C}_{\mathbb{B}_1U} + \mathcal{C}_{\mathbb{B}_2U}\right)(kl) = \left(\mathcal{C}_{\mathbb{B}_1U} \cup \mathcal{C}_{\mathbb{B}_2U}\right)(kl)$ if $kl \in \mathbb{E}_1 \cap \mathbb{E}_2$

$\left(\mathcal{U}_{\mathbb{B}_1U} + \mathcal{U}_{\mathbb{B}_2U}\right)(kl) = \left(\mathcal{U}_{\mathbb{B}_1U} \cup \mathcal{U}_{\mathbb{B}_2U}\right)(kl)$ if $kl \in \mathbb{E}_1 \cap \mathbb{E}_2$

$\left(\mathcal{F}_{\mathbb{B}_1U} + \mathcal{F}_{\mathbb{B}_2U}\right)(kl) = \left(\mathcal{F}_{\mathbb{B}_1U} \cup \mathcal{F}_{\mathbb{B}_2U}\right)(kl)$ if $kl \in \mathbb{E}_1 \cap \mathbb{E}_2$

[3] $\left(\mathcal{T}_{\mathbb{B}_1L} + \mathcal{T}_{\mathbb{B}_2L}\right)(kl) = \min\left(\mathcal{T}_{\mathbb{A}_1L}(k), \mathcal{T}_{\mathbb{A}_2L}(l)\right)$ if $kl \in \mathbb{E}$

$\left(\mathcal{C}_{\mathbb{B}_1L} + \mathcal{C}_{\mathbb{B}_2L}\right)(kl) = \min\left(\mathcal{C}_{\mathbb{A}_1L}(k), \mathcal{C}_{\mathbb{A}_2L}(l)\right)$ if $kl \in \mathbb{E}$

$\left(\mathcal{U}_{\mathbb{B}_1L} + \mathcal{U}_{\mathbb{B}_2L}\right)(kl) = \max\left(\mathcal{C}_{\mathbb{A}_1L}(k), \mathcal{U}_{\mathbb{A}_2L}(l)\right)$ if $kl \in \mathbb{E}$

$\left(\mathcal{F}_{\mathbb{B}_1L} + \mathcal{F}_{\mathbb{B}_2L}\right)(kl) = \max\left(\mathcal{F}_{\mathbb{A}_1L}(k), \mathcal{F}_{\mathbb{A}_2L}(l)\right)$ if $kl \in \mathbb{E}$

$$\left(\mathcal{T}_{\mathbb{B}_1U} + \mathcal{T}_{\mathbb{B}_2U}\right)(kl) = \min\left(\mathcal{T}_{\mathbb{A}_1U}(k), \mathcal{T}_{\mathbb{A}_2U}(l)\right) \text{ if } kl \in \mathbb{E}$$

$$\left(\mathcal{C}_{\mathbb{B}_1U} + \mathcal{C}_{\mathbb{B}_2U}\right)(kl) = \min\left(\mathcal{C}_{\mathbb{A}_1U}(k), \mathcal{C}_{\mathbb{A}_2U}(l)\right) \text{ if } kl \in \mathbb{E}$$

$$\left(\mathcal{U}_{\mathbb{B}_1U} + \mathcal{U}_{\mathbb{B}_2U}\right)(kl) = \max\left(\mathcal{C}_{\mathbb{A}_1U}(k), \mathcal{U}_{\mathbb{A}_2U}(l)\right) \text{ if } kl \in \mathbb{E}$$

$$\left(\mathcal{F}_{\mathbb{B}_1U} + \mathcal{F}_{\mathbb{B}_2U}\right)(kl) = \max\left(\mathcal{F}_{\mathbb{A}_1U}(k), \mathcal{F}_{\mathbb{A}_2U}(l)\right) \text{ if } kl \in \mathbb{E}$$

where $\mathbb{E}$ is the set of all edges joining the vertices of $\mathbb{V}_1$ and $\mathbb{V}_2$.

Proposition 6.4.18 The joined $\mathcal{G}_1 + \mathcal{G}_2 = \left(\mathbb{A}_1 \cup \mathbb{A}_2, \mathbb{B}_1 \cup \mathbb{B}_2\right)$ of two interval quadripartitioned single-valued neutrosophic graphs $\mathcal{G}_1 = \left(\mathbb{A}_1, \mathbb{B}_1\right)$ and $\mathcal{G}_2 = \left(\mathbb{A}_2, \mathbb{B}_2\right)$ is an interval quadripartitioned single-valued neutrosophic graph.

6.5 APPLICATION

Interval QSVNSs for Climatic Characterization in Toronto

In Toronto, winters are cold, windy and dry; summers are pleasant, and it is partly cloudy all year. Throughout the year, the climate usually ranges from 33 $^\circ F$ to 73 $^\circ F$.

Mathematically, we represent these situations as interval-valued quadripartitioned single-valued neutrosophic sets. The warm season lasts June 2 to September 18, with an average daily maximum temperature of over 68 $^\circ F$. The hottest day of the year is July 20, with an average high of 73 $^\circ F$ and a low of 63 $^\circ F$. The cold season spans December 3 to March 16, with an average daily high

Month/temperature in $^\circ F$	*Min. temp.*	*Min. temp.*	*Avg. temp.*
January	33.9	41.8	37.2
February	33.5	43.7	37.7
March	36	47.5	40.9
April	40	53.8	46.1
May	46.1	61	53.2
June	51.8	65.9	58.4
July	56.8	72.2	64.2
August	57.7	72.7	64.7
September	53.1	66.7	58.7
October	45.7	56.6	50.1
November	38.5	46.3	42
December	33.7	40.8	36.9

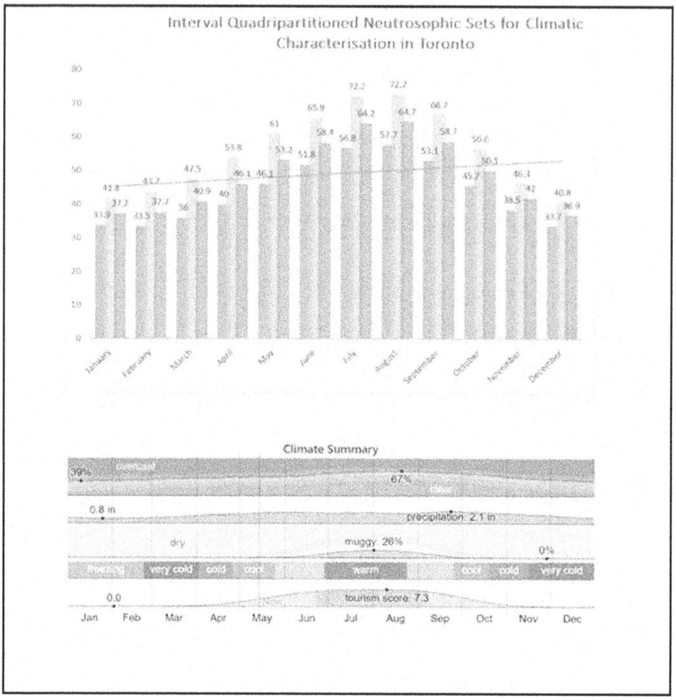

Figure 6.4 Toronto climate summary for a year.

temperature below 39 $^\circ F$. The coldest day of the year is January 29, with an average low of 33 $^\circ F$.

Let A denote the temperature of variant weather conditions in a year in Toronto. Represent Y as an interval quadripartitioned single-valued neutrosophic set within A and A and $\mathcal{T}_{YU}$ as the lower and upper values of the truth membership function, under which the summer season in A falls between May and July, when the temperature rises to the peak of $73^\circ F$. $\mathcal{C}_{YL}$and $\mathcal{C}_{YU}$ are the lower and upper values of the contradiction function, and they represent the warm and cool seasons in A falling from August to October. Thus the temperature ranges from $45^\circ F$ to $72^\circ F$. $\mathcal{U}_{YL}$ and $\mathcal{U}_{YU}$ represent the lower and upper values of the ignorance function and denote the cold season in A. The temperature in February to April ranges from $33^\circ F$ to $53^\circ F$., and $\mathcal{F}_{YL}$ and $\mathcal{F}_{YU}$ represent the lower and upper values of the false membership function; they characterize the winter season with snow in November to January.

For the sake of brevity, take the temperature belonging to set Y as the percentage of neutrosophic values:

$$A = \{[0.46, 0.73], [0.45, 0.72], [0.33, 0.53], [0.38, 0.42]\}.$$

Now, the set A fulfills the interval QSVNS:

$$\mathcal{T}_{YL} \leq \mathcal{T}_{YU}, \mathcal{C}_{YL} \leq \mathcal{C}_{YU}, \mathcal{U}_{YL} \leq \mathcal{U}_{YU}, \mathcal{F}_{YL} \leq \mathcal{F}_{YU} \text{ and also}$$
$$0 \leq \mathcal{T}_{YL} + \mathcal{C}_{YL} + \mathcal{U}_{YL} + \mathcal{F}_{YL} = 1.62 \leq 4, \text{ and}$$
$$0 \leq \mathcal{T}_{YU} + \mathcal{C}_{YU} + \mathcal{U}_{YU} + \mathcal{F}_{YU} = 2.4 \leq 4.$$

In short, interval QSVNSs can be used to understand climate change data.

6.6 CONCLUSIONS

In this work, we developed interval QSVNSs and applied them to their graphs. We investigated the Cartesian product, union, join and composition of interval quadripartitioned single-valued neutrosophic graphs. Finally, we proposed an example model for characterizing the weather in Toronto based on temperature and seasonal variation data. This makes it easier for decision-makers to adapt their opinions to their specific domains of expertise. As a result, the developed sets, graphs, and operations offer sufficient capability to overcome the related dependence on imprecise data. In the future, this research will extend to the algebraic isomorphism properties of the developed methods and be applied to decision-making concepts.

Conflicts of Interest: The authors declare no conflicts of interest.

REFERENCES

1. Zadeh LA, The concept of a linguistic and application to approximate reasoning, *Information Sciences*, 8 (1975), 199–249.
2. Turksen I, Interval valued fuzzy sets based on normal forms, *Fuzzy Sets and Systems*, 20 (1986), 191–210.
3. Wang H, Smarandache F, Zhang Y and Sunderraman R, Single-valued neutrosophic sets, *Multispace and Multistructure*, 4 (2010), 410–413.
4. Smarandache F, Neutrosophic graphs, in his book symbolic neutrosophic theory. Bruxelles, EuropaNova.
5. Bhattacharya S, Neutrosophic informationfusion applied to the options market, *Investment Management and Financial Innovations*, 855(1) (2005), 139–145.
6. Aggarwal S, Biswas R and Ansari AQ, Neutrosophic modeling and control, *Computer and Communication Technology* (2010), 718–723.
7. Guo Y and Cheng HD, New neutrosophic approach to image segmentation, *Pattern Recognition*, 42 (2009), 587–595.

8. Ye J and Fu J, Multi-period medical diagnosis method using a single valued neutrosophic similarity measure based on tangent function, *Computer Methods and Programs in Biomedicine*, 123 (2016), 142–149.
9. Ye J, Multicriteria decision-making method using the correlation coefficient under single valued neutrosophic environment, *International Journal of General Systems*, 42(4) (2013), 386–394.
10. Akram M and Shahzadi S, Neutrosophic soft graphs with application, *Journal of Intelligent & Fuzzy Systems*, 32(1) (2017), 841–858.
11. Ali M and Smarandache F, Complex neutrosophic set, *Neural Computing & Applications*, 28 (2017), 1817–1834.
12. Hussain SS, Hussain RJ, Jun YB and Smarandache F, Neutrosophic bipolar vague set and its application to neutrosophic bipolar vague graphs, *Neutrosophic Sets and Systems*, 28 (2019), 69–86.
13. Hussain SS, Hussain RJ and Smarandache F, On neutrosophic vague graphs, *Neutrosophic Sets and Systems*, 28 (2019), 245–258.
14. Hussain SS, Broumi S, Jun YB and Durga N, Intuitionistic bipolar neutrosophic set and its application to intuitionistic bipolar neutrosophic graphs, *Annals of Communication in Mathematics*, 2(2) (2019), 121–140.
15. Hussain SS, Hussain RJ and Smarandache F, Domination number in neutrosophic soft graphs, *Neutrosophic Sets and Systems*, 28(1) (2019), 228–244.
16. Liang RX, Wang JQ and Li L, Multi-criteria group decision-making method based on interdependent inputs of single-valued trapezoidal neutrosophic information, *Neural Computing & Applications*, 30 (2018), 241–260.
17. Sahin R and Liu P, Correlation coefficient of single-valued neutrosophic hesitant fuzzy sets and its applications in decision making, *Neural Computing & Applications*, 28 (2017), 1387–1395.
18. Chatterjee R, Majumdar P and Samanta SK, On some similarity measures and entropy on quadripartitioned single valued neutrosophic sets, *Journal of Intelligent & Fuzzy Systems*, 30(4) (2016), 2475–2485.
19. Belnap ND, A useful four-valued logic. In Modern uses of multiple-valued logic (pp. 5–37). Springer, Dordrecht (1977).
20. Smarandache F, *n*–Valued refined neutrosophic logic and its applications to physics, *Neutrosophic Computing and Machine Learning*, 2 (2018), 3–8.
21. Akram M and Shahzadi G, Operations on single-valued neutrosophic graphs, *Journal of Uncertain Systems*, 11(1) (2017), 1–26.
22. Akram M and Nasir M, Concepts of interval-valued neutrosophic graphs, *International Journal of Algebra and Statistics*, 6 (2017), 22–41.
23. Sinha K and Majumdar P, Bipolar quadripartitioned single valued neutrosophic sets, *Proyecciones (Antofagasta)*, 39(6) (2020), 1597–1614.
24. Hussain SS, Rosyida I, Rashmanlou H and Mofidnakhaei F, Interval intuitionistic neutrosophic sets with its applications to interval intuitionistic neutrosophic graphs and climatic analysis, *Journal of Computational and Applied Mathematics*, 40(121) (2021). https://doi.org/10.1007/s40314-021-01504-8
25. Ye J, Song J and Du S, Correlation coefficients of consistency neutrosophic sets regarding neutrosophic multi-valued sets and their multi-attribute decision-making method, *International Journal of Fuzzy Systems*, (2020), 1–8.

26. Awang A, Aizam NAH, Ab Ghani AT, Othman M and Abdullah L, A normalized weighted Bonferroni mean aggregation operator considering Shapley fuzzy measure under Interval-valued neutrosophic environment for decision-making, *International Journal of Fuzzy Systems*, 22(1) (2020), 321–336.
27. Nafei A, Javadpour A, Nasseri H and Yuan W, Optimized score function and its application in group multiattribute decision making based on fuzzy neutrosophic sets, *International Journal of Intelligent Systems*, 36(12) (2021), 7522–7543.
28. Chutia R, Gogoi MK, Firozja MA and Smarandache F, Ordering single-valued neutrosophic numbers based on flexibility parameters and its reasonable properties, *International Journal of Intelligent Systems*, 36(4) 2021, 1831–1850.
29. Wang H, Smarandache F, Zhang YQ and Sunderraman R, Interval neutrosophic sets and logic: Theory and applications in computing. Hexis, Phoenix (2005).
30. Smarandache F, Neutrosophy, neutrosophic probability, set, and logic (4th ed., 105 pp). American Press Institute, Rehoboth (1998). http://fs.gallup.unm.edu/eBookneutrosophics4.pdf.

Chapter 7

Neutrosophic Submodule of Direct Sum

R. Binu

7.1 INTRODUCTION

Numerous applications in complex environments frequently involve uncertainty, fuzziness, and indeterminacy. Florentin Smarandache devised analytical neutrosophic sets to describe information's fuzziness, uncertainty, and imprecision. Abstract algebra was one of the first few areas of research using the idea of neutrosophic sets among the diverse branches of applied and pure mathematics. This chapter focuses on applying neutrosophic algebraic structures and operations to the study of classical algebraic structures, particularly the *R*-module.

The definition of neutrosophic submodules *P* and *Q* is further developed in this work in order to create neutrosophic submodules of $P+Q$. We construct and analyze the neutrosophic submodule of the direct sum $M \oplus N$ and examine its associated results. Additionally, several algebraic results of the neutrosophic submodule's direct sum of a nonempty arbitrary family of submodules are examined. Vidan Cetkin [1] consolidated the neutrosophic set theory and algebraic structures, creating neutrosophic subgroups and neutrosophic submodules. The basic features of single-valued neutrosophic submodules of an *R*-module (classical module) are studied by Cetkin and Olgun N. Neutrosophic submodule is one of the generalizations of the algebraic structure "module" that supplements the classic structure by assigning three diverse level graded features of each module component.

7.2 PRELIMINARIES

This section presents some of the preliminary definitions and results that will be basic for clearly understanding the next sections.

Definition 7.1.1 *[2] Let A and B be submodules of an R-module M. The sum of A and B, denoted as a set, is given as*

$$A+B=\{x+y : x \in A, y \in B\},$$

which is also the smallest submodule that contains both A and B.

DOI: 10.1201/9781003487104-7

Theorem 7.1.1 *[2] The intersection of any nonempty collection of submodules of an R-module is a submodule.*

Remark 7.1.1 *[3] "If $t_P, i_P, f_P : X \to [0,1]$, then P is known as single-valued neutrosophic set (SVNS).*

Remark 7.1.2 *This paper considers only SVNS, which we call neutrosophic sets.*

Remark 7.1.3 *U^X denotes the set of all neutrosophic subsets of X or the neutrosophic power set of X.*

Definition 7.1.2 *[4] Let P and Q be neutrosophic sets of an R-Module M. Then their sum $P+Q$ is a neutrosophic set of M, defined as follows:*

$$P+Q(x) = \left\{x, t_{P+Q}(x), i_{P+Q}(x), f_{P+Q}(x) : x \in M\right\} \text{ where}$$
$$t_{P+Q}(x) = \vee\left\{t_P(y) \wedge t_Q(z) \mid x = y+z, y, z \in M\right\}$$
$$i_{P+Q}(x) = \vee\left\{i_P(y) \wedge i_Q(z) \mid x = y+z, y, z \in M\right\}$$
$$f_{P+Q}(x) = \wedge f_P(y) \vee f_Q(z) \mid x = y+z, y, z \in M\big\}.$$

Definition 7.1.3 *[1] Let P be a neutrosophic set of an R-module M and $r \in R$. Define neutrosophic set $rP = \left\{x, t_{rP}(x), i_{rP}(x), f_{rP}(x) : x \in M\right\}$ of M as follows: $t_{rP}(x) =$ $\begin{cases} \vee\{t_P(y)\} & \text{if } y \in M, x = ry \\ 0 & \text{otherwise} \end{cases}$; $i_{rP}(x) = \begin{cases} \vee\{i_P(y)\} & \text{if } y \in M, x = ry \\ 0 & \text{otherwise} \end{cases}$ and $f_{rP}(x) =$*

$$\begin{cases} \wedge\{f_P(y)\} & \text{if } y \in M, x = ry \\ 0 & \text{otherwise} \end{cases}$$

Definition 7.1.4 [1] Let M be an R module. Let $P \in U^M$ where U^M denotes the neutrosophic power set of R-module M. Then a neutrosophic subset $P = \left\{x, t_P(x), i_P(x), f_p(x) : x \in M\right\}$ in M is called a neutrosophic submodule of M if it satisfies the following:

$$1\ t._P(0) = 1, i_P(0) = 1, f_P(0) = 0$$
$$2\ t._P(x+y) \geq t_P(x) \wedge t_P(y)$$
$$i_P(x+y) \geq i_P(x) \wedge i_P(y)$$
$$f_P(x+y) \leq f_P(x) \vee f_P(y), \forall x, y \in M$$
$$3\ t._P(rx) \geq t_P(x), i_P(rx) \geq i_P(x), f_P(rx) \leq f_P(x), \forall x \in M, \forall r \in R.$$

Remark 7.1.4 *The set of all neutrosophic submodules of R-module M represented by U(M).*

Definition 7.1.5 *[5, 6]*

For any neutrosophic subset $P = \left\{\left(x, t_P(x), i_P(x), f_P(x)\right) : x \in X\right\}$ of X, the support P^* of the neutrosophic set P can be defined as

$$P^* = \left\{x \in X, t_P(x) > 0, i_P(x) > 0, f_P(x) < 1\right\}.$$

Proposition 7.1.1 *[6] Let* $P, Q \in U^X$. *If* $P \subseteq Q$, *then* $P^* \subseteq Q^*$.

Proposition 7.1.2 *If* $P, Q \in U^M$, *then* $\forall x, y \in M, r, s \in R$

$$1\ t_{\cdot(rP+sQ)}(rx+sy) \geq t_P(x) \wedge t_Q(y)$$
$$2\ i_{\cdot(rP+sQ)}(rx+sy) \geq i_P(x) \wedge i_Q(y)$$
$$3\ f_{(rP+sQ)}(rx+sy) \leq f_P(x) \wedge f_Q(y)$$

Definition 7.1.6 [6] Let $P_i, i \in J$ *be an arbitrary nonempty family of* U^M *where* $P_i = \left\{x, t_{P_i}(x), i_{P_i}(x), f_{P_i}(x) : x \in M\right\}$ *for each* $i \in J$. *Then*

$$\sum_{i \in J} P_i = \left\{x, t_{\sum_{i\in J} P_i}(x), i_{\sum_{i\in J} P_i} P_i(x), f_{\sum_{i\in J} P_i}(x) : x \in M\right\} \text{ where}$$

$$t_{\sum_{i\in J} P_i}(x) = \vee\left\{\wedge_{i \in J} t_{P_i}(x_i) : x_i \in M, \sum_{i \in J} x_i = x\right\} \forall x \in M$$

$$i_{\sum_{i\in J} P_i}(x) = \vee\left\{\wedge_{i \in J} i_{P_i}(x_i) : x_i \in M, \sum_{i \in J} x_i = x\right\} \forall x \in M$$

$$f_{\sum_{i\in J} P_i}(x) = \wedge\left\{\vee_{i \in J} f_{P_i}(x_i) : x_i \in M, \sum_{i \in J} x_i = x\right\} \forall x \in M$$

in $\sum_{i \in J} x_i$ *at most finitely* x_i*'s are not equal to zero.*

Proposition 7.1.3 *[6] Let* $P_i, i \in J$ *be an arbitrary nonempty family of* U^M, *then* $r\left(\bigcup_{i \in J} P_i\right) = \bigcup_{i \in J}(rP_i)$ *for* $r \in R$.

Definition 7.1.7 *[5] For any* $x \in X$, *the neutrosophic point* $\widehat{N}_{\{x\}}$ *is defined as* $\widehat{N}_{\{x\}}(s) = \left\{s, t_{\widehat{N}_{\{x\}}}(s), i_{\widehat{N}_{\{x\}}}(s), f_{\widehat{N}_{\{x\}}}(s) : s \in X\right\}$ *where*

$$\widehat{N}_{\{x\}}(s) = \begin{cases} (1,1,0) & x = s \\ (0,0,1) & x \neq s \end{cases}.$$

Definition 7.1.8 *[7] If P,Q and* $S \in U(M)$, *then P is said to be the direct sum of neutrosophic submodules Q and S, and we write* $P = Q \oplus S$, *if*

$$1P = Q + S$$
$$2Q \cap S = \widehat{N}_{\{0\}}.$$

Definition 7.1.9 *[7] Let* $P_i \in U(M) \forall i \in J$. *Then we say that P is the direct sum of* $\{P_i : i \in J\}$ *denoted by* $\oplus_{i \in J} P_i$ *if*

1. $P = \sum_{i \in J} P_i$
2. $P_j \cap \sum_{i \in J - \{j\}} P_i = \widehat{N}_{\{0\}} \forall j \in j$

Theorem 7.1.2 *[7] If P,Q and* $S \in U(M)$*, such that* $P = Q \oplus S$*. Then* $P^* = Q^* \oplus S^*$.

Remark 7.1.5 *Let* $P, Q\ S \in U(M)$*. If* $P^* = Q^* \oplus S^* \nRightarrow P = Q \oplus S$*. That is, the converse of the theorem 7.1.2 need not be true.*

7.3 NEUTROSOPHIC SUBMODULE OF DIRECT SUM $M \oplus N$

In this section, we discuss the construction of the neutrosophic submodule of the direct sum $M \oplus N$ and analyze the results associated with it. We also investigate the direct sum of the nonempty arbitrary family of neutrosophic submodules.

Definition 7.2.1 *Let M and N be R-modules. Let* $P = \{m, t_P(m), i_P(m), f_P(m) : m \in M\} \in U(M)$ *and* $Q = \{n, t_Q(n), i_Q(n), f_Q(n) : n \in N\} \in U(N)$*. Consider the direct sum* $M \oplus N$*. We extend the definitions of P and Q to* $M \oplus N$ *to get the neutrosophic set P′ and Q′ in* $M \oplus N$ *such that*

$$P' = \{(m,n), t_{P'}(m,n), i_{P'}(m,n), f_{P'}(m,n) : (m,n) \in M \oplus N\}$$
$$Q' = \{(m,n), t_{Q'}(m,n), i_{Q'}(m,n), f_{Q'}(m,n) : (m,n) \in M \oplus N\}$$

where

$$t_{P'}(m,n) = \begin{cases} t_P(m) & \text{if } n = 0 \\ 0 & \text{if } n \neq 0 \end{cases},$$
$$i_{P'}(m,n) = \begin{cases} i_P(m) & \text{if } n = 0 \\ 0 & \text{if } n \neq 0 \end{cases},$$
$$f_{P'}(m,n) = \begin{cases} f_P(m) & \text{if } n = 0 \\ 1 & \text{if } n \neq 0 \end{cases}$$
$$t_{Q'}(m,n) = \begin{cases} t_Q(n) & \text{if } m = 0 \\ 0 & \text{if } m \neq 0 \end{cases},$$
$$i_{Q'}(m,n) = \begin{cases} i_Q(n) & \text{if } m = 0 \\ 0 & \text{if } m \neq 0 \end{cases},$$
$$f_{Q'}(m,n) = \begin{cases} f_Q(n) & \text{if } m = 0 \\ 1 & \text{if } m \neq 0 \end{cases} \quad \forall (m,n) \in M \oplus N$$

Theorem 7.2.1 *The neutrosophic sets* P' *and* $Q' \in U(M \oplus N)$.

Proof *Consider the neutrosophic set*
$P' = \{(m,n), t_{P'}(m,n), i_{P'}(m,n), f_{P'}(m,n) : (m,n) \in M \oplus N\}$ *defined as*

$$t_{P'}(m,n)=\begin{cases} t_P(m) & \text{if } n=0 \\ 0 & \text{if } n\neq 0 \end{cases},\quad i_{P'}(m,n)=\begin{cases} i_P(m) & \text{if } n=0 \\ 0 & \text{if } n\neq 0 \end{cases},$$

$$f_{P'}(m,n)=\begin{cases} f_P(m) & \text{if } n=0 \\ 1 & \text{if } n\neq 0 \end{cases}$$

Then clearly $t_{p'}(0,0)=1, i_{P'}(0,0)=1$ *and* $f_{P'}(0,0)=0$.

For $(m_1,n_1),(m_2,n_2)\in M\oplus N$

$$t_{P'}\big((m_1,n_1)+(m_2,n_2)\big)=t_{P'}(m_1+m_2,n_1+n_2)$$
$$=\begin{cases} t_P(m_1+m_2) & \text{if } n_1+n_2=0 \\ 0 & \text{if } n_1+n_2\neq 0 \end{cases}$$

Case 1: If $n_1+n_2=0$, *then either* $n_1=0,\ n_2=0$ *or* $n_1\neq 0, n_2=-n_1\neq 0$.
If $n_1=n_2=0$, *then*

$$\begin{aligned} t_{P'}\big((m_1,n_1)+(m_2,n_2)\big) &= t_P(m_1+m_2) \\ &\geq t_P(m_1)\wedge t_P(m_2) \\ &= t_{P'}(m_1,n_1)\wedge t_{P'}(m_2,n_2). \end{aligned}$$

If $n_1\neq 0, n_2=-n_1\neq 0$ *then*

$$\begin{aligned} t_{P'}\big((m_1,n_1)+(m_2,n_2)\big) &= t_p(m_1+m_2,0) \\ &= t_p(m_1+m_2) \\ &\geq t_P(m_1)\wedge t_P(m_2) \\ &= 0\wedge 0 \\ &= t_{P'}(m_1,n_1)\wedge t_{P'}(m_2,n_2) \end{aligned}$$

Case 2: If $n_1+n_2\neq 0$, *then either* n_1 *or* n_2 *or both* $\neq 0$.
If $n_1\neq 0, n_2\neq 0$ *then,*

$$t_{P'}\big((m_1,n_1)+(m_2,n_2)\big)=0=0\wedge 0=t_{P'}(m_1,n_1)\wedge t_{P'}(m_2,n_2).$$

If $n_1=0, n_2\neq 0$ *then,*

$$t_{P'}\big((m_1,n_1)+(m_2,n_2)\big)=0=t_P(m_1)\wedge 0=t_{P'}(m_1,n_1)\wedge t_{P'}(m_2,n_2).$$

If $n_1\neq 0, n_2=0$. *Same as above.*
$\therefore$ *from these two cases,*

$$t_{P'}\left(\left(m_1,n_1\right)+\left(m_2,n_2\right)\right) \geq t_{P'}\left(m_1,n_1\right) \wedge t_{P'}\left(m_2,n_2\right).$$

Similarly, we can prove

$$i_{P'}\left(\left(m_1,n_1\right)+\left(m_2,n_2\right)\right) \geq i_{P'}\left(m_1,n_1\right) \wedge i_{P'}\left(m_2,n_2\right)$$
$$f_{P'}\left(\left(m_1,n_1\right)+\left(m_2,n_2\right)\right) \leq f_{P'}\left(m_1,n_1\right) \vee f_{P'}\left(m_2,n_2\right)$$

Now for any $r \in R, (m,n) \in M \oplus N$,

$$t_{P'}(r(m,n)) = t_{P'}(rm,rn) = \begin{cases} t_P(rm) & \text{if } rn = 0 \\ 0 & \text{if } rn \neq 0 \end{cases}$$

Case 1: If $rn = 0$, *then,*

$$t_{P'}(r(m,n)) = t_P(rm) \geq t_P(m) = t_{P'}(m,n).$$

Case 2: If $rn \neq 0 \Rightarrow n \neq 0$, *then,*

$$t_{P'}(r(m,n)) = t_{P'}(rm,rn) = 0 = t_{P'}(m,n).$$

$\therefore$ *from these two cases,* $t_{P'}(r(m,n)) \geq t_{P'}(m,n)$.
Similarly, we can prove $i_{P'}(r(m,n)) \geq i_{P'}(m,n)$ *and* $f_{P'}(r(m,n)) \leq f_{P'}(m,n)$.
Hence, $P' \in U(M \oplus N)$. *Similarly, we can prove that* $Q' \in U(M \oplus N)$.

Corollary 7.2.1.1 *If* $P', Q' \in U(M \oplus N)$, *then their sum* $P' + Q' \in U(M \oplus N)$.

Remark 7.2.1 $(P' \cap Q')(m,n) = \{(m,n), t_{P'\cap Q'}(m,n), i_{P'\cap Q'}(m,n), f_{P'\cap Q'}(m,n) :$ $(m,n) \in M \oplus N\}$, *then from the definition of* P' *and* Q',

$$t_{P'\cap Q'}(m,n) = \begin{cases} 1 & \text{if } (m,n) = 0 \\ 0 & \text{if } (m,n) \neq 0 \end{cases}, \quad i_{P'\cap Q'}(m,n) = \begin{cases} 1 & \text{if } (m,n) = 0 \\ 0 & \text{if } (m,n) \neq 0 \end{cases}$$

$$f_{P'\cap Q'}(m,n) = \begin{cases} 0 & \text{if } (m,n) = 0 \\ 1 & \text{if } (m,n) \neq 0 \end{cases}$$

$$\therefore P' \cap Q' = \widehat{N}_{\{0\}}$$

$\Rightarrow P' + Q'$ *is a direct sum and is denoted by* $P \oplus Q \in U(M \oplus N)$.

Proposition 7.2.1 *If* $P \oplus Q \in U(M \oplus N)$, *then*

1. $t_{P\oplus Q}(m,n) = t_P(m) \wedge t_Q(n)$
2. $i_{P\oplus Q}(m,n) = i_P(m) \wedge i_Q(n)$
3. $f_{P\oplus Q}(m,n) = f_P(m) \vee f_Q(n)$.

Proof *1. We have*

$$\begin{aligned}(P\oplus Q(m,n)) &= \{(m,n), t_{P\oplus Q}(m,n), i_{P\oplus Q}(m,n), f_{P\oplus Q}(m,n) : (m,n)\in M\oplus N\} \\ &= P' + Q'(m,n) \\ &= \{(m,n), t_{P'+Q'}(m,n), i_{P'+Q'}(m,n), f_{P'+Q'}(m,n) : (m,n)\in M\oplus N\}\end{aligned}$$

Now

$$\begin{aligned}t_{P\oplus Q}(m,n) &= t_{P'+Q'}(m,n) \\ &= \vee\{t_{P'}(m_1,n_1)\wedge t_{Q'}(m_2,n_2) : (m,n) = (m_1,n_1)+(m_2,n_2) : \\ &\quad (m_1,n_1),(m_2,n_2)\in M\oplus Q\} \\ &= t_{P'}(m,0)\wedge t_{Q'}(0,n) \\ &= t_P(m)\wedge t_Q(n)\end{aligned}$$

Similarly we get 2 and 3:

$$i_{P\oplus Q}(m,n) = i_P(m)\wedge i_Q(n), f_{P\oplus Q}(m,n) = f_P(m)\vee f_Q(n).$$

Theorem 7.2.2 *If* $P\in U(M)$ *and* $Q\in U(N)$, *then* $(P\oplus Q)^* = P^*\oplus Q^*$.

Proof *We have* $P\oplus Q = P' + Q' \in U(M\oplus N)$. *Let* $(m,n)\in (P\oplus Q)^*$

$$\begin{aligned}&\Rightarrow t_{(P\oplus Q)^*}(m,n) > 0, i_{(P\oplus Q)^*}(m,n) > 0 \text{ and } f_{(P\oplus Q)^*}(m,n) < 1 \\ &\Rightarrow t_P(m)\wedge t_Q(n) > 0, i_P(m)\wedge i_Q(n) > 0 \text{ and } f_P(m)\vee f_Q(n) < 1 \\ &\Rightarrow t_P(m) > 0, t_Q(n) > 0, i_P(m) > 0\ i_Q(n) > 0 \text{ and } f_Q(m) < 1, f_Q(n) < 1 \\ &\Rightarrow m\in P^* \text{ and } n\in Q^* \\ &\Rightarrow (m,n)\in P^* + Q^*\end{aligned}$$

Now $x\in P^*\cap Q^*$

$$\begin{aligned}&\Rightarrow x\in P^* \text{ and } x\in Q^* \\ &\Rightarrow t_P(x) > 0, i_P(x) > 0, f_P(x) < 1 \text{ and } t_Q(x) > 0, i_Q(x) > 0, f_Q(x) < 1 \\ &\Rightarrow t_P(x)\wedge t_Q(x) > 0, i_P(x)\wedge i_Q(x) > 0 \text{ and } f_P(x)\vee f_Q(x) < 1 \\ &\Rightarrow t_{P\cap Q}(x) > 0, i_{P\cap Q}(x) > 0 \text{ and } f_{P\cap Q}(x) < 1 \\ &\Rightarrow t_{P\cap Q}(x) = 1, i_{P\cap Q}(x) = 1 \text{ and } f_{P\cap Q}(x) = 0 \\ &\quad \left[\text{Since } P\oplus Q \text{ is the direct sum, hence } P\cap Q = \widehat{N}_{\{0\}}\right] \\ &\Rightarrow x = 0 \\ &\Rightarrow P^*\cap Q^* = \{0\}\end{aligned}$$

Hence $P^* + Q^*$ *is the direct sum, denoted as* $P^* \oplus Q^*$, *which is a submodule of* $M \oplus N$.

Hence,

$$(P \oplus Q)^* \subseteq P^* \oplus Q^*. \tag{7.1}$$

Let $(m,n) \in P^* \oplus Q^*$

$$\begin{aligned}
&\Rightarrow m \in P^* \text{ and } n \in Q^* \\
&\Rightarrow t_P(m) > 0, t_Q(n) > 0, i_P(m) > 0 i_Q(n) > 0 \text{ and } f_P(m) < 1, f_Q(n) < 1 \\
&\Rightarrow t_P(m) \wedge t_Q(n) > 0, i_P(m) \wedge i_Q(n) > 0 \text{ and } f_P(m) \vee f_Q(n) < 1 \\
&\Rightarrow t_{P\oplus Q}(m,n) > 0, i_{P\oplus Q}(m,n) > 0 \text{ and } f_{P\oplus Q}(m,n) 1 \\
&\Rightarrow (m,n) \in (P \oplus Q)^*
\end{aligned}$$

Hence

$$P^* \oplus Q^* \subseteq (P \oplus Q)^* \tag{7.2}$$

$\therefore$ *from equations (7.1) and (7.2), we get* $(P \oplus Q)^* = P^* \oplus Q^*$.

Corollary 7.2.2.1 *If* $P, Q \in U(M)$, *then* $(P \oplus Q)^* = P^* \oplus Q^*$.

Theorem 7.2.3 *Let* $P_i, i \in J$ *be an arbitrary nonempty family of neutrosophic submodule of an R-module M where* $\sum_{i \in J} P_i$ *is the direct sum of* $\oplus_{i\in J} P_i$ *and* $Q \in U(M)$. *If*

$Q \cap \sum_{i\in J} P_i = \widehat{N}_{\{0\}}$, *then* $Q + \sum_{i\in J} P_i$ *is the direct sum* $Q \oplus (\oplus_{i\in J} P_i)$.

Proof *Given that* $\sum_{i\in J} P_i$ *is a direct sum,* $P_j \cap \left(\sum_{i\in J-\{j\}} P_i\right) = \widehat{N}_{\{0\}} \forall j \in J$.

Also given that $Q \cap \sum_{i\in J} P_i = \widehat{N}_{\{0\}}$,

for $x \in M, j \in J$, *the neutosophic components of* $\left(P_j \cap \left(Q + \sum_{J-\{j\}} P_i\right)\right)(x)$ *are*

$$\left\{x, t_{P_j \cap \left(Q + \sum_{i\in J-\{j\}} P_i\right)}(x), i_{P_j \cap \left(Q + \sum_{J-\{j\}} P_i\right)}(x), f_{P_j \cap \left(Q + \sum_{J-\{j\}} P_i\right)}(x)\right\}.$$

Now consider

$$\begin{aligned}
t_{P_j \cap \left(Q + \sum_{i\in J-\{j\}} P_i\right)}(x) &= t_{P_j}(x) \wedge t_{Q + \sum_{i\in J-\{j\}} P_i}(x) \\
&= t_{P_j}(x) \wedge \vee\Big\{t_Q(y) \wedge \left(\wedge_i t_{Pi}(x_i)\right) \\
&\qquad : x = y + \sum_i x_i, i \in J - \{j\}, y, x_i \in M\Big\}
\end{aligned}$$

$$= t_{P_j}\left(y+\sum x_i\right)\wedge\vee\Big\{t_Q(y)\wedge\left(\wedge_i t_{P_i}(x_i)\right)$$

$$: x = y+\sum_i x_i, i\in J-\{j\}, y, x_i \in M\Big\}$$

$$=\left(\vee\left\{t_{P_j}(y)\wedge\left(\wedge_i t_{P_j}(x_i)\right): x = y+\sum_i x_i, i\in J-\{j\}, y, x_i\in M\right\}\right)$$

$$\wedge\left(\vee\left\{t_Q(y)\wedge\left(\wedge_i t_{P_i}(x_i)\right): x = y+\sum_i x_i, i\in J-\{j\}, y, x_i\in M\right\}\right)$$

$$=\vee\Big\{\left(t_{P_j}(y)\wedge(\wedge_i t_{P_j}(x_i)\right)\wedge\left(t_Q(y)\wedge\left(\wedge_i t_{P_i}(x_i)\right)\right):$$

$$x = y+\sum_i x_i, i\in J-\{j\}, y, x_i\in M\Big\}$$

$$=\vee\Big\{\left(t_{P_j}(y)\wedge t_Q(y)\right)\wedge\wedge_i\left(t_{P_j}(x_i)\wedge t_{P_i}(x_i)\right):$$

$$x = y+\sum_i x_i, i\in J-\{j\}, y, x_i\in M\Big\}$$

$$=\vee\Big\{t_{P_j\cap Q}(y)\wedge\wedge_i\left(t_{P_j\cap P_i}(x_i)\right): x = y+\sum_i x_i$$

$$i\in J-\{j\}, y, x_i\in M\Big\}$$

$$=\begin{cases}1 & \textit{if } y=0, x_i=0\,\forall i\in J-\{j\}\\ 0 & \textit{if } y\neq 0 \text{ or } x_i\neq 0 \textit{ for some } i\in J-\{j\}\end{cases}$$

$$\left[\textit{Since } P_j\cap Q=\widehat{N}_{\{0\}}\forall j\in J \textit{ and } P_j\cap P_i=\widehat{N}_{\{0\}}\forall i\in J-\{j\}\right]$$

$$=\begin{cases}1 & \text{if } x=0\\ 0 & \text{if } x\neq 0\end{cases}\quad\left[\textit{since } x = y+\sum_i x_i, i\in J-\{j\}, y, x_i\in M\right]$$

$$= t_{\widehat{N}_{\{0\}}}(x)$$

Similarly we can show that

$$i_{P_j\cap\left(Q+\sum_{i\in J-\{j\}}P_i\right)}(x)=\begin{cases}1 & \textit{if } x=0\\ 0 & \textit{if } x\neq 0\end{cases} = i_{\widehat{N}_{\{0\}}}(x)$$

$$f_{P_j\cap\left(Q+\sum_{i\in J-\{j\}}P_i\right)}(x)=\begin{cases}0 & \textit{if } x=0\\ 1 & \textit{if } x\neq 0\end{cases} = f_{\widehat{N}_{\{0\}}}(x)$$

$$\Rightarrow P_j \cap \left(Q + \sum_{J-\{j\}} P_i\right) = \widehat{N}_{\{0\}} \forall j \in J$$

$\therefore Q + \sum_{i \in J-\{j\}} P_i$ *is the direct sum* $Q \oplus (\oplus_{i \in J} P_i)$

Thus the theorem is proved.

7.4 CONCLUSION

The primary goal of this study was to infuse the contemporary speculations of neutrosophic sets and module structure into conventional algebraic structures. We investigated and improved the construction of neutrosophic submodules from the direct sum of submodules in the arena of abstract algebra. The present study should lead to explorations of the concept of tensor product of neutrosophic submodule, injective and projective neutrosophic submodules of an *R*-module, a semi-simple neutrosophic submodule of an *R*-module, and a quasi neutrosophic submodule of an *R*-module.

REFERENCES

1. Cetkin, V., Varol, B. P., & Aygün, H. (2017). On neutrosophic submodules of a module. Hacettepe Journal of Mathematics and Statistics, 46(5), 791–799.
2. Dummit, D. S., & Foote, R. M. (2004). Abstract algebra (Vol. 3). Hoboken: Wiley.
3. Smarandache, F. (1999). A unifying field in Logics: Neutrosophic Logic. In Philosophy (pp. 1–141). American Research Press.
4. Smarandache, F. (2010). Neutrosophic set–a generalization of the intuitionistic fuzzy set. Journal of Defense Resources Management (JoDRM), 1(1), 107–116.
5. Binu, R., & Isaac, P. (2020). Neutrosophic quotient submodules and homomorphisms. Punjab University Journal of Mathematics, 52(1), 33–45.
6. Binu, R., & Isaac, P. (2021). Some characterizations of neutrosophic submodules of an-module. Applied Mathematics and Nonlinear Sciences, 6(1), 359–372.
7. Babak, D., & Ulcay, V. (2022). Neutrosophic algebraic structures and their applications. Neutrosophic Algebraic Structures and Their Applications, 167.

Chapter 8

Quota-Harvesting Logistic Growth Model

A Neutrosophic Differential Equation Approach

Ashish Acharya, Nikhilesh Sil, Animesh Mahata, Subrata Paul, Manajat Ali Biswas, Supriya Mukherjee, Said Broumi, and Banamali Roy

8.1 INTRODUCTION

It is common knowledge that fuzzy intuitionistic sets are unable to incorporate indeterminacy. Uses of neutrosophic fuzzy set theory have been developed to address this challenge. In this chapter, we formulate a logistic model with quota harvesting in a neutrosophic fuzzy environment. We investigate the suggested model, in which all parameters are assumed to be TrSVNNs of type 3. We also test equilibrium along with conducting a stability analysis. The nature of neutroshopic solutions are reflected in this chapter, and we elucidate the concept of strong and weak neutrosophic solutions. Using MATLAB®, we perform numerical simulations to validate the model's entire set of findings.

Harvesting is one of the most important biological phenomena predator–prey systems for managing renewable resources. Generally, harvesting of prey populations is used in fishery and forestry [1–3]. In the last few decades, many researchers have devoted their time to investigate the dynamical behavior of predator–prey systems with harvesting in the management of renewable resources [4–17]. The exploitation of renewable resources and its bioeconomic analysis is a major concern. Clark discussed the socioeconomic exploitation of renewable resources discussed by Clark [1, 18], and many other researchers have described the dynamics of multispecies harvesting models [19–23]. Chaudhuri, Saha Ray and Kar et al. [24, 25] studied a nonselective harvesting model, although there is a problem with nonselective harvesting with logistic growth [1]. This type of model has maximum sustainable yield (MSY). If the harvesting of a population exceeds its MSY (i.e., the population is overexploited), then the population will become extinct, and if the harvesting rate is smaller than its MSY then the decreased harvested species can be recovered [26].

There are two types of harvesting policy: constant-quota harvesting and constant-effort harvesting. In constant-quota harvesting, a constant number of individuals in the population are harvested per unit of time, and in constant-effort harvesting, fixed effort is applied per unit of time to catch the animals in a population [27].

 DOI: 10.1201/9781003487104-8

Although harvesting does not follow a particular rule, we can say that many more species of fish are harvested in the warmer seasons than in cold seasons [28].

In the field of ecology, many situations are uncertain and ambiguous in nature. Therefore, it is very difficult to model some real-life scenarios with the help of classical mathematics. To handle this type of uncertainty in nature, in 1965, L.A. Zadeh [29] discovered fuzzy set theory (FST), which extends classical set theory with a membership function. Many researchers have examined using FST to handle uncertainty [30–33, 35–37].

Some ambiguous situations cannot be handled with the membership function. Following the development of FST, Atanassov [38–41] developed intuitionistic fuzzy set theory (IFST) which generalized FST and consists of degree of membership and a nonmembership function [34, 42–49]. However, neither of these two sets can address indeterminacy.

To handle undetermined, insufficient, and inconsistent data, Smarandache and other researchers developed neutrosophic set theory (NST) [50–52], which is a further generalization of FST and IFST. The single-valued neutrosophic set was developed in [53], and the simplified neutrosophic set was given in [54]. Peng et al. [55, 56] discuss ideas on new operations and aggregation operators. Researchers also developed triangular neutrosophic sets [57], bipolar neutrosophic sets [58–61], and multivalued neutrosophic sets [62].

Saini et al. [63] developed single-valued trapezoidal neutrosophic numbers and applied them to a transportation problem. Chakraborty et al. [64] developed cylindrical neutrosophic single-valued numbers in different perspectivse and applied them to solving a networking problem and finding a minimal spanning tree. Haque et al. [65] described the exponential operational law in a trapezoidal neutrosophic environment.

Banik et al. [66] developed a multicriteria group decision-making strategy in a pentagonal neutrosophic environment. Biswas et al. [67] discussed different types of Gaussian quadrature rules for the numeric integration of a neutrosophic valued function. Moi et al. [68] developed a neutrosophic differential equation. Moi et al. [69] investigated a numeric method for solving the neutrosophic Fredholm integral equation. Parikh and Sahni [70] used Sumudu transform to solve an ordinary differential equation in a mechanical spring mass system with neutrosophic initial conditions, etc.

In short, we can say that many researchers have developed and studied neutrosophic numbers and applied them in different fields of engineering. However, to date, no actions have been taken to describe ecological problems with the help of neutrosophic numbers. In this chapter, we describe and analyzed an ecological model of harvesting with the help of neutrosophic set theory.

8.2 MOTIVATION AND NOVELTY

Multiple factors both natural and related to human activities affect the parameters of a biological model, and this can lead to vagueness, impreciseness, or indeterminacy in parameters. To consider situations of uncertainty, researchers have examined approaches like interval differential equation (IDE), fuzzy differential equation (FDE), and intuitionistic fuzzy differential equation (IFDE). In IDE, the parameters are considered to belong to certain intervals, whereas in FDE, the parameters are assigned certain membership values. IFDE considers the membership as well as nonmembership values of the parameters. However, to deal with indeterminant parameter values, we need neutrosophic differential equation (NDE).

8.3 STRUCTURE OF THE CHAPTER

8.3.1 Prerequisite Concept

Definition 8.3.1.1 [26]: Let U be a universe set. A neutrosophic set $B_{n\tilde{e}}$ on U defined as

$B_{n\tilde{e}} = \{< T_{n\tilde{e}}(x), I_{n\tilde{e}}(x), F_{n\tilde{e}}(x) >: x \in U\}$, where $T_{n\tilde{e}}(x), I_{n\tilde{e}}(x), F_{n\tilde{e}}(x): U \to {}^{-}]0,1[^{+}$ represents the degree of membership, degree of indeterministic and degree of nonmembership, respectively, such that $0 \le T_{n\tilde{e}}(x) + I_{n\tilde{e}}(x) + F_{n\tilde{e}}(x) \le 3^{+}, x \in U$.

Definition 8.3.1.2 [26]: (α, β, γ) cuts: The (α, β, γ) cut in a neutrosophic set is defined by $K_{\alpha,\beta,\gamma}$ where $\alpha, \beta, \gamma \in [0,1]$ is a fixed number such that $\alpha + \beta + \gamma \le 3$ is defined as $K_{\alpha,\beta,\gamma} = \{< T_{n\tilde{e}}(x), I_{n\tilde{e}}(x), F_{n\tilde{e}}(x) >: x \in U, T_{n\tilde{e}}(x) \ge \alpha, I_{n\tilde{e}}(x) \le \beta, F_{n\tilde{e}}(x) \le \gamma\}$.

Definition 8.3.1.3 [26]: A neutrosophic set $B_{n\tilde{e}}$ defined on the universal set of real number R is said to be a neutrosophic number if it satisfies the following conditions:

$B_{n\tilde{e}}$ is normal if there exists $x_0 \in R$ such that $T_{n\tilde{e}}(x_0) = 1, I_{n\tilde{e}}(x_0) = F_{n\tilde{e}}(x_0) = 0$, $B_{n\tilde{e}}$ is convex for the truth function $T_{n\tilde{e}}(x)$. That is, $T_{n\tilde{e}}(\lambda x_1 + (1 - \lambda x_1)) \ge \min(T_{n\tilde{e}}(x_1), T_{n\tilde{e}}(x_2))$, $\forall x_1, x_2 \in R, \lambda \in [0,1]$, $F_{n\tilde{e}}(\lambda x_1 + (1 - \lambda x_1)) \ge \max(F_{n\tilde{e}}(x_1), F_{n\tilde{e}}(x_2))$, $\forall x_1, x_2 \in R, \lambda \in [0,1]$.

Definition 8.3.1.4 [57]: A triangular single-valued neutrosophic number of type 3 (TrSVNN 3) is defined as $A_{n\tilde{u}} = (r_1, r_2, r_3; p_{nu}, q_{nu}, l_{nu})$, with truth, indeterminancy, and falsity membership defined as follows:

$$T_{n\tilde{u}}(x) = \begin{cases} p_{nu}(x - r_1)/(r_2 - r_1), r_1 \le x < r_2 \\ p_{nu}, x = r_2 \\ p_{nu}(r_3 - x)/(r_3 - r_2), r_2 < x \le r_3 \\ 0, \; otherwise \end{cases}$$

$$I_{n\tilde{u}}(x) = \begin{cases} \{r_2 - x + q_{nu}(x - r_1)\}/(r_2 - r_1), r_1 \le x < r_2 \\ q_{nu}, x = r_2 \\ \{x - r_2 + q_{nu}(r_3 - x)\}/(r_3 - r_2), r_2 < x \le r_3 \\ 1, \; otherwise \end{cases}$$

$$F_{n\tilde{u}}(x)=\begin{cases}\{r_2-x+l_{nu}(x-r_1)\}/(r_2-r_1), r_1\le x<r_2\\ l_{nu}, x=r_2\\ \{x-r_2+l_{nu}(r_3-x)\}/(r_3-r_2),\ r_2<x\le r_3\\ 1,\ \ otherwise\end{cases}$$

where $0\le T_{n\tilde{e}}(x)+I_{n\tilde{e}}(x)+F_{n\tilde{e}}(x)\le 1, x\in A_{n\tilde{u}}$.
The parametric form of TrSVNN type 3 is

$$A_{n\tilde{u}}=[T_{nu1}(\alpha),T_{nu2}(\alpha),I_{nu1}(\beta),I_{nu2}(\beta),F_{nu1}(\gamma),F_{nu2}(\gamma)], \text{ where}$$
$$T_{nu1}(\alpha)=r_1+\alpha/p_{nu}(r_2-r_1),$$
$$T_{nu2}(\alpha)=r_3-\alpha/p_{nu}(r_3-r_2),$$
$$I_{nu1}(\beta)=\{r_2-q_{nu}r_1-\beta(r_2-r_1)\}/(1-q_{nu}),$$
$$I_{nu2}(\beta)=\{r_2-q_{nu}r_3+\beta(r_3-r_2)\}/(1-q_{nu}),$$
$$F_{nu1}(\gamma)=\{r_2-l_{nu}r_1-\beta(r_2-r_1)\}/(1-l_{nu}),$$
$$F_{nu2}(\gamma)=\{r_2-l_{nu}r_3+\beta(r_3-r_2)\}/(1-l_{nu}).$$

Here, $0<\alpha\le p_{nu},\ q_{nu}<\beta\le 1, l_{nu}<\gamma\le 1,\ \ 0<\alpha+\beta+\gamma\le 1$.

Definition 8.3.1.5 [70]: Hukuhara difference on neutrosophic function: Let E^*be the set of all neutrosophic functions $\tilde{S},\tilde{r}\in E^*$. If there exists a neutrosophic number, $\tilde{v}\in E^*$ and $\tilde{v}$ satisfy $\tilde{s}=\tilde{v}+\tilde{r}$, and then $\tilde{v}$ is defined to be a Hukuhara difference of $\tilde{s},\tilde{r}$, denoted by $\tilde{v}=\tilde{s}\Theta\tilde{r}$.

Definition 8.3.1.6 [70]: Hukuhara derivative on neutrosophic function: Let $gh_{n\tilde{u}}:(c,d)\to N(R)$ be a neutrosophic single-valued function and $k_0,k_0+h,k_0-h\in(c,d)$. We say that $g_{n\tilde{u}}$ is Hukuhara differentiable at k_0 if there exists an element $g'h_{n\tilde{u}}\in N(R)$ for all $h>0$.

8.3.2 Model Formulation

The quota-harvesting logistic growth model [72] is given by

$$\mathrm{dP}(\mathrm{t})/\mathrm{dt}=\mathrm{rP}(\mathrm{t})(1-\mathrm{P}(\mathrm{t})/\mathrm{K})-\mathrm{q_h EP}(\mathrm{t}). \tag{8.1}$$

where $\mathrm{P}=\mathrm{P}(\mathrm{t})$ is the population density, r is the intrinsic growth rate of the prey population, K is the carrying capacity, q_h is the catchability of the species, and E is the harvesting effort. Now we modify with the help of a neutrosophic differential equation as follows.
We consider crisp quota-harvesting model (8.1) in a neutrosophic environment as

$$\mathrm{d\tilde{P}}(\mathrm{t})/\mathrm{dt}=\mathrm{r\tilde{P}}(\mathrm{t})(1-\mathrm{\tilde{P}}(\mathrm{t})/\mathrm{K})-\mathrm{q_h E\tilde{P}}(\mathrm{t}). \tag{8.2}$$

with the initial condition, $\tilde{\mathrm{P}}(\mathrm{t}_0)=\tilde{\mathrm{P}}_0$.

8.3.3 Analysis of the Quota-Harvesting Logistic Growth Model in the Neutrosophic Environment

We consider that the initial population of the system (8.2) $\tilde{P}_0$ is a neutrosophic number. The following many cases arise if we apply the GN-derivative to model (8.2):

(a) $\tilde{P}(t)$ is GN-derivative of type 1;
(b) $\tilde{P}(t)$ is GN-derivative of type 2;
(c) $\tilde{P}(t)$ is GN-derivative of type 3;
(d) $\tilde{P}(t)$ is GN-derivative of type 4;
(e) $\tilde{P}(t)$ is GN-derivative of type 5;
(f) $\tilde{P}(t)$ is GN-derivative of type 6;
(g) $\tilde{P}(t)$ is GN-derivative of type 7;
(h) $\tilde{P}(t)$ is GN-derivative of type 8.

In this chapter, to study the dynamic behaviour of population model (8.2) in a neutrosophic environment, we took two cases:

(a) $\tilde{P}(t)$ is GN-derivative of type 1 and
(b) $\tilde{P}(t)$ is GN-derivative of type 2.

8.4 WHEN $\tilde{P}(t)$ IS GN-DERIVATIVE OF TYPE 1

The quota-harvesting logistic growth model (8.2) is transformed into the system of differential equation as

$$\begin{aligned}
dP_{1L}(t,\alpha)/dt &= rP_{1L}(t,\alpha)\left(1-P_{1R}(t,\alpha)/K\right)-q_hEP_{1R}(t,\alpha),\\
dP_{1R}(t,\alpha)/dt &= rP_{1R}(t,\alpha)\left(1-P_{1L}(t,\alpha)/K\right)-q_hEP_{1L}(t,\alpha),\\
dP_{2L}(t,\alpha)/dt &= rP_{2L}(t,\beta)\left(1-P_{2R}(t,\beta)/K\right)-q_hEP_{2R}(t,\beta),\\
dP_{2R}(t,\beta)/dt &= rP_{2R}(t,\beta)\left(1-P_{2L}(t,\beta)/K\right)-q_hEP_{2L}(t,\beta),\\
dP_{3L}(t,\gamma)/dt &= rP_{3L}(t,\gamma)\left(1-P_{3R}(t,\gamma)/K\right)-q_hEP_{3R}(t,\gamma),\\
dP_{3R}(t,\gamma)/dt &= rP_{3R}(t,\gamma)\left(1-P_{3L}(t,\gamma)/K\right)-q_hEP_{3L}(t,\gamma).
\end{aligned} \tag{8.3}$$

With initial condition $P_{1L}(t_0,\alpha)=P_{01L}(\alpha)$, $P_{1R}(t_0,\alpha)=P_{01R}(\alpha)$, $P_{2L}(t_0,\beta) = P_{02L}(\beta)$, $P_{2R}(t_0,\beta)=P_{02R}(\beta)$, $P_{3L}(t_0,\gamma)=P_{03L}(\gamma)$, $P_{3R}(t_0,\gamma)=P_{03R}(\gamma)$.

The $(\alpha,\beta,\gamma)-$ cut of $\tilde{P}(t)$ is given by $\tilde{P}_{\alpha,\beta,\gamma} =< P_{1L}(t,\alpha),\ P_{1R}(t,\alpha);\ P_{2L}(t,\alpha),\ P_{2R}(t,\beta);\ P_{3L}(t,\gamma),\ P_{3R}(t,\gamma) >$.

8.4.1 Equilibrium Points

(i) **Trivial equilibrium:** The transformed system (8.3) has a trivial equilibrium point $E^1_{00}(P_{1L}(t,\alpha), P_{1R}(t,\alpha); P_{2L}(t,\alpha), P_{2R}(t,\beta); P_{3L}(t,\gamma), P_{3R}(t,\gamma))$ where $P_{1L}(t,\alpha) = P_{1R}(t,\alpha) = P_{2L}(t,\alpha) = P_{2R}(t,\beta) = P_{3L}(t,\gamma) = P_{3R}(t,\gamma) = 0$.

(ii) **Coexistence equilibrium:** The transformed model (8.3) has the coexistence equilibrium point $E^*_{11c}(P^*_{1L}(t,\alpha) = P^*_{1R}(t,\alpha) = P^*_{2L}(t,\alpha) = P^*_{2R}(t,\beta) = P^*_{3L}(t,\gamma) = P^*_{3R}(t,\gamma) = K(r - q_h E)/r$.

The coexistence equilibrium point is feasible if $r > q_h E$.

8.4.2 Stability Analysis

The variational matrix of the transformed system (8.3) is given by

$$V^1_P = \begin{pmatrix} x_1 & x_2 & 0 & 0 & 0 & 0 \\ x_3 & x_4 & 0 & 0 & 0 & 0 \\ 0 & 0 & x_5 & x_6 & 0 & 0 \\ 0 & 0 & x_7 & x_8 & 0 & 0 \\ 0 & 0 & 0 & 0 & x_9 & x_{10} \\ 0 & 0 & 0 & 0 & x_{11} & x_{12} \end{pmatrix}.$$

where $x_1 = r(1 - P_{1R}(t,\alpha)/dt), x_2 = -(rP_{1R}(t,\alpha)/K + q_h E), x_3 = -(rP_{1L}(t,\alpha)/K + q_h E)$,

$x_4 = r(1 - P_{1L}(t,\alpha)/dt), x_5 = r(1 - P_{2R}(t,\beta)/dt), x_6 = -(rP_{2R}(t,\beta)/K + q_h E)$,
$x_7 = -(rP_{2L}(t,\beta)/K + q_h E), x_8 = r(1 - P_{2L}(t,\beta)/dt), x_9 = r(1 - P_{3R}(t,\gamma)/dt)$,
$x_{10} = -(rP_{3R}(t,\gamma)/K + q_h E), x_{11} = -(rP_{3L}(t,\alpha)/K + q_h E), x_{12} = r(1 - P_{3L}(t,\alpha)/dt)$.

Theorem 1. The transformed system (8.3) is unstable at E^1_{00}.

Proof. The variational matrix at E^1_{00} becomes

$$V^1_{00} = \begin{pmatrix} r & -q_h E & 0 & 0 & 0 & 0 \\ -q_h E & r & 0 & 0 & 0 & 0 \\ 0 & 0 & r & -q_h E & 0 & 0 \\ 0 & 0 & -q_h E & r & 0 & 0 \\ 0 & 0 & 0 & 0 & r & -q_h E \\ 0 & 0 & 0 & 0 & -q_h E & r \end{pmatrix}.$$

Let μ_1 be the eigenvalue of V^1_{00}; then, the characteristic equation of V^1_{00} is given by $\left[(\mu_1 - r)^2 - (q_h E)^2\right]^3 = 0$.

The eigenvalues of V_{00}^{1} are $(r+q_h E),(r-q_h E),(r+q_h E),(r-q_h E),(r+q_h E)$, $(r-q_h E)$. Since the eigenvalues are positive, the trivial equilibrium point E_{11c}^{*} of the system (8.3) is unstable.

Theorem 2. The transformed system (8.3) is unstable at E_{11c}^{*}.

Proof. The variational matrix at E_{11c}^{*} is given by

$$V_{11}^{c} = \begin{pmatrix} q_h E & -r & 0 & 0 & 0 & 0 \\ -r & q_h E & 0 & 0 & 0 & 0 \\ 0 & 0 & q_h E & -r & 0 & 0 \\ 0 & 0 & -r & q_h E & 0 & 0 \\ 0 & 0 & 0 & 0 & q_h E & -r \\ 0 & 0 & 0 & 0 & -r & q_h E \end{pmatrix}.$$

If μ_2 is the eigenvalue of V_{11}^{c}, then the characteristics equation of V_{11}^{c} becomes $\left[(\mu_2 - q_h E)^2 - (r)^2\right]^3 = 0.$

The eigenvalues are $(q_h E + r)$, $(q_h E - r)$, $(q_h E + r)$, $(q_h E - r)$, $(q_h E + r)$, $(q_h E - r)$. Here, all theeigenvalues are positive. Therefore, the coexistence of equilibrium point E_{11c}^{*} of the transformed system (8.3) is unstable.

8.5 WHEN $\tilde{P}(t)$ IS GN-DERIVATIVE OF TYPE 2

The quota-harvesting logistic growth model (8.2) is transformed into the system of differential equation as

$$\begin{aligned} dP_{1L}(t,\alpha)/dt &= rP_{1R}(t,\alpha)\left(1-P_{1L}(t,\alpha)/K\right)-q_h EP_{1L}(t,\alpha), \\ dP_{1R}(t,\alpha)/dt &= rP_{1L}(t,\alpha)\left(1-P_{1R}(t,\alpha)/K\right)-q_h EP_{1R}(t,\alpha), \\ dP_{2L}(t,\beta)/dt &= rP_{2R}(t,\beta)\left(1-P_{2L}(t,\beta)/K\right)-q_h EP_{2L}(t,\beta), \\ dP_{2R}(t,\beta)/dt &= rP_{2L}(t,\beta)\left(1-P_{2R}(t,\beta)/K\right)-q_h EP_{2R}(t,\beta), \\ dP_{3L}(t,\gamma)/dt &= rP_{3R}(t,\gamma)\left(1-P_{3L}(t,\gamma)/K\right)-q_h EP_{3L}(t,\gamma), \\ dP_{3R}(t,\gamma)/dt &= rP_{3L}(t,\gamma)\left(1-P_{3R}(t,\gamma)/K\right)-q_h EP_{3R}(t,\gamma). \end{aligned} \tag{8.4}$$

With initial condition $P_{1L}(t_0,\alpha) = P_{01L}(\alpha)$, $P_{1R}(t_0,\alpha) = P_{01R}(\alpha)$, $P_{2L}(t_0,\beta) = P_{02L}(\beta)$, $P_{2R}(t_0,\beta) = P_{02R}(\beta)$, $P_{3L}(t_0,\gamma) = P_{03L}(\gamma)$, $P_{3R}(t_0,\gamma) = P_{03R}(\gamma)$.

The $(\alpha,\beta,\gamma)-$cut of $\tilde{P}(t)$ is given by

$\tilde{P}_{\alpha,\beta,\gamma} =< P_{1L}(t,\alpha), P_{1R}(t,\alpha); P_{2L}(t,\alpha), P_{2R}(t,\beta); P_{3L}(t,\gamma), P_{3R}(t,\gamma) >.$

8.5.1 Equilibrium Points

(i) **Trivial equilibrium:** The transformed system (8.4) has the trivial equilibrium point $E_{00}^{2}(P_{1L}(t,\alpha), P_{1R}(t,\alpha); P_{2L}(t,\alpha), P_{2R}(t,\beta); P_{3L}(t,\gamma), P_{3R}(t,\gamma))$, where $P_{1L}(t,\pm) = P_{1R}(t,\alpha) = P_{2L}(t,\alpha) = P_{2R}(t,\beta) = P_{3L}(t,\gamma) = P_{3R}(t,\gamma) = 0$.

(ii) **Coexistence equilibrium:** The transformed model (8.4) has the coexistence equilibrium point $E_{22c}^{*}(P^{*}_{1L}(t,\alpha) = P^{*}_{1R}(t,\alpha) = P^{*}_{2L}(t,\alpha) = P^{*}_{2R}(t,\beta) = P^{*}_{3L}(t,\gamma) = P^{*}_{3R}(t,\gamma) = K(r - q_h E)/r$.

The coexistence equilibrium point is feasible if $r > q_h E$.

8.5.1.1 *Stability Analysis*

Theorem 3. The transformed system (8.4) is locally asymptotically stable (LAS) at E_{00}^{2} when $q_h E > r$.

Proof. The variational matrix of transformed system (8.4) at E_{00}^{2} is given by

$$V_{00}^{2} = \begin{pmatrix} -q_h E & r & 0 & 0 & 0 & 0 \\ r & -q_h E & 0 & 0 & 0 & 0 \\ 0 & 0 & -q_h E & r & 0 & 0 \\ 0 & 0 & r & -q_h E & 0 & 0 \\ 0 & 0 & 0 & 0 & -q_h E & r \\ 0 & 0 & 0 & 0 & r & -q_h E \end{pmatrix}.$$

If μ_3 is the eigenvalue of V_{00}^{2}, then the characteristic equation of V_{00}^{2} becomes $[(\mu_3 + q_h E)^2 - r^2]^3 = 0$. Here, the eigenvalues are $-(r + q_h E), -(q_h E - r), -(r + q_h E), -(q_h E - r), -(r + q_h E), -(q_h E - r)$. Second, fourth, and sixth eigenvalues are negative if $q_h E > r$.

Therefore, system (8.4) is LAS at trivial equilibrium point E_{00}^{2} when $q_h E > r$.

Theorem 4. Transformed system (8.4) is LAS at E_{22c}^{*} when $r > q_h E$.

Proof. The variational matrix at E_{22c}^{*} of system (8.4) becomes

$$V_{22}^{c} = \begin{pmatrix} -r & q_h E & 0 & 0 & 0 & 0 \\ q_h E & -r & 0 & 0 & 0 & 0 \\ 0 & 0 & -r & q_h E & 0 & 0 \\ 0 & 0 & q_h E & -r & 0 & 0 \\ 0 & 0 & 0 & 0 & -r & q_h E \\ 0 & 0 & 0 & 0 & q_h E & -r \end{pmatrix}.$$

If μ_4 is the eigenvalue of V_{22}^{c}, then the characteristic equation of V_{22}^{c} is given by $[(\mu_4 + r)^2 - (q_h E)^2]^3 = 0$.

Table 8.1 The System Parameters

Parameters	*Values*	*Source*
K	100	[71]
E	0.005	[72]
R	0.5	[72]
q_h	10	[72]

The eigenvalues of V_{22}^{c} are $-(r+q_h E), -(r-q_h E), -(r+q_h E), -(r-q_h E), -(r+q_h E), -(r-q_h E)$.

The second, fourth, and sixth eigenvalues are negative if $r > q_h E$.

Therefore, transformed system (8.4) is LAS at the coexistence of equilibrium point E_{22c}^{*} when $r > q_h E$.

8.5.2 Numeric Simulation

In this section, we numerically verify and validate all the results and theorems of transformed systems (8.3) and (8.4) using MATLAB® software. First, we considered that the initial values of (8.3) and (8.4) were TrSVNN 3s, and we set the various collections of parameter values as follows.

8.6 WHEN $\tilde{P}(t)$ IS GN-DERIVATIVE OF TYPE I

The parameters we used in proposed system (8.3) are reported in Table 8.1; the initial population is an TrSVNN 3 as follows: $\tilde{P}(t_0) = \langle 10, 20, 30; 0.5, 0.4, 0, 0.6 \rangle$. The $(\alpha, \beta, \gamma)-$cut of $\tilde{P}(t_0)$ is given by

$$P_{1L}(\pm) = 10 + 20\alpha;\ P_{1R}(\alpha) = 30 - 20\alpha, P_{2L}(\beta) = (15 - 10\beta)/0.6;\ P_{2R}(\beta) = (5 - 10\beta)/0.6, P_{3L}(\gamma) = (14 - 10\gamma)/0.4;\ P_{3R}(\gamma) = (2 + 10\gamma)/0.4.$$

Figures 8.1–8.4 portray the values in Table 8.1 over the $(\alpha, \beta, \gamma)-$cuts of $\tilde{P}(t_0)$ for $\alpha = 0, \beta = 0.4, \gamma = 0.6$; $\alpha = 0.2, \beta = 0.6, \gamma = 0.8$; $\alpha = 0.4, \beta = 0.8, \gamma = 0.9$; $\alpha = 0.5, \beta = 1, \gamma = 1$; and $t \in [0,3]$. The figures display time-series solutions of (8.3) in a neutrosophic fuzzy environment and show that $P_{1L}(t,\alpha) \le P_{1R}(t,\alpha)$, $P_{2L}(t,\beta) \le P_{2R}(t,\beta)$, and $P_{3L}(t,\gamma) \le P_{3R}(t,\gamma)$. We verified that all solutions of (8.3) are neutrosophic solutions where $\pm \in [0,1]$, $\beta \in [0.4,1]$, $\gamma \in [0.6,1]$ for $t \in [0,3]$.

From Table 8.2, we see that $P_{1L}(t,\alpha)$ is increasing, $P_{1R}(t,\alpha)$ is decreasing, $P_{2L}(t,\beta)$ is decreasing, $P_{2R}(t,\beta)$ is increasing, $P_{3L}(t,\gamma)$ is decreasing, and $P_{3R}(t,\gamma)$ is increasing for $\alpha \in [0,1], \beta \in [0.4,1], \gamma \in [0.6,1]$ for $t = 2$. Hence, $\tilde{P}(t)$ gives strong neutrosophic solutions to transformed model (8.3); see Figure 8.5.

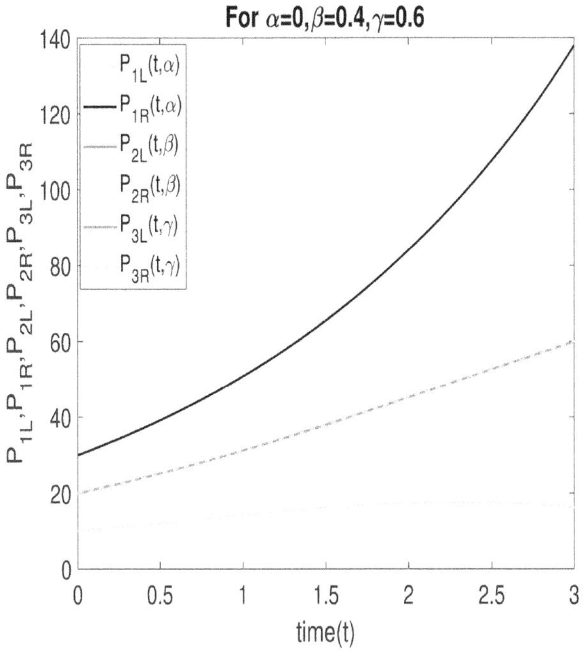

Figure 8.1 Neutrosophic solution for $\alpha = 0$, $\beta = 0.4, \gamma = 0.6$.

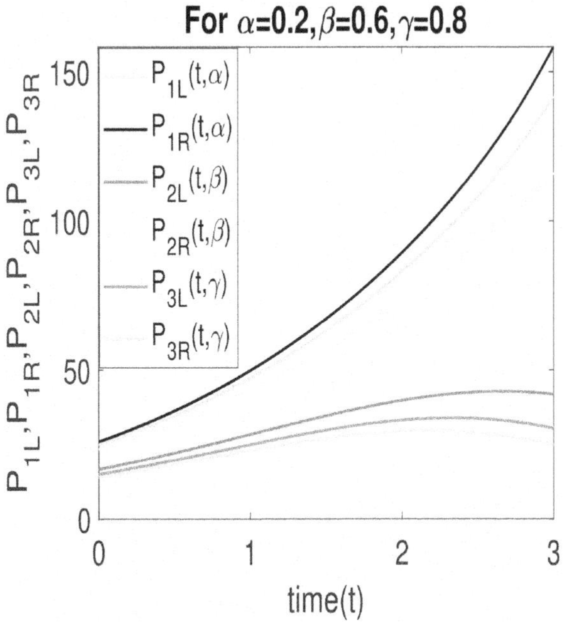

Figure 8.2 Neutrosophic solution for $\alpha = 0.2$, $\beta = 0.6$, $\gamma = 0.8$.

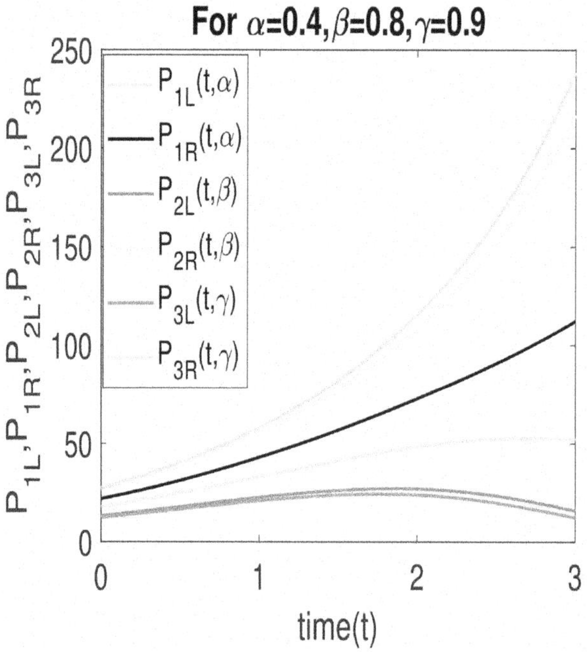

Figure 8.3 Neutrosophic solution for $\alpha = 0.4$, $\beta = 0.8$, $\gamma = 0.9$.

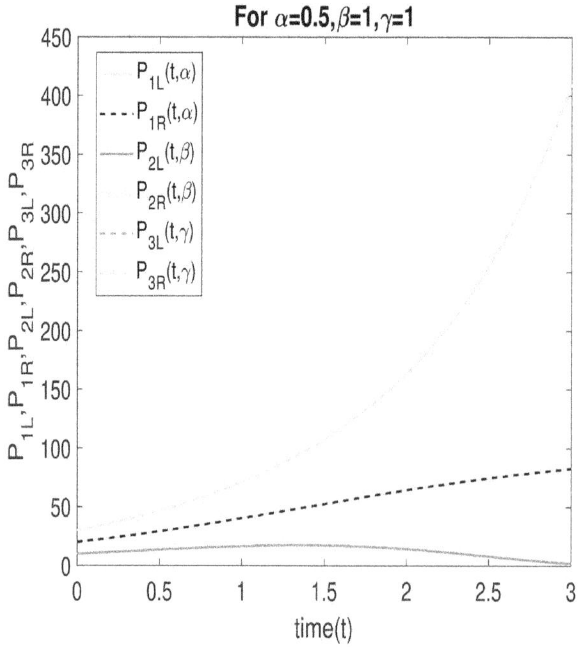

Figure 8.4 Neutrosophic solution for $\alpha = 0.5, \beta = 1, \gamma = 1$.

Table 8.2 Neutrosophic Solutions of System (8.3) for $t = 2$

α	$P_{1L}(t,\alpha)$	$P_{1R}(t,\alpha)$	β	$P_{2L}(t,\beta)$	$P_{2R}(t,\beta)$	γ	$P_{3L}(t,\gamma)$	$P_{3R}(t,\gamma)$
0	17.3524	84.0209						
0.1	22.0629	75.3977						
0.2	27.1961	67.1972						
0.3	32.7610	59.4284						
0.4	38.7641	52.0978	0.4	45.2092	45.2092			
0.5	45.2092	45.2092	0.5	39.7655	50.8880			
			0.6	34.6190	56.8863	0.6	45.2092	45.2092
			0.7	29.7691	63.2035	0.7	40.5294	60.8867
			0.8	25.2134	69.8372	0.8	33.3440	83.0729
			0.9	20.9489	76.7844	0.9	24.2901	115.3985
			1	16.9708	84.0405	1	14.8190	163.1924

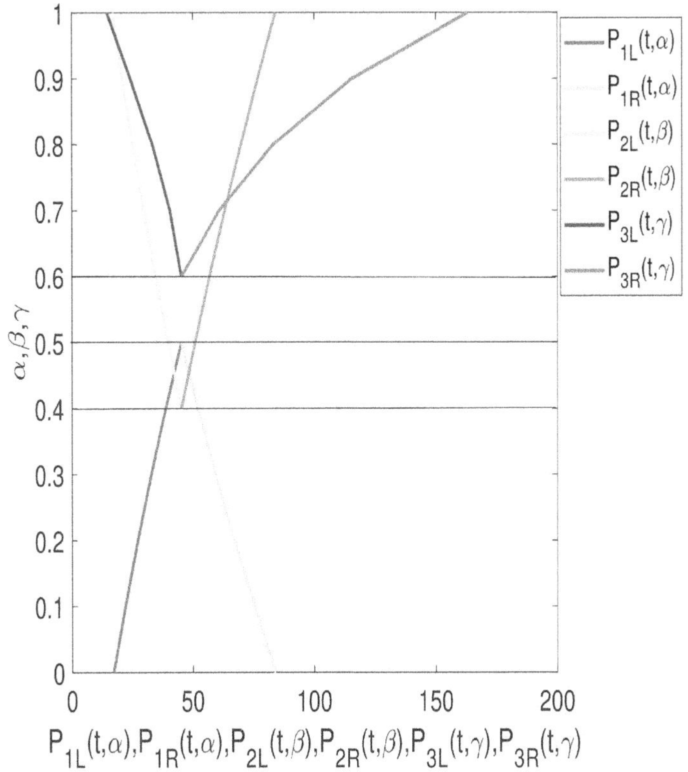

Figure 8.5 Truth, indetermnancy and falsity-membership functions for neutrosophic solutions of (8.3) for $\alpha \in [0,1], \beta \in [0.4,1], \gamma \in [0.6,1]$ at $t = 2$.

8.7 WHEN $\tilde{P}(t)$ IS GN-DERIVATIVE OF TYPE 2

In this part, we plot Figures 8.6–8.9 using the parameters reported in Table 8.3 and the same initial population of the proposed model $\tilde{P}(t_0)$ that is considered as TrS-VNN Type 3. We described the parametric form, or $(\alpha,\beta,\gamma)-$, of $\tilde{P}(t_0)$ in subsection 8.5.1 considering all the parameter values used in the transformed system (8.4) reported in Table 8.3 and under the same initial conditions in that subsection.

In Figures 8.6–8.9, we see that $P_{1L}(t,\alpha)\le P_{1R}(t,\alpha)$, $P_{2L}(t,\beta)\le P_{2R}(t,\beta)$, $P_{3L}(t,\gamma)\le P_{3R}(t,\gamma)$ for $t\in[0,15]$. Obviously, the Figures 8.6–8.9 supports that the system (8.4) is LAS (locally asymptotically stable) at E^*_{22c}.

From Table 8.4, we see that $P_{1L}(t,\alpha)$ is increasing, $P_{1R}(t,\alpha)$ is decreasing; $P_{2L}(t,\beta)$ is decreasing, $P_{2R}(t,\beta)$ is increasing; $P_{3L}(t,\gamma)$ is decreasing, and $P_{3R}(t,\gamma)$ is increasing for $\alpha\in[0,1],\beta\in[0.4,1],\gamma\in[0.6,1]$ for $t=2$. Hence, $\tilde{P}(t)$ gives a strong neutrosophic solution of transformed model (8.4).

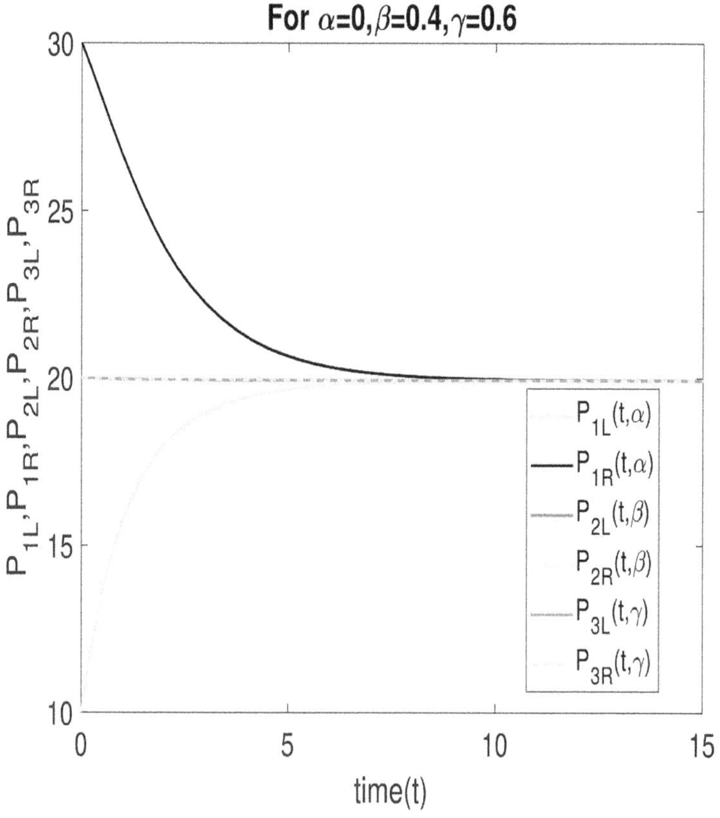

Figure 8.6 Neutrosophic solution for $\alpha=0,\beta=0.4,\gamma=0.6$.

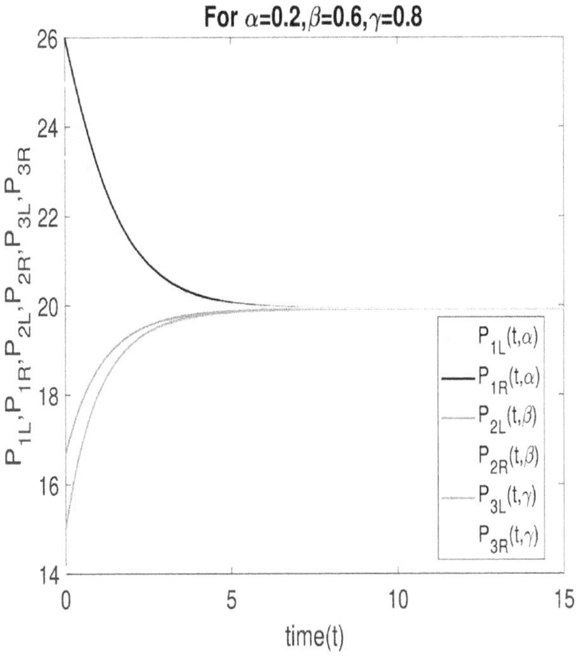

Figure 8.7 Neutrosophic solution for $\alpha = 0.2, \beta = 0.6, \gamma = 0.8$.

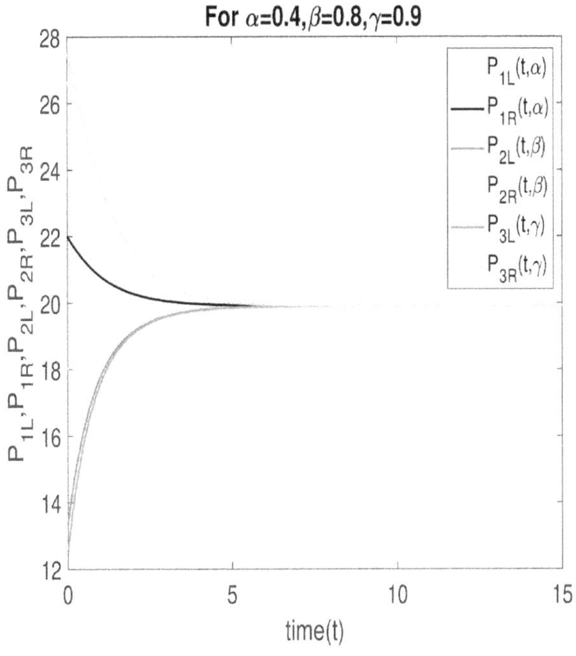

Figure 8.8 Neutrosophic solution for $\alpha = 0.4, \beta = 0.8, \gamma = 0.9$.

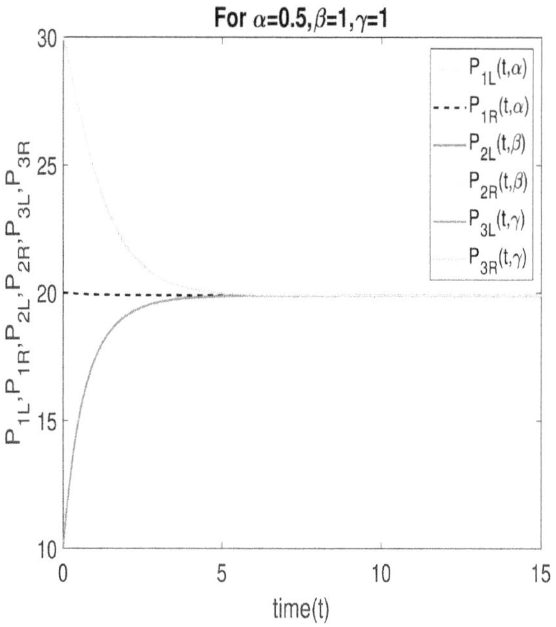

Figure 8.9 Neutrosophic solution for $\alpha = 0.5, \beta = 1, \gamma = 1$.

Table 8.3 System Parameters for TrSVNN 3

Parameters	*Values*	*Source*
K	20	[Assumed]
E	0.005	[72]
R	0.5	[72]
q_h	10	[72]

Table 8.4 Neutrosophic solution of the system (8.4) for t = 2.

α	$P_{1L}(t,\alpha)$	$P_{1R}(t,\alpha)$	β	$P_{2L}(t,\beta)$	$P_{2R}(t,\beta)$	γ	$P_{3L}(t,\gamma)$	$P_{3R}(t,\gamma)$
0	15.7076	26.6619						
0.1	16.3704	25.1338						
0.2	17.1257	23.6983						
0.3	17.9762	22.3579						
0.4	18.9239	21.1148	0.4	19.9549	19.9549			
0.5	19.9699	19.9699	0.5	19.0843	20.9091			
			0.6	18.2679	21.9157	0.6	19.9249	19.9249
			0.7	17.5200	22.9889	0.7	18.7705	21.2411
			0.8	16.8395	24.1278	0.8	18.0017	22.4725
			0.9	16.2251	25.3309	0.9	17.5646	23.6327
			1	15.6750	26.5965	1	17.3970	24.7179

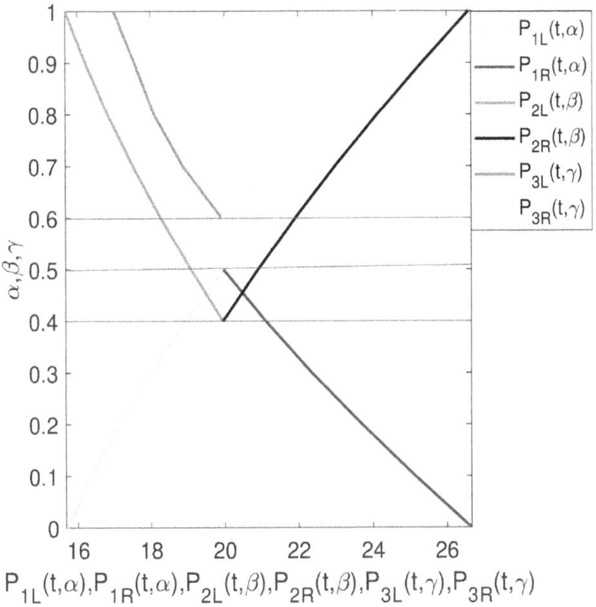

Figure 8.10 Truth, indetermnancy and falsity-membership functions for the neutrosphic solutions of (8.4) for $\alpha \in [0,1], \beta \in [0.4,1], \gamma \in [0.6,1]$ at $t = 2$.

8.8 CONCLUSION

A robust mathematical tool always plays a pivotal role in describing real-world problems in the diverse field of biology. Dealing with different model systems with the help of differential equations has added an extra dimension in recent decades, but due to parameters involving uncertainty in data, mathematical modeling is challenging today. To overcome the complexities, we envisaged a harvesting logistic growth model using neutrosophic differential equations.

In this chapter, we took initial populations as TrSVNN of type 3 to advance the study logistic model. We applied generalized Hukuhara differentiability to neutrosophic functions in the proposed model and converted the neutrosophic harvesting logistic model into different systems of classical ODEs in different (α, β, γ)-cuts form. We also performed Neutrosophic stability analysis. We explored all neutrosophic solution results for when P(t) is GN-derivative of type 1 and GN-derivative type 2, and we found strong neutrosophic solutions in the various cases. MATLAB® software assisted us immensely in verifying the mathematical outcomes of neutrosophic harvesting logistic model numerically and graphically in an elegant way. To conclude, the procedure we demonstrated in this chapter on a very simple ecological model can be applied to models using soft sets.

REFERENCES

1. Clark, C.W. Mathematical bioeconomics: the optimal management resources. *John Wiley & Sons* (1976).
2. Christensen, V. Managing fisheries involving predator and prey species. *Reviews in Fish Biology and Fisheries* 6 (1996): 417–442.
3. Hill, S.L., Murphy, E.J., Reid, K., Trathan, P.N. and Constable, A.J. Modelling Southern ocean ecosystems: Krill, the food-web, and the impacts of harvesting. *Biological Reviews* 81(4) (2006): 81–608.
4. Beddington, J.R. and Cooke, J.G. Harvesting from a prey-predator complex. *Ecological Modelling* 14(3) (1982): 155–177.
5. Beddington, J.R. and May, R.M. Maximum sustainable yields in systems subject to harvesting at more than one trophic level. *Mathematical Biosciences* 51(3) (1980): 261–281.
6. Brauer, F. and Soudack, A.C. Stability regions and transition phenomena for harvested predator-prey systems. *Journal of Mathematical Biology* 7(4) (1979): 319–337.
7. Brauer, F. and Soudack, A.C. Stability regions in predator-prey systems with constant-rate prey harvesting. *Journal of Mathematical Biology* 8 (1979): 55–71.
8. Brauer, F. and Soudack, A.C. Coexistence properties of some predator-prey systems under constant rate harvesting and stocking. *Journal of Mathematical Biology* 12 (1982): 101–114.
9. Dai, G. and Tang, M. Coexistence region and global dynamics of a harvested predator-preysystem. *SIAM Journal on Applied Mathematics* 58(1) (1998): 193–210.
10. Etoua, R.M. and Rousseau, C. Bifurcation analysis of a generalized Gause model with prey harvesting and a generalized Holling response function of type III. *Journal of Differential Equations* 249(9) (2010): 2316–2356.
11. Hogarth, W.L., Norbury, J., Cunning, I. and Sommers, K. Stability of a predator-prey model with harvesting. *Ecological Modeling* 62(1) (1992): 83–106.
12. Huang, J., Gong, Y. and Ruan, S. Bifurcation analysis in a predator-preymodel with constant-yield predator harvesting. *Discrete and Continuous Dynamical Systems-B* 18(8) (2013): 2101–2121.
13. Paul, S., Mahata, A., Mukherjee, S., et al. Study of fractional order SEIR epidemic model and effect of vaccination on the spread of COVID-19. *International Journal of Applied and Computational Mathematics* 8 (2022): 237.
14. Leard, B., Lewis, C. and Rebaza, J. Dynamics of ratio-dependent predatorprey models with non constant harvesting. *Discrete and Continuous Dynamical Systems* 1(2) (2008): 303–315.
15. Myerscough, M.R., Gray, B.F., Hogarth, W.L. and Norbury, J. An analysis of an ordinary differential equation model for a two-species predator-prey system with harvesting and stocking. *Journal of Mathematical Biology* 30 (1992): 389–411.
16. Xiao, D. and Ruan, S. Bogdanov-Takens bifurcations in predator-prey systems with constant rate harvesting. *Fields Institute Communications* 21 (1999): 493–506.
17. Mahata, A., Matia, S. N., Roy, B., Alam, S. and Sinha, H. The behaviour of logistic equation in fuzzy environment: Fuzzy differential equation approach. *International Journal of Hybrid Intelligence* 2(1) (2021): 26–46.
18. Clark, C.W. Bioeconomic modelling and fisheries management. *Wiley-Interscience* (1985).

19. Chaudhuri, K. A bioeconomic model of harvesting a multispecies fishery. *Eco-logical Modeling* 32(4) (1986): 267–279.
20. Chaudhuri, K. Dynamic optimization of combined harvesting of a two-species fishery. *Ecological Modelling* 41(1) (1988): 17–25.
21. Mesterton-Gibbons, M. On the optimal policy for combined harvesting of independent species. *Natural Resource Modeling* 2(1) (1987): 109–134.
22. Kar, T.K. and Chaudhuri, K.S. On non-selective harvesting of a multispecies fishery. *International Journal of Mathematical Education in Science and Technology* 33(4) (2002): 543–556.
23. Kar, T.K. and Chaudhuri, K.S. On non-selective harvesting of two competing fish species in the presence of toxicity. *Ecological Modelling* 161(1) (2003): 125–137.
24. Chaudhuri, K.S. and Ray, S.S. On the combined harvesting of a prey-predator system. *Journal of Biological Systems* 4(3) (1996): 373–389.
25. Kar, T.K., Pahari, U.K. and Chaudhuri, K.S. Management of a prey-predator fishery based on continuous fishing effort. *Journal of Biological Systems* 12(3) (2004): 301–313.
26. Sumathi, I.R. and Priya, V.M. A new perspective on neutrosophic differential equation. *International Journal of Engineering and Technology* 7(4) (2018): 422–425.
27. Chen, J., Huang, J., Ruan, S. and Wang, J. Bifurcations of invariant tori in predator-prey models with seasonal prey harvesting. *SIAM Journal on Applied Mathematics* 73(5) (2013): 1876–1905.
28. Hirsch, M.W., Smale, S. and Devaney, R.L. Differential equations, dynamical systems, and an introduction to chaos. *Academic Press* (2012).
29. Zadeh, L.A. Fuzzy sets. *Information and Control* 8(3) (1965): 338–353.
30. Dubois, D. and Prade, H. Operations on fuzzy numbers. *International Journal of Systems Science* 9(6) (1978): 613–626.
31. Allahviranloo, T., Kiani, N.A. and Motamedi, N. Solving fuzzy differential equations by differential transformation method. *Information Sciences* 179(7) (2009): 956–966.
32. Altaie, S.A., Jameel, A.F. and Saaban, A. Homotopy perturbation method approximate analytical solution of fuzzy partial differential equation. *IAENG Inter-national Journal of Applied Mathematics* 49(1) (2019): 22–28.
33. Jameel, A.F., Amen, S.G., Saaban, A., Man, N.H. and Alipiah, F.M. Homotopy perturbation method for solving linear fuzzy delay differential equations using double parametric approach. *Stat* 8 (2020): 551–558.
34. Ettoussi, R., Melliani, S., Elomari, M. and Chadli, L.S. Solution of intu- itionistic fuzzy differential equations by successive approximations method. *Notes on Intuitionistic Fuzzy Sets* 21(2) (2015): 51–62.
35. Salahshour, S., Ahmadian, A. and Mahata, A. The behavior of logistic equation with alley effect in fuzzy environment: Fuzzy differential equation approach. *International Journal of Computational Mathematics* 4 (2018).
36. Mahata, A., Roy, B., Mondal, S. P. and Alam, S. Application of ordinary differential equation in glucose-insulin regulatory system modeling in fuzzy environment. *Ecological Genetics and Genomics* 3 (2017): 60–66.
37. Mahata, A., Mondal, S. P., Alam, S., Chakraborty, A., De, S. K. and Goswami, A. Mathematical model for diabetes in fuzzy environment with stability analysis. *Journal of Intelligent and Fuzzy Systems* 36(3) (2019): 2923–2932.

38. Atanassov, K.T. and Atanassov, K.T. Intuitionistic fuzzy sets. *Physica-Verlag HD* (1999): 1–137.
39. Atanassov, K.T. Intuitionistic fuzzy sets VII ITKR's session. *Sofia*, 1 (1983).
40. Atanassov, K.T. Operators over interval valued intuitionistic fuzzy sets. *Fuzzy Sets and Systems* 64(2) (1994): 159–174.
41. Atanassov, K. Intuitionistic fuzzy sets. *Fuzzy Sets and Systems* 20(1) (1986): 87–96.
42. Melliani, S., Elomari, M., Chadli, L.S. and Ettoussi, R. Intuitionistic fuzzy metric space. *Notes on Intuitionistic Fuzzy Sets* 21(1) (2015): 43–53.
43. Melliani, S., Elomari, M., Atraoui, M. and Chadli, L.S. Intuitionistic fuzzy differential equation with nonlocal condition. *Notes on Intuitionistic Fuzzy Sets* 21(4) (2015): 58–68.
44. Wan, S.P., Lin, L.L. and Dong, J.Y. MAGDM based on triangular Atanassov's intuitionistic fuzzy information aggregation. *Neural Computing and Applications* 28 (2017): 2687–2702.
45. Wan, S.P., Wang, F. and Dong, J.Y. Theory and method of intuitionistic fuzzy preference relation group decision making (2019).
46. Xu, J., Dong, J.Y., Wan, S.P. and Gao, J. Multiple attribute decision making with triangular intuitionistic fuzzy numbers based on zero-sum game approach. *Iranian Journal of Fuzzy Systems* 16(3) (2019): 97–112.
47. Wan, S. and Dong, J. Decision making theories and methods based on intervalvalued intuitionistic fuzzy sets. *Springer Nature* (2020).
48. Liu, F. and Yuan, X. Fuzzy number intuitionistic fuzzy set. *Fuzzy Systems and Mathematics* 21(1) (2007): 88–91.
49. Ye, J. Prioritized aggregation operators of trapezoidal intuitionistic fuzzy sets and their application to multicriteria decision-making. *Neural Computing and Applications* 25 (2014): 1447–1454.
50. Smarandache, F. Neutrosophy: Neutrosophic probability, set, and logic: Analytic synthesis & synthetic analysis. *American Research Press* (1998).
51. Smarandache, F. First international conference on neutrosophy, neutrosophic logic, set, probability and statistics. *Florentin Smarandache* 4 (2001).
52. Smarandache, F. May. Neutrosophic set-a generalization of the intuitionistic fuzzy set. *International Journal of Pure and Applied Mathematics* 24(3) (2005): 287–297.
53. Wang, H., Smarandache, F., Zhang, Y. and Sunderraman, R. Single valued neutrosophic sets. *Infinite Study* 12 (2010).
54. Ye, J. Single-valued neutrosophic minimum spanning tree and its clustering method. *Journal of Intelligent Systems* 23(3) (2014): 311–324.
55. Peng, J.J., Wang, J.Q., Wu, X.H., Zhang, H.Y. and Chen, X.H. The fuzzy cross entropy intuitionistic indeterminacy fuzzy sets and their application in multi-criteria decision making. *International Journal of Systems Science* 46(13) (2014): 2335–2350.
56. Peng, J.J., Wang, J.Q., Wang, J., Zhang, H.Y. and Chen, X.H. Simplified neutrosophic sets and their applications in multi-criteria group decision-making problems. *International Journal of Systems Science* 47(10) (2016): 2342–2358.
57. Chakraborty, A., Mondal, S., Ahmadian, A., et al. Different forms of triangular neutrosophic numbers, de-neutrosophication techniques, and their applications. *Symmetry* 10 (2018): 327.
58. Chakraborty, A., Mondal, S., Ahmadian, A., et al. Disjunctive representation of triangular bipolar neutrosophic numbers, de-bipolarization technique and application in multi-criteria decision-making problems. *Symmetry* 11(7) (2019): 932.

59. Deli, I., Ali, M. and Smarandache, F. Bipolar neutrosophic sets and their application based on multi-criteria decision making problems. In 2015 International conference on advanced mechatronic systems (ICAMechS), Beijing, China (2015): 249–254.
60. Lee, K.M. Bipolar-valued fuzzy sets and their operations. In Proceedings of international conference on intelligent technologies, Bangkok, Thailand (2000): 307–312.
61. Kang, M.K. and Kang, J.G. Bipolar fuzzy set theory applied to sub-semigroups with operators in semigroups. *The Pure and Applied Mathematics* 19(1) (2012): 23–35.
62. Smarandache, F. Neutrosophic perspectives: Triplets, duplets, multisets, hybrid operators, modal logic, hedge algebras. And applications. *Pons Publishing House* (2017).
63. Saini, R.K., Sangal, A. and Manisha, M. Application of single valued trapezoidal neutrosophic numbers in transportation problem. *Neutrosophic Sets and Systems* 35(1) (2020): 33.
64. Chakraborty, A., Mondal, S.P., Alam, S. and Mahata, A. Cylindrical neutrosophic single-valued number and its application in networking problem, multicriterion group decision-making problem and graph theory. *CAAI Transactions on Intelligence Technology* 5(2) (2020): 68–77.
65. Haque, T.S., Chakraborty, A., Mondal, S.P. and Alam, S. New exponential operational law for measuring pollution attributes in mega-cities based on MCGDM problem with trapezoidal neutrosophic data. *Journal of Ambient Intelligence and Humanized Computing* 13(12) (2022): 5591–5608.
66. Banik, B., Alam, S. and Chakraborty, A. Comparative study between GRA and MEREC technique on an agricultural-based MCGDM problem in pentagonal neutrosophic environment. *International Journal of Environmental Science and Technology* (2023): 1–16.
67. Biswas, S., Moi, S. and Sarkar, S.P. Numerical integration of neutrosophic valued function by Gaussian quadrature methods. *Arabian Journal of Mathematics* 11(2) (2022): 189–211.
68. Moi, S., Biswas, S. and Pal, S. Second-order neutrosophic boundary-value problem. *Complex & Intelligent Systems* 7 (2021): 1079–1098.
69. Moi, S., Biswas, S. and Sarkar, S.P. An efficient method for solving neutrosophic Fredholm integral equations of second kind. *Granular Computing* 8(1) (2023): 1–22.
70. Parikh, M. and Sahni, M. Sumudu transform for solving second order ordinary differential equation under neutrosophic initial conditions. *Infinite Study* 38 (2020).
71. Paul, S., Mondal, S.P. and Bhattacharya, P. Discussion on fuzzy quota harvesting model in fuzzy environment: Fuzzy differential equation approach. *Modeling Earth Systems and Environment* 2 (2016): 70.
72. Paul, S., Mondal, S.P. and Bhattacharya, P. Discussion on proportional harvesting model in fuzzy environment: fuzzy differential equation approach. *International Journal of Applied and Computational Mathematics* (2016).

Chapter 9

Pentapartitioned Neutrosophic Weighted Geometric Bonferroni Mean Operators in Multiple-Attribute Decision Making

R. Radha, M. Kishorekumar, S. Krishnaprakash, Said Broumi, and S. A. Edalatpanah

9.1 INTRODUCTION

In 1968, Zadeh [1] introduced the concept of a fuzzy set, which has many applications in medical diagnosis, signal processing, image processing, robotics, decision-making, business, economy, data mining, etc. Based on this concept, many studies have been conducted in general theoretical areas and for different applications. Moreover, the Bonferroni mean, first introduced by Bonferroni [2], is one of the aggregation operators. It is composed of arithmetic and the product and has desirable properties that capture the interrelationships among arguments.

The Bonferroni mean has been identified as a useful averaging aggregation function with the potential for interesting applications in fuzzy systems and multicriteria decision making and has attracted much attention from researchers. Xu and Yager [3] extended the Bonferroni mean to accommodate the intuitionistic fuzzy environment and developed an intuitionistic fuzzy Bonferroni mean. The aggregation operators are powerful tools for multi-attribute group decision-making problems. In this paper, we applied the generalized and weighted geometric Bonferroni mean operators to a pentapartitioned neutrosophic environment and investigated their properties in detail. Finally, we give a practical example for groundnut yield (GY) to verify the developed approach and demonstrate its practicality and effectiveness.

Wang's single-valued neutrosophic sets are still being used to improve pentapartitioned neutrosophic sets. Rama Malik et al. (2020) proposed the idea of pentapartitioned single-valued neutrosophic sets (PSVNSs) that relies on Belnap's five valued logic and Smarandache's five numerical valued logic. In PSVNSs, indeterminacy is split into three functions: contradicition (both true and false), ignorance (neither true nor false), and unknown, which gives the PSVNS five parts, T, C, U, F, and G, that additionally lie within the nonnormal unit interval [0, 1].

DOI: 10.1201/9781003487104-9

9.2 PRELIMINARIES

Definition 9.2.1 [2] Let $x, y \geq 0$, and $u_i \left(i = 1,2,3,\ldots\ldots,n\right)$ be a collection of non-negative numbers. If

$$B^{x.y}\left(u_1,u_2,\ldots\ldots\ldots,u_n\right) = \left(\frac{1}{n\left(n-1\right)}\sum_{\substack{i,j=1\\ i\neq j}}^{n} u_i^x u_j^y\right)^{\frac{1}{x+y}},$$

then $B^{x.y}$ is called the Bonferroni mean.

Definition 9.2.2.[4] Let $x, y \geq 0$, and $u_i \left(i = 1,2,3,\ldots\ldots,n\right)$ be a collection of non-negative numbers. If

$$GB^{x.y}\left(u_1,u_2,\ldots\ldots\ldots,u_n\right) = \frac{1}{x+y}\prod_{\substack{i,j=1\\ i\neq j}}^{n}\left(xu_i + yu_j\right)^{\frac{1}{n(n-1)}},$$

then $GB^{x,y}$ is called the geometric Bonferroni mean.

Definition 9.2.3. Let P be a nonempty set. A pentapartitioned neutrosophic set A over P characterizes each element p in P a truth membership function t, a contradiction membership function c, an ignorance membership function g, an unknown membership function n and a false membership function f, such that for each p in P

$$t + c + g + n + f \leq 5.$$

9.3 PENTAPARTITIONED NEUTROSOPHIC GEOMETRIC BONFERRONI MEAN

Definition 9.3.1 Let $x, y \geq 0$ and $u_i \left(i = 1,2,\ldots\ldots\ldots,n\right)$ be a collection of nonnegative numbers and $W = \left(w_1, w_2,\ldots\ldots\ldots\ldots, w_n\right)$ is the weight vector of $u_i \left(i = 1,2,\ldots\ldots,n\right)$ and satisfies $w_i \geq 0, \sum_{i=1}^{n} w_i = 1$, if

$$PNPBM^{x,y}\left(u_1,u_2,\ldots\ldots\ldots,u_n\right) = \frac{1}{x+y}\left(\prod_{\substack{i,j=1\\ i\neq j}}^{n}\left(xu_i \oplus yu_j\right)^{\frac{1}{n(n-1)}}\right)$$

Then $PNBM^{x,y}$ is called the pentapartitioned neutrosophic geometric Bonferroni mean (PNGBM).

Theorem 9.3.2 The aggregated value using the PNGBM operator is also a pentapartitioned neutrosophic number (PNN) in which
PNB$M^{x,y}$ $(u_1, u_2, \ldots$

$$PNGBM^{x,y}\left(u_1,u_2,\ldots\ldots\ldots,u_n\right)=\left(\begin{array}{c}
\left(1-\left(\prod_{\substack{i,j=1\\ i\neq j}}^{n}\left(1-t_{u_i}\right)^x\left(1-t_{u_j}\right)^y\right)^{\frac{1}{n(n-1)}}\right)^{\frac{1}{x+y}}, \\
\left(1-\left(\prod_{\substack{i,j=1\\ i\neq j}}^{n}\left(1-c_{u_i}\right)^x\left(1-c_{u_j}\right)^y\right)^{\frac{1}{n(n-1)}}\right)^{\frac{1}{x+y}}, \\
1-\left(1-\prod_{\substack{i,j=1\\ i\neq j}}^{n}\left(1-\left(g_{u_i}\right)^x\left(g_{u_j}^{U}\right)^y\right)^{\frac{1}{n(n-1)}}\right)^{\frac{1}{x+y}} \\
1-\left(1-\prod_{\substack{i,j=1\\ i\neq j}}^{n}\left(1-\left(n_{u_i}\right)^x\left(n_{u_j}\right)^y\right)^{\frac{1}{n(n-1)}}\right)^{\frac{1}{x+y}}, \\
1-\left(1-\prod_{\substack{i,j=1\\ i\neq j}}^{n}\left(1-\left(f_{u_i}\right)^x\left(f_{u_i}\right)^y\right)^{\frac{1}{n(n-1)}}\right)^{\frac{1}{x+y}}
\end{array}\right)$$

where $x, y \geq 0$ and $I = [0,1]$.

Proof. By PNN operational laws, we have

$$xu_i=\left(\begin{array}{c}
1-\left(1-t_{u_i}\right)^x, \\
1-\left(1-c_{u_i}\right)^x, \\
(g_{u_i})^x, \\
(n_{u_i})^x, \\
(f_{u_i})^x,
\end{array}\right) \quad and \quad yu_j=\left(\begin{array}{c}
1-\left(1-t_{u_j}\right)^y, \\
1-\left(1-c_{u_j}\right)^y, \\
(g_{u_j})^y, \\
(n_{u_j})^y, \\
(f_{u_j})^y,
\end{array}\right)$$

Now,

$$xu_i \oplus yu_j = \begin{pmatrix} 1-\left(1-t_{u_i}\right)^x\left(1-t_{u_j}\right)^y, \\ 1-\left(1-c_{u_i}\right)^x\left(1-c_{u_j}\right)^y, \\ \left(g_{u_i}\right)^x\left(g_{u_j}\right)^y, \\ \left(n_{u_i}\right)^x\left(n_{u_j}\right)^y, \\ \left(f_{u_i}\right)^x\left(f_{u_j}\right)^y \end{pmatrix}$$

and

$$\left(xu_i \oplus yu_j\right)^{\frac{1}{n(n-1)}} = \begin{pmatrix} \left(1-\left(1-t_{u_i}\right)^x\left(1-t_{u_i}\right)^y\right)^{\frac{1}{n(n-1)}}, \\ \left(1-\left(1-c_{u_i}\right)^x\left(1-c_{u_j}\right)^y\right)^{\frac{1}{n(n-1)}}, \\ 1-\left(1-\left(g_{u_i}\right)^x\left(g_{u_j}\right)^y\right)^{\frac{1}{n(n-1)}}, \\ 1-\left(1-\left(n_{u_i}\right)^x\left(n_{u_j}\right)^y\right)^{\frac{1}{n(n-1)}}, \\ 1-\left(1-\left(f_{u_i}\right)^x\left(f_{u_j}\right)^y\right)^{\frac{1}{n(n-1)}} \end{pmatrix}$$

Therefore,

$$\prod_{\substack{i,j=1 \\ i\neq j}}^{n}\left(xu_i \oplus yu_j\right)^{\frac{1}{n(n-1)}} = \begin{pmatrix} \prod_{\substack{i,j=1 \\ i\neq j}}^{n}\left(1-\left(1-t_{u_i}\right)^x\left(1-t_{u_j}\right)^y\right)^{\frac{1}{n(n-1)}}, \\ \prod_{\substack{i,j=1 \\ i\neq j}}^{n}\left(1-\left(1-c_{u_i}\right)^x\left(1-c_{u_j}\right)^y\right)^{\frac{1}{n(n-1)}}, \\ 1-\prod_{\substack{i,j=1 \\ i\neq j}}^{n}\left(1-\left(g_{u_i}\right)^x\left(g_{u_j}\right)^y\right)^{\frac{1}{n(n-1)}}, \\ 1-\prod_{\substack{i,j=1 \\ i\neq j}}^{n}\left(1-\left(n_{u_i}\right)^x\left(n_{u_j}\right)^y\right)^{\frac{1}{n(n-1)}}, \\ 1-\prod_{\substack{i,j=1 \\ i\neq j}}^{n}\left(1-\left(g_{u_i}\right)^x\left(g_{u_j}\right)^y\right)^{\frac{1}{n(n-1)}} \end{pmatrix}$$

Hence, $PNGBM^{x,y}(u_1,u_2,\ldots\ldots\ldots,u_n) = \frac{1}{x+y}\left(\prod_{\substack{i,j=1\\ i\neq j}}^{n}\left(xu_i \oplus yu_j\right)^{\frac{1}{n(n-1)}}\right)$

$$= \left(\begin{array}{c} 1-\left(1-\prod_{\substack{i,j=1\\ i\neq j}}^{n}\left(1-\left(1-t_{u_i}\right)^x\left(1-t_{u_j}\right)^y\right)^{\frac{1}{n(n-1)}}\right)^{\frac{1}{x+y}}, \\ 1-\left(1-\prod_{\substack{i,j=1\\ i\neq j}}^{n}\left(1-\left(1-c_{u_i}\right)^x\left(1-c_{u_j}\right)^y\right)^{\frac{1}{n(n-1)}}\right)^{\frac{1}{x+y}}, \\ \left(1-\prod_{\substack{i,j=1\\ i\neq j}}^{n}\left(1-\left(g_{u_i}\right)^x\left(g_{u_j}\right)^y\right)^{\frac{1}{n(n-1)}}\right)^{\frac{1}{x+y}}, \\ \left(1-\prod_{\substack{i,j=1\\ i\neq j}}^{n}\left(1-\left(n_{u_i}\right)^x\left(n_{u_j}\right)^y\right)^{\frac{1}{n(n-1)}}\right)^{\frac{1}{x+y}}, \\ \left(1-\prod_{\substack{i,j=1\\ i\neq j}}^{n}\left(1-\left(f_{u_i}\right)^x\left(f_{u_j}\right)^y\right)^{\frac{1}{n(n-1)}}\right)^{\frac{1}{x+y}} \end{array}\right).$$

The theorem is proved.

It can be easily proven that the PNGBM operator has the following properties.

Property 1. (Idempotency property) If all $u_j\left(j=1,2,\ldots\ldots,n\right)$ are equal, *i.e.*,$u_j = u$ for all $j, u \geq 0, v \geq 0$ do not take the value 0 simultaneously, then $PNGBM^{x,y}(u_1,u_2,\ldots\ldots\ldots,u_n) = u$.

Property 2. (Boundary property) Let $u_j\left(j=1,2,\ldots\ldots,n\right)$ be a collection of PNNs, and let $u^- = min_j u_j, u^+ = max_j u_j$; then $u^- \leq PNGBM^{x,y}(u_1,u_2,\ldots\ldots\ldots,u_n) \leq u^+$.

Property 3. (Monotonicity property) Let $j\left(j=1,2,\ldots\ldots,n\right)$ and $u'_j\left(j=1,2,\ldots\ldots,n\right)$ be a two set of PNNs; if $u_j \leq u'_j$, for all j, then $PNGBM^{x,y}(u_1,u_2,\ldots\ldots\ldots,u_n) \leq PNGBM^{x,y}(u'_1,u'_2,\ldots\ldots\ldots,u'_n)$.

Property 4. (Permutation property) Let $u_j\left(j=1,2,\ldots\ldots,n\right)$ be a collection of PNNs. Then

$$PNGBM^{x,y}(u_1,u_2,\ldots\ldots\ldots,u_n) = PNBM^{x,y}\left(\tilde{\dot{u}}_1, \tilde{\dot{u}}_2,\ldots\ldots\ldots,\tilde{\dot{u}}_n\right)$$

where $\left(\tilde{\dot{u}}_1, \tilde{\dot{u}}_2, \ldots\ldots\ldots, \tilde{\dot{u}}_n\right)$ is a permutation of $(u_1, u_2, \ldots\ldots\ldots, u_n)$

The following are some particular cases of the PNGBM that can be obtained by giving different values to the parameters x and y:

Case 1. If $x = y = \frac{1}{2}$, then PNGBM is reduced to

$$PNGBM^{\frac{1}{2},\frac{1}{2}}(u_1, u_2, \ldots\ldots\ldots, u_n)$$

$$= \left(\begin{array}{c} 1-\left(1-\prod_{\substack{i,j=1 \\ i\neq j}}^{n}\left(1-\sqrt{\left(1-t_{u_i}\right)\left(1-t_{u_j}\right)}\right)^{\frac{1}{n(n-1)}}\right), \\ 1-\left(1-\prod_{\substack{i,j=1 \\ i\neq j}}^{n}\left(1-\sqrt{\left(1-c_{u_i}\right)\left(1-c_{u_j}\right)}\right)^{\frac{1}{n(n-1)}}\right), \\ \left(1-\prod_{\substack{i,j=1 \\ i\neq j}}^{n}\left(1-\sqrt{\left(g_{u_i}\right)\left(g_{u_j}\right)}\right)^{\frac{1}{n(n-1)}}\right), \\ \left(1-\prod_{\substack{i,j=1 \\ i\neq j}}^{n}\left(1-\sqrt{\left(n_{u_i}\right)\left(n_{u_j}\right)}\right)^{\frac{1}{n(n-1)}}\right), \\ \left(1-\prod_{\substack{i,j=1 \\ i\neq j}}^{n}\left(1-\sqrt{\left(f_{u_i}\right)\left(f_{u_j}\right)}\right)^{\frac{1}{n(n-1)}}\right) \end{array} \right)$$

which we call the pentapartitioned neutrosophic geometric Bonferroni mean.
Case 2. If $y \to 0$, then PNGBM is reduced to $\lim_{y\to 0} PNGBM^{x,y}(u_1, u_2, \ldots\ldots\ldots, u_n)$

$$= \begin{pmatrix} 1-\left(1-\prod_{\substack{i,j=1\\ i\neq j}}^{n}\left(1-\left(1-t_{u_i}\right)^{x}\right)^{\frac{1}{n}}\right)^{\frac{1}{x}}, \\ 1-\left(1-\prod_{\substack{i,j=1\\ i\neq j}}^{n}\left(1-\left(1-c_{u_i}\right)^{x}\right)^{\frac{1}{n}}\right)^{\frac{1}{x}}, \\ \left(1-\prod_{\substack{i,j=1\\ i\neq j}}^{n}\left(1-\left(g_{u_i}\right)^{x}\right)^{\frac{1}{n}}\right)^{\frac{1}{x}}, \\ \left(1-\prod_{\substack{i,j=1\\ i\neq j}}^{n}\left(1-\left(n_{u_i}\right)^{x}\right)^{\frac{1}{n}}\right)^{\frac{1}{x}} \\ \left(1-\prod_{\substack{i,j=1\\ i\neq j}}^{n}\left(1-\left(f_{u_i}\right)^{x}\right)^{\frac{1}{n}}\right)^{\frac{1}{x}} \end{pmatrix} = PNGBM^{x,0}\left(u_1,u_2,\ldots\ldots\ldots,u_n\right)$$

which is the PNGBM.

Case 3. If $x = 2, y \rightarrow 0$, then PNGBM is reduced to

$$PNGHM^{2,0}\left(u_1,u_2,\ldots\ldots\ldots,u_n\right) = \begin{pmatrix} 1-\left(1-\prod_{\substack{i,j=1\\ i\neq j}}^{n}\left(1-\left(1-t_{u_i}\right)^{2}\right)^{\frac{1}{n}}\right)^{\frac{1}{2}}, \\ 1-\left(1-\prod_{\substack{i,j=1\\ i\neq j}}^{n}\left(1-\left(1-c_{u_i}\right)^{2}\right)^{\frac{1}{n}}\right)^{\frac{1}{2}}, \\ \left(1-\prod_{\substack{i,j=1\\ i\neq j}}^{n}\left(\left(1-g_{u_i}\right)^{2}\right)^{\frac{1}{n}}\right)^{\frac{1}{2}}, \\ \left(1-\prod_{\substack{i,j=1\\ i\neq j}}^{n}\left(\left(1-n_{u_i}\right)^{2}\right)^{\frac{1}{n}}\right)^{\frac{1}{2}}, \\ \left(1-\prod_{\substack{i,j=1\\ i\neq j}}^{n}\left(\left(1-f_{u_i}\right)^{2}\right)^{\frac{1}{n}}\right)^{\frac{1}{2}} \end{pmatrix}$$

This is called the pentapartitioned neutrosophic square geometric mean.

Case 4. If $x = 1, y \to 0$, then PNGBM is reduced to the pentapartitioned neutrosophic average:

$$PNGBM^{1,0}\left(u_1, u_2, \ldots\ldots\ldots, u_n\right)$$

$$= \begin{pmatrix} \prod_{\substack{i,j=1 \\ i \neq j}}^{n} \left(t_{u_i}\right)^{\frac{1}{n}}, \\ \prod_{\substack{i,j=1 \\ i \neq j}}^{n} \left(c_{u_i}\right)^{\frac{1}{n}}, \\ 1 - \prod_{\substack{i,j=1 \\ i \neq j}}^{n} \left(1 - g_{u_i}\right)^{\frac{1}{n}}, \\ 1 - \prod_{\substack{i,j=1 \\ i \neq j}}^{n} \left(1 - n_{u_i}\right)^{\frac{1}{n}}, \\ 1 - \prod_{\substack{i,j=1 \\ i \neq j}}^{n} \left(1 - u_{u_i}\right)^{\frac{1}{n}} \end{pmatrix}$$

Case 5. If $x \to \infty,\ y = 0$, then PNPGBM is reduced to

$$\lim_{x \to \infty} PNPGBM^{x,0}\left(u_1, u_2, \ldots\ldots\ldots, u_n\right) = \max_i u_i.$$

Case 6. If $x \to 0, y = 0$, then PNPGBM is reduced to

$$\lim_{x \to 0} PNPGBM^{x,0}\left(u_1, u_2, \ldots\ldots\ldots, u_n\right) = \lim_{x \to 0} \left(\frac{1}{n} \bigotimes_{i=1}^{n} \left(u_i\right)^x \right)^{\frac{1}{x}} = \bigotimes_{i=1}^{n} \left(u_i\right)^{\frac{1}{n}} = \left(\bigotimes_{i=1}^{n} u_i \right)^{\frac{1}{n}}$$

which is the pentapartitioned neutrosophic geometric mean operator.

Case 7. If $x = y = 1$, then PNGBM is reduced to

$$PNPGBM^{1,1}\left(u_1,u_2,\ldots\ldots\ldots,u_n\right)=\left(\begin{array}{c}\left(1-\prod_{\substack{i,j=1\\i\neq j}}^{n}\left(1-\left(1-t_{u_i}\right)\left(1-t_{u_j}\right)\right)^{\frac{1}{n(n-1)}}\right)^{\frac{1}{2}},\\ \left(1-\prod_{\substack{i,j=1\\i\neq j}}^{n}\left(1-\left(1-c_{u_i}\right)\left(1-c_{u_j}\right)\right)^{\frac{1}{n(n-1)}}\right)^{\frac{1}{2}},\\ \left(1-\prod_{\substack{i,j=1\\i\neq j}}^{n}\left(1-g_{u_i}g_{u_j}\right)^{\frac{1}{n(n-1)}}\right)^{\frac{1}{2}},\\ \left(1-\prod_{\substack{i,j=1\\i\neq j}}^{n}\left(1-n_{u_i}n_{u_j}\right)^{\frac{1}{n(n-1)}}\right)^{\frac{1}{2}},\\ \left(1-\prod_{\substack{i,j=1\\i\neq j}}^{n}\left(1-f_{u_i}f_{u_j}\right)^{\frac{1}{n(n-1)}}\right)^{\frac{1}{2}}\end{array}\right)$$

This is called a pentapartitioned neutrosophic interrelated square geometric mean.

9.4 PENTAPARTIITONED NEUTROSOPHIC WEIGHTED GEOMETRIC BONFERRONI MEAN

Definition 9.4.1. Let $x, y \geq 0$, and $u_j\left(j=1,2,\ldots\ldots,n\right)$ be a collection of PNNs; $w=\left(w_1,w_2,\ldots\ldots\ldots\ldots,w_n\right)^T$ is the weight vector of $u_j\left(j=1,2,\ldots\ldots,n\right)$, where w_j indicates the importance degree of u_j satisfying $w_j>0, j=1,2,3,\ldots\ldots\ldots,n$ and $\sum_{j=1}^{n} w_j=1$. Then the aggregated value using the PNWGBM is also an PNN, and $\left(u_1,u_2,\ldots\ldots\ldots,u_n\right)=$

$$\frac{1}{x+y}\left(\prod_{\substack{i,j=1\\i\neq j}}^{n}\left(xu_i^{w_i}\oplus yu_j^{w_j}\right)^{\frac{1}{n(n-1)}}\right).$$

Theorem 9.4.2. The aggregated value using the PNWGBM operator is also a PNN, where

$$PNWGBM^{x,y}\left(u_1,u_2,\ldots\ldots\ldots,u_n\right)$$

$$=\left(\begin{array}{l}
1-\left(1-\prod_{\substack{i,j=1\\ i\neq j}}^{n}\left(1-\left(1-\left(t_{u_i}\right)^{w_i}\right)^{x}\left(1-\left(t_{u_j}\right)^{w_j}\right)^{y}\right)^{\frac{1}{n(n-1)}}\right)^{\frac{1}{x+y}}, \\
1-\left(1-\prod_{\substack{i,j=1\\ i\neq j}}^{n}\left(1-\left(1-\left(c_{u_i}\right)^{w_i}\right)^{x}\left(1-\left(c_{u_j}\right)^{w_j}\right)^{y}\right)^{\frac{1}{n(n-1)}}\right)^{\frac{1}{x+y}}, \\
\left(1-\prod_{\substack{i,j=1\\ i\neq j}}^{n}\left(1-\left(1-\left(1-g_{u_i}\right)^{w_i}\right)^{x}\left(1-\left(1-g_{u_i}\right)^{w_j}\right)^{y}\right)^{\frac{1}{n(n-1)}}\right)^{\frac{1}{x+y}}, \\
\left(1-\prod_{\substack{i,j=1\\ i\neq j}}^{n}\left(1-\left(1-\left(1-n_{u_i}\right)^{w_i}\right)^{x}\left(1-\left(1-n_{u_j}\right)^{w_j}\right)^{y}\right)^{\frac{1}{n(n-1)}}\right)^{\frac{1}{x+y}}, \\
\left(1-\prod_{\substack{i,j=1\\ i\neq j}}^{n}\left(1-\left(1-\left(1-f_{u_i}\right)^{w_i}\right)^{x}\left(1-\left(1-f_{u_i}\right)^{w_j}\right)^{y}\right)^{\frac{1}{n(n-1)}}\right)^{\frac{1}{x+y}}
\end{array}\right).$$

9.5 AN APPROACH TO PNWGBM IN MULTI-ATTRIBUTE DECISION-MAKING

Table 9.1 lists the linguistic terms we apply in the model, and Table 9.2 presents the linguistic terms with the specific problem attributes. A set of alternatives in GY $\{U_1,U_2,U_3,U_4\}$ needs to be estimated from lists of lands. GY has a set of finite

Table 9.1 Linguistic Terms

Linguistic Term		*t*	*c*	*g*	*n*	*f*
Very Very High	VVH	1.0	1.0	0.00	0	0.0
Very High	VH	0.912	0.823	0.190	0.289	0.378
High	H	0.823	0.734	0.289	0.378	0.467
Medium High	MH	0.734	0.645	0.378	0.467	0.556
Medium	M	0.645	0.556	0.467	0.556	0.645
Medium Low	ML	0.556	0.467	0.556	0.645	0.734
Low	L	0.467	0.378	0.645	0.734	0.823
Very Low	VL	0.378	0.289	0.734	0.823	0.912
Very Very Low	VVL	0.000	0.000	0.823	0.912	1.000

Table 9.2 Linguistic Terms and Problem Attributes

	Soil (V_1)	*Water (V_2)*	*Fertilizer (V_3)*	*Environment (V_4)*
U1	VH	VH	H	M
U2	VVH	H	L	MH
U3	H	VVH	H	M
U4	M	H	M	MH

attributes denoted as $\{V_1, V_2, V_3, V_4\}$ in which V_1 represents soil preparation, V_2 represents the water system, V_3 represents fertilizer usage, and V_4 represents the land environment. The aim is to select the best option GY. Each attribute is associated with a weight given the weight vector: $w = (0.35, 0.35, 0.20, 0.10)^T$. The procedure of the GY is elaborated as follows.

Step 1. According to Table 9.3, aggregate all PNNs α_{ij} $(j = 1, 2, \dots\dots\dots, n)$ by using the PNWGBM operator to derive the overall PNN u_i $(i = 1, 2, 3, 4)$ of the alternative U_i. The aggregating results are shown in Table 9.4.

Step 2. According to the aggregating results shown in Table 9.4, the score functions of the GY systems are shown in Table 9.5.

Step 3. According to the comparison formulas of the score functions, the ordering of the GY systems are shown in Table 9.5.

Step 4. Finally, we obtained the ranking of the score functions.

Table 9.3 The Pentapartiitoned Neutrosophic Decision Matrix

	V_1					V_2				
	t	*c*	*g*	*n*	*f*	*t*	*c*	*g*	*n*	*f*
U2	0.912	0.823	0.190	0.289	0.378	0.912	0.823	0.190	0.289	0.378
U3	1.000	1.000	0.000	0.000	0.000	0.823	0.734	0.289	0.378	0.467
U1	0.823	0.734	0.289	0.378	0.467	1.000	1.000	0.000	0.000	0.000
U4	0.645	0.556	0.467	0.556	0.645	0.823	0.734	0.289	0.378	0.467
			V3					V4		
U1	0.823	0.734	0.289	0.378	0.467	0.645	0.556	0.467	0.556	0.645
U2	0.467	0.378	0.645	0.734	0.823	0.734	0.645	0.378	0.467	0.556
U3	0.823	0.734	0.289	0.378	0.467	0.645	0.556	0.467	0.556	0.645
U4	0.645	0.556	0.467	0.556	0.645	0.734	0.645	0.378	0.467	0.556

Table 9.4 The Aggregated Results of the Pentapartiitoned Neutrosophic Weighted Geometric Bonferroni Mean Operators

	$p = 1$ $q = 1$	$p = 1$ $q = 5$
U1	(0.3751,0.2948,0.7167,0.7749,0.8184)	(0.4577,0.3628,0.6489,0.7192,0.7705)
U2	(0.3904,0.3411,0.6703,0.7114,0.7495)	(0.5369,0.5064,0.5012,0.5314,0.5670)
U3	(0.4373,0.3822,0.6303,0.6743,0.7139)	(0.6781,0.6490,0.3578,0.3828,0.4070)
U4	(0.2588,0.2090,0.8024,0.8420,0.8766)	(0.3207,0.2613,0.7520,0.7984,0.8387)
	p = 1 q = 10	p = 5 q = 1
U1	(0.5054,0.4010,0.6114,0.6893,0.7457)	(0.4733,0.3752,0.6370,0.7100,0.7635)
U2	(0.7033,0.6855,0.3194,0.3406,0.3676)	(0.6989,0.6699,0.3370,0.3625,0.3876)
U3	(0.8082,0.7896,0.2149,0.2316,0.2481)	(0.6505,0.6209,0.3860,0.4112,0.4357)
U4	(0.3727,0.3041,0.7113,0.7643,0.8099)	(0.3292,0.2676,0.7462,0.7944,0.8360)
	p = 10 q = 1	p = 5 q = 5
U1	(0.5149,0.4085,0.6043,0.6837,0.7415)	(0.4822,0.3826,0.6293,0.7036,0.7574)
U2	(0.8220,0.8035,0.2009,0.2178,0.2347)	(0.5642,0.5094,0.5035,0.5508,0.5956)
U3	(0.7876,0.7695,0.2348,0.2515,0.2684)	(0.5701,0.5140,0.4991,0.5471,0.5925)
U4	(0.3742,0.3048,0.7107,0.7641,0.8099)	(0.3141,0.2557,0.7576,0.8047,0.8462)

Table 9.5 Scores Obtained from the PNGHM Operator and the Rankings of Alternatives

		U1	*U2*	*U3*	*U4*	*RANK*
p = 1	q = 1	0.5929	0.5829	0.5356	0.718	U4 > U1 > U2 > U3
p = 1	q = 5	0.5027	0.4351	0.3045	0.6523	U4 > U1 > U2 > U3
p = 1	q = 10	0.4506	0.2746	0.1798	0.5971	U4 > U1 > U2 > U3
p = 5	q = 1	0.4854	0.2829	0.3319	0.643	U4 > U1 > U3 > U2
p = 10	q = 1	0.4401	0.1655	0.1998	0.5955	U4 > U1 > U3 > U2
p = 5	p = 5	0.4759	0.4039	0.3976	0.6577	U4 > U1 > U2 > U3

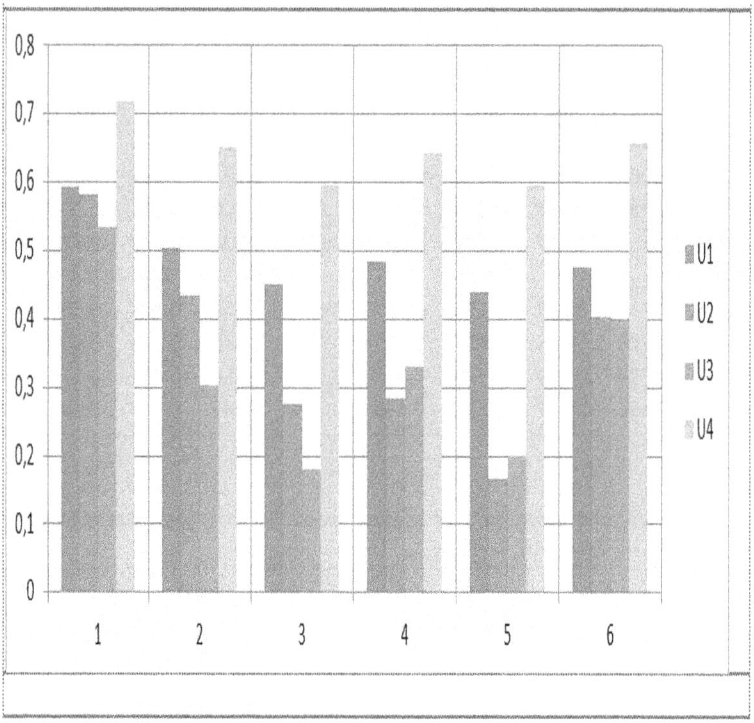

Figure 9.1 Scores for alternative U obtained from the PNWGBM operator ($p \in (0, 10]$, $q \in (0, 10]$).

9.6 CONCLUSION

In this paper, we studied multiple-attribute decision-making problems based on PNWGBM operators. First, we reviewed certain fundamental pentapartiitoned neutrosophic set notions and discussed their fundamental properties. We furthermore discussed pentapartiitoned neutrosophic geometric Bonferroni mean operators, and we used these new pentapartiitoned neutrosophic weighted geometric Bonferroni mean operators to solve problems involving multiple attributes. Finally, we gave a real example for evaluating these operators to show the applicability and validity of the novel method. We compared the pentapartiitoned neutrosophic weighted geometric Bonferroni mean operator with two basic aggregation operators. Research on these novel operators is both interesting and important.

Funding: There was no outside funding.

Declarations:

Ethics Approval: No human or animal subjects were involved in this study.

Informed Consent: There were no human participants in this study.

Conflict of interest: The authors declare that they have no conflicts of interest.

REFERENCES

[1] Z. Xu, R. R. Yager, Intuitionistic fuzzy Bonferroni means, IEEE Transactions on Systems, Man, and Cybernetics, Part B (Cybernetics), 41(2) (2011), 568–578. http://dx.doi.org/10.1109/TSMCB.2010.2072918

[2] C. Bonferroni, Sullemedie multiple di potenze, Bolletino Mathematica Italiana, 5 (1950), 267–270. http://eudml.org/doc/196058

[3] Rama Malik, Surpathi Pranamik, Pentapartitioned neutrosophic sets and its properties, Neutrosophic Sets and Stystems, 36 (2020), 184–192.

[4] Meimei Xia, Zeshui Xu, Bin Zhu, Geometric bonferroni means with their application in multi-criteria decision making, Knowledge-Based Systems, 40 (2013), 88–100. https://doi.org/10.1016/j.knosys.2012.11.013

[5] L. A. Zadeh, Fuzzy sets, Information and Control, 8 (1965), 338–353. http://dx.doi.org/10.1016/S0019-9958(65)90241-X

Chapter 10

A Novel Technique of Neutrosophic Mayor–Torrens Hybrid Aggregation Operator in MCDM

Augus Kurian and I. R. Sumathi

10.1 INTRODUCTION

In real life, uncertainty plays a pivotal role in daily activities. Zadeh [1] proposed the notion of fuzzy sets to deal with uncertainty, but the fuzzy set could not express the negation of statements in detail. Atanassov [2] extended the fuzzy set to an intuitionistic fuzzy set such that the degree of non-membership is independent of the degree of membership. Later, Samarandache [3, 4] proposed a branch of philosophy that studied the scope, origin and nature of neutralities.

A neutrosophic set provides a more comprehensive framework for handling problems with inconsistent, ambiguous and inaccurate information. Neutrosophic sets play a significant role in multicriteria decision-making (MCDM) problems. Aggregation operators were developed to deal with decision-making problems systematically. Ordered weighted averaging aggregation operators, also known as orands, were introduced by Yagar [5] as an application for MCDM problems. This operator provides a fuzzy aggregation between the two extremes, but these operators weight only either the intuitionistic fuzzy values or the ordered positions of the intuitionistic fuzzy values.

To overcome the difficulty, Xu [6] combined the operators defined by Yagar and developed the hybrid aggregation operator. The effectiveness of aggregation operators was studied in [7]. Hybrid intuitionistic fuzzy aggregation operators have been used in human resource management [8], diagnostic granule systems [9] and enterprise resource planning [10]. Hybrid aggregation operators for Pythagorean-resistant fuzzy information [11] showed effectiveness and flexibility in decision-making problems.

The aggregation operators were extended to the neutrosophic system [12], and an approach was developed for solving MCDM problems. Basic operational laws of neutrosophic single-valued sets were defined in detail to equip operators capable of dealing with practical applications such as decision-making [13]. Different aggregation operators such as Hamacher [14], logarithmic [15] and dombi [16, 17] were defined to make MCDM problems easy, and their effectiveness was examined in hydrogen power plants [18], solar power plants [19] and traffic control management [20, 21].

 DOI: 10.1201/9781003487104-10

Decision-making problems in neutrosophic sets [22–25] using other operators can be extended to aggregation operators to improve their efficiency. Mayor–Torrens norms were defined in classical set theory [26]. In this paper, we define neutrosophic Mayor–Torrens aggregation operators and apply them to a MCDM in the education sector. We propose the neutrosophic Mayor–Torrens t-norm and t-conorms and their aggregation operators. This method is compared with existing approaches to show its capability and effectiveness.

This chapter is organized as follows. In Section 10.2, we give basic preliminaries. In Section 10.3, we define neutrosophic Mayor–Torrens t-norms and t-conorms and also discuss related theorems. The neutrosophic Mayor–Torrens operating laws are discussed in Section 10.4 and verify some fundamental properties. We define the neutrosophic Mayor–Torrens aggregation operators in Section 10.5 and discuss related theorems. In Section 10.6, we give the algorithm for MCDM problems to show the effectiveness and easiness of the aggregator operators. Finally, we compare our proposed method with existing methods to show its efficacy in MCDM problems.

10.2 PRELIMINARIES

A few fundamental terms are defined in this section.

Definition 10.1 *[4] A single-valued neutrosophic set (SVNS) S in* $\mathcal{N}$ *(universal set) is defined as* $S = \{\langle x, \theta(x), \iota(x), \phi(x)\rangle / x \in \mathcal{N}\}$ *where* θ, ι, ϕ *are functions from* $\mathcal{N}$ *to [0,1]. Furthermore,* $\forall x \in \mathcal{N}$, $0 \le \theta(x) + \iota(x) + \phi(x) \le 3$.

Definition 10.2 *[16] A function* $\mathbb{A}$ *from* $[0,1] \times [0,1]$ *to* $[0,1]$ *is called a t-norm if it satisfies the following four conditions:*

1. $\mathbb{A}(0, b) = b$ *and* $\mathbb{A}(1, b) = 1$,
2. $\mathbb{A}(a, b) = \mathbb{A}(b, a) \forall$ *a and b*,
3. $\mathbb{A}(a, \mathbb{A}(b, c)) = \mathbb{A}(\mathbb{A}(a, b), c) \forall$ *a, b and c*,
4. $a \le a^*$ *and* $b \le b^*$ *then* $\mathbb{A}(a, b) \le \mathbb{A}(a^*, b^*)$.

Definition 10.3 *[16] A function* $\tilde{\mathbb{A}}$ *from* $[0,1] \times [0,1]$ *to* $[0,1]$ *is called a t-conorm if it satisfies the following four conditions:*

1. $\tilde{\mathbb{A}}(a,0) = 0$ *and* $\tilde{\mathbb{A}}(a,1) = a$,
2. $\tilde{\mathbb{A}}(a,b) = \tilde{\mathbb{A}}(b,a) \forall$ *a and b*,
3. $\tilde{\mathbb{A}}(a, \tilde{\mathbb{A}}(b,c)) = \tilde{\mathbb{A}}(\tilde{\mathbb{A}}(a,b), c) \forall$ *a,b and c*,
4. $a \le a^*$ *and* $b \le b^*$ $\tilde{\mathbb{A}}(a,b) \le \tilde{\mathbb{A}}(a^*, b^*)$.

Definition 10.4 *[16] Let* $\mathcal{N}_1 = (\theta_1, \iota_1, \phi_1)$, $\mathcal{N}_2 = (\theta_2, \iota_2, \phi_2) \in SVNS(\mathcal{N})$ *Then the neutrosophic t-norm is denoted by S, and S is a function from* $(]^-0,1^+[\times]^-0,1^+[\times]^-0,1^+[)^2$ *to* $]^-0,1^+[\times]^-0,1^+[\times]^-0,1^+[$ *such that* $S(\mathcal{N}_1, \mathcal{N}_2) = (\mathbb{A}(\theta_1, \theta_2), (\tilde{\mathbb{A}}(\iota_1, \iota_2), \tilde{\mathbb{A}}(\phi_1, \phi_2))$.

Definition 10.5 *[16] Let* $\mathcal{N}_1 = (\theta_1, \iota_1, \phi_1)$, $\mathcal{N}_2 = (\theta_2, \iota_2, \phi_2) \in SVNS(\mathcal{N})$. *Then neutrosophic t-conorm* $\tilde{S}$ *is a function from* $(]^{-}0,1^{+}[\times]^{-}0,1^{+}[\times]^{-}0,1^{+}[)^2$ *to* $]^{-}0,1^{+}[\times]^{-}0,1^{+}[\times]^{-}0,1^{+}[$ *such that* $\tilde{S}(\mathcal{N}_1,\mathcal{N}_2) = (\tilde{\mathbb{A}}(\theta_1,\theta_2), \mathbb{A}(\iota_1,\iota_2), \mathbb{A}(\phi_1,\phi_2))$.

Definition 10.6 *[16] Let* $\mathcal{N}_1 = (\theta_1, \iota_1, \phi_1)$, $\mathcal{N}_2 = (\theta_2, \iota_2, \phi_2) \in SVNS(\mathcal{N})$. *Then the neutrosophic t-norm denoted by* $S(\mathcal{N}_1,\mathcal{N}_2) = (\mathbb{A}(\theta_1,\theta_2), \tilde{\mathbb{A}}(\iota_1,\iota_2), \tilde{\mathbb{A}}(\phi_1,\phi_2))$ *is continous if* $\mathbb{A}(\theta_1,\theta_2)$, $\widetilde{\mathbb{A}}(\iota_1,\iota_2)$ *and* $\widetilde{\mathbb{A}}(\phi_1,\phi_2)$ *are continous functions.*

Definition 10.7 *[16] Let* $\mathcal{N}_1 = (\theta_1, \iota_1, \phi_1)$, $\mathcal{N}_2 = (\theta_2, \iota_2, \phi_2) \in SVNS(\mathcal{N})$. *Then the neutrosophic t-conorm denoted by* $\tilde{S}(\mathcal{N}_1,\mathcal{N}_2) = (\tilde{\mathbb{A}}(\theta_1,\theta_2), \mathbb{A}(\iota_1,\iota_2), \mathbb{A}(\phi_1,\phi_2))$ *is continous if* $\widetilde{\mathbb{A}}(\theta_1,\theta_2), \mathbb{A}(\iota_1,\iota_2)$ *and* $\mathbb{A}(\phi_1,\phi_2)$ *are continous functions.*

Definition 10.8 *[16] Let* $\mathcal{N} = (\theta, \iota, \phi) \in SVNS(\mathcal{N})$. *Then the neutrosophic t-norm denoted by* $S(\mathcal{N},\mathcal{N}) = (\mathbb{A}(\theta,\theta), \tilde{\mathbb{A}}(\iota,\iota), \tilde{\mathbb{A}}(\phi,\phi))$ *is Archimedean if* $\mathbb{A}(\theta,\theta) > \theta, \widetilde{\mathbb{A}}(\iota,\iota) < \iota$ *and* $\widetilde{\mathbb{A}}(\phi,\phi) < \phi$.

Definition 10.9 *[16] Let* $\mathcal{N} = (\theta, \iota, \phi) \in SVNS(\mathcal{N})$. *Then the neutrosophic t-conorm denoted by* $\tilde{S}(\mathcal{N}, \mathcal{N}) = (\tilde{\mathbb{A}}(\theta,\theta), \mathbb{A}(\iota,\iota), \mathbb{A}(\phi, \phi))$ *is Archimedean if* $\widetilde{\mathbb{A}}(\theta,\theta) < \theta, \mathbb{A}(\iota,\iota) > \iota$ *and* $\mathbb{A}(\phi,\phi) > \phi$.

Definition 10.10 *[19] The score function of a SVNS(*$\mathcal{N}$*) is defined as* $\tilde{S}(S) = \theta(x) - \iota(x) - \phi(x)$. *The accuracy function of a SVNS(*$\mathcal{N}$*) is defined as* $\tilde{A}(S) = \theta(x) + \iota(x) + \phi(x)$.

Definition 10.11 *[19] Let* $\mathcal{N}_1 = (\theta_1, \iota_1, \phi_1)$, $\mathcal{N}_2 = (\theta_2, \iota_2, \phi_2) \in SVNS(\mathcal{N})$. *Then*

1. $\mathcal{N}_1 < \mathcal{N}_2$ *if* $\tilde{S}(\mathcal{N}_1) < \tilde{S}(\mathcal{N}_2)$ *or* $\tilde{S}(\mathcal{N}_1) = \tilde{S}(\mathcal{N}_2)$ *and* $\tilde{A}(\mathcal{N}_1) < \tilde{A}(\mathcal{N}_2)$
2. $\mathcal{N}_1 = \mathcal{N}_2$ *if* $\tilde{S}(\mathcal{N}_1) = \tilde{S}(\mathcal{N}_2)$ *and* $\tilde{A}(\mathcal{N}_1) = \tilde{A}(\mathcal{N}_2)$.

10.3 NEUTROSOPHIC MAYOR–TORRENS T-NORM AND T-CONORM

Definition 10.12 *Let* $\mathcal{N}_1 = (\theta_1, \iota_1, \phi_1)$, $\mathcal{N}_2 = (\theta_2, \iota_2, \phi_2) \in SVNS(\mathcal{N})$. *The neutrosophic Mayor–Torrens t-norm is defined as follows:*

$$
\begin{aligned}
&M_\lambda^T((\theta_1,\iota_1,\phi_1),(\theta_2,\iota_2,\phi_2)) \\
&= \mathcal{N}_1 \bar{\wedge} \mathcal{N}_2 \\
&= \begin{cases} (\theta,\iota,\phi), \text{ if } \lambda \in (0,1], (\theta_1,\theta_2) \in [0,\lambda]^2 \textit{ and } (\iota_1,\iota_2),(\phi_1,\phi_2) \in [1-\lambda,1]^2 \\ (\tilde{t},\tilde{i},\tilde{f}), \text{ otherwise} \end{cases}
\end{aligned}
$$

where $\theta = max(0, \theta_1 + \theta_2 - \lambda)$, $\iota = min(\iota_1 + \iota_2 + \lambda - 1, 1)$, $\phi = min(\phi_1 + \phi_2 + \lambda - 1, 1)$, $\tilde{t} = min(\theta_1, \theta_2)$, $\tilde{i} = max(\iota_1, \iota_2)$ *and* $\tilde{f} = max(\phi_1, \phi_2)$.

Definition 10.13 *Let* $\mathcal{N}_1 = (\theta_1, \iota_1, \phi_1)$, $\mathcal{N}_2 = (\theta_2, \iota_2, \phi_2) \in SVNS(\mathcal{N})$. *The neutrosophic Mayor–Torrens t-conorm is defined as follows:* $M_\lambda^S((\theta_1, \iota_1, \phi_1), (\theta_2, \iota_2, \phi_2)) =$

$$\mathcal{N}_1 \veebar \mathcal{N}_2 = \begin{cases} (\theta, \iota, \phi), \text{ if } \lambda \in (0,1], (\theta_1, \theta_2) \in [1-\lambda, 1]^2 \text{ and } (\iota_1, \iota_2), (\phi_1, \phi_2) \in [0, \lambda]^2 \\ (\tilde{t}, \tilde{i}, \tilde{f}), \text{ otherwise} \end{cases}$$

where $\theta = min(\theta_1 + \theta_2 + \lambda - 1, 1)$, $\iota = max(0, \iota_1 + \iota_2 - \lambda)$, $\phi = max(0, \phi_1 + \phi_2 - \lambda)$, $\tilde{t} = max(\theta_1, \theta_2)$, $\tilde{i} = min(\iota_1, \iota_2)$ *and* $\tilde{f} = min(\phi_1, \phi_2)$.

Remark 10.1 $M^C(\theta_1, \iota_1, \phi_1) = (1-\theta_1, 1-\iota_1, 1-\phi_1)$ *gives the neutrosophic Mayor–Torrens complement* (M^C); *there exist many other complements. This neutrosophic complement is best suited for calculations and applications defined in this paper.*

Theorem 10.1 *The neutrosophic Mayor–Torrens t-norm and t-conorm are continuous.*

Proof *If* $\lambda = 0$, *then* $\mathcal{N}_1 \bar{\wedge} \mathcal{N}_2 = (min(\theta_1, \theta_2), max(\iota_1, \iota_2), max(\phi_1, \phi_2))$ *and* $\mathcal{N}_1 \veebar \mathcal{N}_2 = (max(\theta_1, \theta_2), min(\iota_1, \iota_2), min(\phi_1, \phi_2))$. *The functions in each of the coordinates are continuous. If* $\lambda = 1$, *then* $\mathcal{N}_1 \bar{\wedge} \mathcal{N}_2 = (max(\theta_1 + \theta_2 - 1), min(\iota_1 + \iota_2, 1), min(\phi_1 + \phi_2 + \lambda - 1, 1))$ *and* $\mathcal{N}_1 \veebar \mathcal{N}_2 = (min(\theta_1 + \theta_2 + \lambda - 1, 1), max(0, \iota_1 + \iota_2 - \lambda), max(0, \phi_1 + \phi_2 - \lambda))$. *The functions in each of the coordinates are continuous. If* $\lambda \in (0,1)$, *the problematic line segments are* $y = \lambda (0 \leq x \leq \lambda)$ *and* $x = \lambda (0 \leq y \leq \lambda)$. *In this line segment,* $max\ (x + y + \lambda, 0) = min(x, y)$, *so the truth part is continuous. Similarly, we can prove indeterminacy and falsity. Therefore, the neutrosophic Mayor–Torrens t-norm is continuous. A similar proof follows for neutrosophic Mayor–Torrens t-conorm.*

Theorem 10.2 *The neutrosophic Mayor–Torrens t-norm and t-conorm are Archimedean at* $\lambda = 1$.

Proof *First, we will prove the sub-idempotency for the Archimedian neutrosophic Mayor–Torrens t-norm. All other parts are trivial.* $\mathcal{N}_1 \bar{\wedge} \mathcal{N}_2 = (2\theta - 1, 2\iota, 2\phi) < (\theta, \iota, \phi)$. *Since* $2\theta - 1 < \theta$, $2\iota > \iota$ *and* $2\phi > \phi$. *A similar proof follows for the neutrosophic Mayor–Torrens t-conorm.*

10.4 THE OPERATING LAWS FOR NEUTROSOPHIC MAYOR–TORRENS

Definition 10.14 *Let* $\mathcal{N}_1 = (\theta_1, \iota_1, \phi_1)$, $\mathcal{N}_2 = (\theta_2, \iota_2, \phi_2) \in SVNS(\mathcal{N})$ *where* $\omega > 0$.

The neutrosophic Mayor–Torrens operating laws are defined as follows:

1. $\mathcal{N}_1 \barwedge \mathcal{N}_2$

$$= \begin{cases} (\theta,\iota,\phi), \text{if } \lambda \in (0,1], (\theta_1,\theta_2) \in [0,\lambda]^2 \textit{ and } (\iota_1,\iota_2),(\phi_1,\phi_2) \in [1-\lambda,1]^2 \\ (\tilde{t},\tilde{i},\tilde{f}), \text{otherwise} \end{cases}$$

where $\theta = max(0, \theta_1 + \theta_2 - \lambda)$, $\iota = min(\iota_1 + \iota_2 + \lambda - 1, 1)$, $\phi = min(\phi_1 + \phi_2 + \lambda - 1, 1)$, $\tilde{t} = min(\theta_1, \theta_2)$, $\tilde{i} = max(\iota_1, \iota_2)$ *and* $\tilde{f} = max(\phi_1, \phi_2)$.

2. $\mathcal{N}_1 \veebar \mathcal{N}_2$

$$= \begin{cases} (\theta,\iota,\phi), \text{if } \lambda \in (0,1], (\theta_1,\theta_2) \in [1-\lambda,1]^2 \textit{ and } (\iota_1,\iota_2),(\phi_1,\phi_2) \in [0,\lambda]^2 \\ (\tilde{t},\tilde{i},\tilde{f}), \text{otherwise} \end{cases}$$

where $\theta = min(\theta_1 + \theta_2 + \lambda - 1, 1)$, $\iota = max(0, \iota_1 + \iota_2 - \lambda)$, $\phi = max(0, \phi_1 + \phi_2 - \lambda)$, $\tilde{t} = max(\theta_1, \theta_2)$, $\tilde{i} = min(\iota_1, \iota_2)$ *and* $\tilde{f} = min(\phi_1, \phi_2)$.

3. $\mathcal{N}_1^{\omega} = \begin{cases} (\theta,\iota,\phi), \text{if } \lambda \in (0,1], \theta_1 \in [0,\lambda] \textit{ and } \iota_1, \phi_1 \in [1-\lambda,1] \\ (\tilde{t},\tilde{i},\tilde{f}), \text{otherwise} \end{cases}$

where $\theta = max(0, \omega\theta_1 + \lambda(1-\omega))$, $\iota = min(\omega\iota_1 + (\omega-1)(\lambda-1), 1)$, $\phi = min(\omega\phi_1 + (\omega-1)(\lambda-1), 1)$, $\tilde{t} = \omega\theta_1, \tilde{i} = \omega\iota_1$ *and* $\tilde{f} = \omega\phi_1$.

4. $\omega(\mathcal{N}_1) = \begin{cases} (\theta,\iota,\phi), \text{if } \lambda \in (0,1], \theta_1 \in [1-\lambda,1] \textit{ and } \iota_1, \phi_1 \in [0,\lambda] \\ (\tilde{t},\tilde{i},\tilde{f}), \text{otherwise} \end{cases}$

where $\theta = min(\omega\theta_1 + (\omega-1)(\lambda-1), 1)$, $\iota = max(0, \omega\iota_1 + \lambda(1-\omega))$, $\phi = max(0, \omega\phi_1 + \lambda(1-\omega))$, $\tilde{t} = \omega\theta_1$, $\tilde{i} = \omega\iota_1$ *and* $\tilde{f} = \omega\phi_1$.

Theorem 10.3 *Let* $\mathcal{N}_1, \mathcal{N}_2 \in SVNS(\mathcal{N})$ *with* $\omega_1, \omega_2 > 0$. *Then*

1. $\mathcal{N}_1 \barwedge \mathcal{N}_2 = \mathcal{N}_2 \barwedge \mathcal{N}_1$
2. $\mathcal{N}_1 \veebar \mathcal{N}_2 = \mathcal{N}_2 \veebar \mathcal{N}_1$
3. $\omega(\mathcal{N}_1 \veebar \mathcal{N}_2) = \omega\mathcal{N}_1 \veebar \omega\mathcal{N}_2$
4. $(\omega_1 + \omega_2)\mathcal{N}_1 = \omega_1\mathcal{N}_1 \veebar \omega_2\mathcal{N}_1$
5. $(\mathcal{N}_1 \barwedge \mathcal{N}_2)^{\omega} = \mathcal{N}_1^{\omega} \barwedge \mathcal{N}_2^{\omega}$
6. $\mathcal{N}_1^{\omega_1} \barwedge \mathcal{N}_1^{\omega_2} = \mathcal{N}_1^{(\omega_1+\omega_2]}$

Proof *proof of (1) and (2) is direct.*

(3) If $\lambda \in (0,1], (\theta_1,\theta_2) \in [1-\lambda,1]^2$ *and* $(\iota_1,\iota_2),(\phi_1,\phi_2) \in [0,\lambda]^2$, *then* $\omega(\mathcal{N}_1 \veebar \mathcal{N}_2) =$ *(t,i,f) where,* $\theta = min(\omega(\theta_1 + \theta_2 + \lambda - 1) + (\omega-1)(\lambda-1), 1)$, $i = max(0, \omega(\iota_1 + \iota_2 - \lambda) + \lambda(1-\omega))$, $\phi = max(0, \omega(\phi_1 + \phi_2 - \lambda) + \lambda(1-\omega)))$.

$= (\tilde{t}, \tilde{i}, \tilde{f})$ *where*

$$\theta = min\left(\omega\theta_1 + (\omega-1)(\lambda-1) + \omega\theta_2 + (\omega-1)(\lambda-1) + \lambda - 1, 1\right),$$

$$\iota = max\left(0, \omega\iota_1 + \lambda(1-\omega) + \omega\iota_2 + \lambda(1-\omega) - \lambda\right),$$

$\phi = max\left(0, \omega\phi_1 + \lambda(1-\omega) + \omega\phi_2 + \lambda(1-\omega) - \lambda\right)$. *We can verify that* $t = \tilde{t}$, $\iota = \tilde{i}$ *and* $\phi = \tilde{f}$. *Therefore,* $\omega(\mathcal{N}_1 \veebar \mathcal{N}_2) = \omega\mathcal{N}_1 \veebar \omega\mathcal{N}_2$. *The other case is obvious.*

(4) If $\lambda \in (0,1], (\theta_1, \theta_2) \in [1-\lambda, 1]^2$ *and* $(\iota_1, \iota_2), (\phi_1, \phi_2) \in [0, \lambda]^2$ *then,* $(\omega_1 + \omega_2)\mathcal{N}_1 = (t, i, f)$ *where* $\theta = min\left((\omega_1 + \omega_2)\theta_1 + (\omega_1 + \omega_2 - 1)(\lambda - 1), 1\right)$, $\iota = max\left(0, (\omega_1 + \omega_2)\iota_1 + \lambda(1 - \omega_1 - \omega_2)\right)$, $\phi = max\left(0, (\omega_1 + \omega_2)\phi_1 + \lambda(1 - \omega_1 - \omega_2)\right)$. $\omega_1\mathcal{N}_1 \veebar \omega_2\mathcal{N}_1 = (\tilde{t}, \tilde{i}, \tilde{f})$ *where* $\theta = min\left(\omega_1\theta_1 + (\omega_1 - 1)(\lambda - 1) + \omega_2\theta_2 + (\omega_2 - 1)\right.$ $\left.(\lambda - 1) + \lambda - 1, 1\right)$, $\iota = max\left(0, \omega_1\iota_1 + \lambda(1-\omega_1) + \omega_2\iota_1 + \lambda(1-\omega_2) - \lambda\right)$, $\phi = max\left(0,\right.$ $\left.\omega_1\phi_1 + \lambda(1-\omega_1) + \omega_2\phi_1 + \lambda(1-\omega_2) - \lambda\right)$. *We can verify that* $t = \tilde{t}$, $\iota = \tilde{i}$ *and* $\phi = \tilde{f}$. *Hence* $(\omega_1 + \omega_2)\mathcal{N}_1 = \omega_1\mathcal{N}_1 \veebar \omega_2\mathcal{N}_1$. *The other case is obvious.*

The proofs of (5) and (6) are the same.

10.5 NEUTROSOPHIC MAYOR–TORRENS AGGREGATION OPERATORS

In this section, we define neutrosophic Mayor–Torrens Aggregation operators and some of their properties.

10.5.1 Neutrosophic Mayor–Torrens Weighted Averaging Aggregation Operators

Definition 10.15 *Let* $\mathcal{N}_j = \left(\theta_j, \iota_j, f_j\right) \in SVNS(\mathcal{N})$ *where* $j \in \mathbb{N}$. *Then, the neutrosophic Mayor–Torrens weighted average aggregation operator for SVNS(*$\mathcal{N}$*) is defined as follows:*

$$SVNMTWA(\mathcal{N}_1, \mathcal{N}_2, \ldots, \mathcal{N}_n) = \omega_1\mathcal{N}_1 \veebar \omega_2\mathcal{N}_2 \veebar \ldots \veebar \omega_n\mathcal{N}_n = \sum\nolimits_{j=1}^{n} \omega_j \mathcal{N}_j,$$

where the weight $\omega_1, \omega_2, \ldots, \omega_n$ *of* $\mathcal{N}_j$ *satisfies* $\omega_j \geq 0$ *and* $\sum\nolimits_{j=1}^{n} \omega_h = 1$.

Theorem 10.4 *Let* $\mathcal{N}_j = \left(\theta_j, \iota_j, f_j\right) \in SVNS(\mathcal{N})$ *where* $j \in \mathbb{N}$ *and the weights* $(\omega_1, \omega_2, \ldots, \omega_n)$ *of* $\mathcal{N}_n$ *are* $\omega_j \geq 0$ *and* $\sum\nolimits_{j=1}^{n} \omega_h = 1$. *The SVNMTWA (single-valued* *neutrosophic Mayor–Torrens weighted average) aggregation operators map from* $\mathcal{N}^n \rightarrow \mathcal{N}$ *such that* $SVNMTWA(\mathcal{N}_1, \mathcal{N}_2, ..., \mathcal{N}_n) =$

$$\begin{cases} (\theta, \iota, f), \text{if } \lambda \in (0,1], (\theta_1, \theta_2, ..., \theta_n) \in [1-\lambda, 1]^n \; and (\iota_1, \iota_2, ..., \iota_n), \\ \qquad (f_1, f_2, ..., f_n) \in [0, \lambda]^n \\ (\tilde{t}, \tilde{i}, \tilde{f}), \text{otherwise} \end{cases}$$

$where\, \theta = min\left(\sum_{j=1}^{n} \omega_j \theta_j, 1\right), \iota = max\left(0, \sum_{j=1}^{n} \omega_j \iota_j\right), f = max\left(0, \sum_{j=1}^{n} \omega_j f_j\right),$
$\tilde{t} = max(\omega_j \theta_j), \tilde{i} = min(\omega_j \iota_j) \; and \; \tilde{f} = min(\omega_j \phi_j).$

Proof *We prove this by the method of induction. For k=2,* $SVNMTWA(\mathcal{N}_1, \mathcal{N}_2) =$

$$\begin{cases} (\theta, \iota, \phi), \text{if } \lambda \in (0,1], (\theta_1, \theta_2) \in [1-\lambda, 1]^2 \; and (\iota_1, \iota_2), (\phi_1, \phi_2) \in [0, \lambda]^2 \\ (\tilde{t}, \tilde{i}, \tilde{f}), \text{otherwise} \end{cases}$$

$where\; \theta = min(\omega_1\theta_1 + \omega_2\theta_2 + (\omega_1 - 1)(\lambda - 1) + (\omega_2 - 1)(\lambda - 1) + (\lambda - 1), 1) =$
$min(\omega_1\theta_1 + \omega_2\theta_2, 1)$

$$\iota = max(0, \omega_1\iota_1 + \omega_2\iota_2 + \lambda(1-\omega_1) + \lambda(1-\omega_2) - \lambda) = max(0, \omega_1\iota_1 + \omega_2\iota_2),$$
$$\phi = max(0, \omega_1\phi_1 + \omega_2\phi_2 + \lambda(1-\omega_1) + \lambda(1-\omega_2) - \lambda) = max(0, \omega_1\phi_1 + \omega_2\phi_2),$$
$$\tilde{t} = max(\omega_1\theta_1, \omega_2\theta_2), \tilde{i} = min(\omega_1\iota_1, \omega_2\iota_2) \; and \; \tilde{f} = min(\omega_1\phi_1, \omega_2\phi_2).$$

Therefore, for k=2, the statement is true. Now suppose that for k = n, the statement is true: $SVNMTWA(\mathcal{N}_1, \mathcal{N}_2, ..., \mathcal{N}_n) =$

$$\begin{cases} (\theta, \iota, \phi), \text{if } \lambda \in (0,1], (\theta_1, \theta_2, ..., \theta_n) \in [1-\lambda, 1]^n \; and (\iota_1, \iota_2, ..., \iota_n), \\ \qquad (\phi_1, \phi_2, ..., \phi_n) \in [0, \lambda]^n \\ (\tilde{t}, \tilde{i}, \tilde{f}), \text{otherwise} \end{cases}$$

$where\; \theta = min\left(\sum_{j=1}^{n} \omega_j \theta_j, 1\right), \; \iota = max\left(0, \sum_{j=1}^{n} \omega_j \iota_j\right), \phi = max\left(0, \sum_{j=1}^{n} \omega_j \phi_j\right),$
$\tilde{t} = max(\omega_j \theta_j), \tilde{i} = min(\omega_j \iota_j) \; and \; \tilde{f} = min(\omega_j \phi_j).$

Now we prove that this statement is true for k=n+1:

$$SVNMTWA\left(\mathcal{N}_1,\mathcal{N}_2,\ldots,\mathcal{N}_n,\mathcal{N}_{n+1}\right)=SVNMTWA\left(\mathcal{N}_1,\mathcal{N}_2,\ldots,\mathcal{N}_n,\mathcal{N}_n\right)\veebar\omega_{n+1}\mathcal{N}_{n+1}$$

$$\begin{cases}(\theta,\iota,\phi),\text{ if } \lambda\in(0,1],(\theta_1,\theta_2,\ldots,\theta_{n+1})\in[1-\lambda,1]^{n+1}\ and\left(\iota_1,\iota_2,\ldots,\iota_{n+1}\right),\\ \qquad(\phi_1,\phi_2,\ldots,\phi_{n+1})\in[0,\lambda]^{n+1}\\ \left(\tilde{\iota},\tilde{i},\tilde{f}\right),\text{ otherwise}\end{cases}$$

where $\theta=min\left(\sum_{j=1}^{n+1}\omega_j\theta_j,1\right)$, $\iota=max\left(0,\sum_{j=1}^{n+1}\omega_j\iota_j\right)$, $\phi=max\left(0,\sum_{j=1}^{n+1}\omega_j\phi_j\right)$, $\tilde{\iota}=max\left(\omega_j\theta_j\right)$, $\tilde{i}=min\left(\omega_j\iota_j\right)$ *and* $\tilde{f}=min\left(\omega_j\phi_j\right)$.

Hence, this statement is true for all n.

Theorem 10.5 *Let* $\mathcal{N}_j=\left(\theta_j,\iota_j,\phi_j\right)\in SVNS(\mathcal{N})$ *where* $\mathcal{N}_j=\mathcal{N}$.

Then $SVNMTWA(\mathcal{N}_1,\mathcal{N}_2,\ldots,\mathcal{N}_n)=\mathcal{N}$.

Proof *Applying theorem 10.4 and simplification in each of the tuples proves the theorem.*

Theorem 10.6 *Let* $\mathcal{N}_j^-=\left(min\left(\theta_j\right),\ max\left(\iota_j\right),max\left(\phi_j\right)\right)$ *and* $\mathcal{N}_j^+=\left(max\left(\theta_j\right),\ min\left(\iota_j\right),min\left(\phi_j\right)\right)\in SVNS(\mathcal{N})$.

Then $\mathcal{N}_j^-\le SVNMTWA(\mathcal{N}_1,\mathcal{N}_2,\ldots,\mathcal{N}_n)\le\mathcal{N}_j^+$.

Proof *Applying theorem 10.4 and simplification in each of the tuples proves the theorem.*

Theorem 10.7 *Let* $\mathcal{N}_j=\left(\theta_j,\iota_j,\phi_j\right)$ *and* $\mathcal{N}_j^+=\left(\theta_j^+,\iota_j^+,\phi_j^+\right)\in SVNS(\mathcal{N})$ *where* $\left(\theta_j,\iota_j,\phi_j\right)\le\left(\theta_j^+,\iota_j^+,\phi_j^+\right)$.

Then $SVNMTWA(\mathcal{N}_1,\mathcal{N}_2,\ldots,\mathcal{N}_n)\le SVNMTWA(\mathcal{N}_1^+,\mathcal{N}_2^+,\ldots,\mathcal{N}_n^+)$.

Proof *This proof is the same as that for theorem 10.6.*

Definition 10.16 *Let* $\mathcal{N}_j=\left(\theta_j,\iota_j,\phi_j\right)\in SVNS(\mathcal{N})$ *where* $j\in\mathbb{N}$. *Then, the neutrosophic Mayor–Torrens ordered weighted average aggregation operator for SVNS(𝒩) is defined as follows:*

$$SVNMTOWA(\mathcal{N}_1,\mathcal{N}_2,\ldots,\mathcal{N}_n)=\omega_1\mathcal{N}_{f(1)}\veebar\omega_2\mathcal{N}_{f(2)}\veebar\ldots\veebar\omega_n\mathcal{N}_{f(n)}=$$

$\sum_{j=1}^{n}\omega_j\mathcal{N}_{f(j)}$, *where* $\mu:1,2,\ldots,n\rightarrow1,2,\ldots,n$ *is the permutation such that* $N_{f(j)}$ *is the* μ *(j)th largest element of the collection of SVNS(𝒩)* N_j *and the weight* $\omega_1,\omega_2,\ldots,\omega_n$ *of* $\mathcal{N}_j$ *satisfies* $\omega_j\ge0$ *and* $\sum_{j=1}^{n}\omega_h=1$.

Theorem 10.8 *Let* $\mathcal{N}_j = \left(\theta_j, \iota_j, \phi_j\right) \in SVNS(\mathcal{N})$ *where* $j \in \mathbb{N}$ *and the weights* $(\omega_1, \omega_2, \ldots, \omega_n)$ *of* $\mathcal{N}_n$ *are* $\omega_j \geq 0$ *and* $\sum_{j=1}^{n} \omega_h = 1$. *The SVNMTOWA aggregation operators are a mapping from* $\mathcal{N}^n \to \mathcal{N}$ *such that* $SVNMTWA\left(\mathcal{N}_1, \mathcal{N}_2, \ldots, \mathcal{N}_n\right) =$

$$\begin{cases} (\theta, \iota, \phi), \text{ if } \lambda \in (0,1], (\theta_1, \theta_2, \ldots, \theta_n) \in [1-\lambda, 1]^n \ and \left(\iota_1, \iota_2, \ldots, \iota_n\right), \\ \qquad (\phi_1, \phi_2, \ldots, \phi_n) \in [0, \lambda]^n \\ \left(\tilde{t}, \tilde{i}, \tilde{f}\right), \text{ otherwise} \end{cases}$$

where $\theta = min\left(\sum_{j=1}^{n} \omega_j t_{f(j)}, 1\right)$, $\iota = max\left(0, \sum_{j=1}^{n} \omega_j i_{f(j)}\right)$, $\phi = max$ $\left(0, \sum_{j=1}^{n} \omega_j f_{f(j)}\right)$, $\tilde{t} = max\left(\omega_j t_{f(j)}\right)$, $\tilde{i} = min\left(\omega_j i_{f(j)}\right)$ *and* $\tilde{f} = min\left(\omega_j f_{f(j)}\right)$.

Proof *The proof is the same as that for theorem 10.4.*

Theorem 10.9 *Let* $\mathcal{N}_j = \left(\theta_j, \iota_j, \phi_j\right) \in SVNS(\mathcal{N})$ *where* $\mathcal{N}_j = \mathcal{N}$.

Then SVNMTOWA $(\mathcal{N}_1, \mathcal{N}_2, \ldots, \mathcal{N}_n) = \mathcal{N}$.

Proof *The proof is the same as that for theorem 10.5.*

Theorem 10.10 $\mathcal{N}_j = \left(\theta_j, \iota_j, \phi_j\right), \mathcal{N}_j^- = \left(min\left(\theta_j\right), max\left(\iota_j\right), max\left(\phi_j\right)\right)$ *and* $\mathcal{N}_j^+ = \left(max\left(\theta_j\right), min\left(\iota_j\right), min\left(\phi_j\right)\right) \in SVNS(\mathcal{N})$.

Then $\mathcal{N}_j^- \leq SVNMTOWA(\mathcal{N}_1, \mathcal{N}_2, \ldots, \mathcal{N}_n) \leq \mathcal{N}_j^+$.

Proof *The proof is the same as that for theorem 10.6.*

Theorem 10.11 *Let* $\mathcal{N}_j = \left(\theta_j, \iota_j, \phi_j\right)$ *and* $\mathcal{N}_j^+ = \left(\theta_j^+, \iota_j^+, \phi_j^+\right) \in SVNS(\mathcal{N})$ *where* $\left(\theta_j, \iota_j, \phi_j\right) \leq \left(\theta_j^+, \iota_j^+, \phi_j^+\right)$.

Then $SVNMTOWA(\mathcal{N}_1, \mathcal{N}_2, \ldots, \mathcal{N}_n) \leq SVNMTOWA(\mathcal{N}_1^+, \mathcal{N}_2^+, \ldots, \mathcal{N}_n^+)$.

Proof *The proof is the same as that for theorem 10.7.*

Definition 10.17 *Let* $\mathcal{N}_j = \left(\theta_j, \iota_j, \phi_j\right) \in SVNS(\mathcal{N})$ *where* $j \in \mathbb{N}$. *Then, the neutrosophic Mayor–Torrens hybrid weighted average aggregation operator for* $SVNS(\mathcal{N})$ *is defined as follows:* $SVNMTHWA(\mathcal{N}_1, \mathcal{N}_2, \ldots, \mathcal{N}_n) = \omega_1 \tilde{\mathcal{N}}_{f(1)} \veebar \omega_2 \tilde{\mathcal{N}}_{f(2)} \veebar \ldots \veebar \omega_n \tilde{\mathcal{N}}_{f(n)} = \sum_{j=1}^{n} \omega_j \tilde{\mathcal{N}}_{f(j)}$, *where* $\tilde{\mathcal{N}}_{f(j)}$ *is the jth largest of the weighted neutrosophic values* $\tilde{\mathcal{N}}_j (\tilde{\mathcal{N}}_j = n\omega_j \mathcal{N}_j)$ *and the weight* $\omega_1, \omega_2, \ldots, \omega_n$ *of* $\mathcal{N}_j$ *satisfies* $\omega_j \geq 0$ *and* $\sum_{j=1}^{n} \omega_h = 1$.

Theorem 10.12 *Let* $\mathcal{N}_j = (\theta_j, \iota_j, \phi_j) \in SVNS(\mathcal{N})$ *where* $j \in \mathbb{N}$ *and the weights* $(\omega_1, \omega_2, ..., \omega_n)$ *of* $\mathcal{N}_n$ *having* $\omega_j \geq 0$ *and* $\sum_{j=1}^{n} \omega_h = 1$. *The SVNMTOWA aggregation operators are a mapping from* $\mathcal{N}^n \to \mathcal{N}$ *such that* $SVNMTWA(\mathcal{N}_1, \mathcal{N}_2, ..., \mathcal{N}_n) =$

$$\begin{cases} (\theta, \iota, \phi), \text{ if } \lambda \in (0,1], (\theta_1, \theta_2, ..., \theta_n) \in [1-\lambda, 1]^n \textit{ and } (\iota_1, \iota_2, ..., \iota_n), \\ \qquad (\phi_1, \phi_2, ..., \phi_n) \in [0, \lambda]^n \\ (\tilde{t}, \tilde{i}, \tilde{f}), \text{ otherwise} \end{cases}$$

where $\theta = min\left(\sum_{j=1}^{n} \omega_j \tilde{t}_{f(j)}, 1\right)$, $\iota = max\left(0, \sum_{j=1}^{n} \omega_j \tilde{i}_{f(j)}\right)$, $\phi = max\left(0, \sum_{j=1}^{n} \omega_j \tilde{f}_{f(j)}\right)$, $\tilde{t} = max\left(\omega_j \tilde{t}_{f(j)}\right)$, $\tilde{i} = min\left(\omega_j \tilde{i}_{f(j)}\right)$ *and* $\tilde{f} = min(\omega_j \tilde{f}_{f(j)})$.

Proof *The proof is the same as that for theorem 10.4.*

Theorem 10.13 *Let* $\mathcal{N}_j = (\theta_j, \iota_j, \phi_j) \in SVNS(\mathcal{N})$ *where* $\mathcal{N}_j = \mathcal{N}$.

Then SVNMTHWA $(\mathcal{N}_1, \mathcal{N}_2, ..., \mathcal{N}_n) = \mathcal{N}$.

Proof *The proof is the same as that for theorem 10.5.*

Theorem 10.14 *Let* $\mathcal{N}_j = (\theta_j, \iota_j, \phi_j)$, $\mathcal{N}_j^- = \left(min(\theta_j), max(\iota_j), max(\phi_j)\right)$ *and* $\mathcal{N}_j^+ = \left(max(\theta_j), min(\iota_j), min(\phi_j)\right) \in SVNS(\mathcal{N})$.

Then $\mathcal{N}_j^- \leq SVNMTHWA(\mathcal{N}_1, \mathcal{N}_2, ..., \mathbb{N}_n) \leq \mathcal{N}_j^+$

Proof *The proof is the same as that for theorem 10.6.*

Theorem 10.15 *Let* $\mathcal{N}_j = (\theta_j, \iota_j, \phi_j)$ *and* $\mathcal{N}_j^+ = (\theta_j^+, \iota_j^+, \phi_j^+) \in SVNS(\mathcal{N})$ *where* $(\theta_j, \iota_j, \phi_j) \leq (\theta_j^+, \iota_j^+, \phi_j^+)$.

Then $SVNMTHWA(\mathcal{N}_1, \mathcal{N}_2, ..., \mathcal{N}_n) \leq SVNMTHWA(\mathcal{N}_1^+, \mathcal{N}_2^+, ..., \mathcal{N}_n^+)$

Proof *The proof is the same as that for theorem 10.7.*

10.5.2 Neutrosophic Mayor–Torrens Weighted Geometric Aggregation Operators

Definition 10.18 *Let* $\mathcal{N}_j = (\theta_j, \iota_j, \phi_j) \in SVNS(\mathcal{N})$ *where* $j \in \mathbb{N}$. *Then, neutrosophic Mayor–Torrens weighted geometric aggregation operator for SVNS*($\mathcal{N}$) *is defined as follows:*

$$SVNMTWG(\mathcal{N}_1, \mathcal{N}_2, ..., \mathcal{N}_n) = \mathcal{N}_1^{\omega_1} \barwedge \mathcal{N}_2^{\omega_2} \barwedge ... \barwedge \mathcal{N}_n^{\omega_n} = \prod_{j=1}^{n} \mathcal{N}_j^{\omega_j},$$

where the weights $\omega_1, \omega_2, \ldots, \omega_n$ of $\mathcal{N}_j$ satisfies $\omega_j \geq 0$ and $\sum_{j=1}^{n} \omega_h = 1$.

Theorem 10.16 *Let $\mathcal{N}_j = (\theta_j, \iota_j, \phi_j) \in SVNS(\mathcal{N})$ where $j \in \mathbb{N}$ and the weights $(\omega_1, \omega_2, \ldots, \omega_n)$ of $\mathcal{N}_n$ having $\omega_j \geq 0$ and $\sum_{j=1}^{n} \omega_h = 1$. The SVNMTWG aggregation operators map from $\mathcal{N}^n \to \mathcal{N}$ such that $SVNMTWG(\mathcal{N}_1, \mathcal{N}_2, \ldots, \mathcal{N}_n) =$*

$$\begin{cases} (\theta, \iota, \phi), \text{ if } \lambda \in (0,1], (\theta_1, \theta_2, \ldots, \theta_n) \in [1-\lambda, 1]^n \; and\, (\iota_1, \iota_2, \ldots, \iota_n), \\ \qquad (\phi_1, \phi_2, \ldots, \phi_n) \in [0, \lambda]^n \\ (\tilde{t}, \tilde{i}, \tilde{f}), \text{ otherwise} \end{cases}$$

where $\theta = max\left(0, \sum_{j=1}^{n} \omega_j \theta_j\right)$, $\iota = min\left(\sum_{j=1}^{n} \omega_j \iota_j, 1\right)$, $\phi = min\left(\sum_{j=1}^{n} \omega_j \phi_j, 1\right)$, $\tilde{t} = min(\omega_j \theta_j)$, $\tilde{i} = max(\omega_j \iota_j)$ and $\tilde{f} = max(\omega_j \phi_j)$.

Proof *The proof is the same as that for theorem 10.4.*

Theorem 10.17 *Let $\mathcal{N}_j = (\theta_j, \iota_j, \phi_j) \in SVNS(\mathcal{N})$ where $\mathcal{N}_j = \mathcal{N}$.*

Then $SVNMTWG(\mathcal{N}_1, \mathcal{N}_2, \ldots, \mathcal{N}_n) = \mathcal{N}$

Proof *Applying theorem 10.16 and simplification in each of the tuples proves the theorem.*

Theorem 10.18 *Let $\mathcal{N}_j = (\theta_j, \iota_j, \phi_j)$, $\mathcal{N}_j^- = (min(\theta_j), max(\iota_j), max(\phi_j))$ and $\mathcal{N}_j^+ = (max(\theta_j), min(\iota_j), min(\phi_j)) \in SVNS(\mathcal{N})$.*

Then $\mathcal{N}_j^- \leq SVNMTWG(\mathcal{N}_1, \mathcal{N}_2, \ldots, \mathcal{N}_n) \leq \mathcal{N}_j^+$

Proof *Applying theorem 10.16 and simplification in each of the tuples proves the theorem.*

Theorem 10.19 *Let $\mathcal{N}_j = (\theta_j, \iota_j, \phi_j)$ and $\mathcal{N}_j^+ = (\theta_j^+, \iota_j^+, \phi_j^+) \in SVNS(\mathcal{N})$ where $(\theta_j, \iota_j, \phi_j) \leq (\theta_j^+, \iota_j^+, \phi_j^+)$.*

Then $SVNMTWG(\mathcal{N}_1, \mathcal{N}_2, \ldots, \mathcal{N}_n) \leq SVNMTWG(\mathcal{N}_1^*, \mathcal{N}_2^+, \ldots, \mathcal{N}_n^+)$

Proof *The proof the same as that for theorem 10.18.*

Definition 10.19 *Let $\mathcal{N}_j = (\theta_j, \iota_j, \phi_j) \in SVNS(\mathcal{N})$ where $j \in \mathbb{N}$. Then, the neutrosophic Mayor−Torrens ordered weighted geometric aggregation operator for $SVNS(\mathcal{N})$ is defined as follows:*

$$SVNMTOWG(\mathcal{N}_1, \mathcal{N}_2, \ldots, \mathcal{N}_n) = \mathcal{N}_{f(1)}^{\omega_1} \barwedge \mathcal{N}_{f(2)}^{\omega_2} \barwedge \ldots \barwedge \mathcal{N}_{f(n)}^{\omega_n} = \prod_{j=1}^{n} \mathcal{N}_{f(j)}^{\omega_j},$$

where $\mu : 1,2,...,n \rightarrow 1,2,...,n$ *is the permutation such that* $\mathcal{N}_{f(j)}$ *is the* μ *(j)th largest element of the collection of SVNS(*$\mathcal{N}$*)* $\mathcal{N}_j$, *the weights* $\omega_1, \omega_2, ..., \omega_n$ *of* $\mathcal{N}_j$ *satisfies* $\omega_j \geq 0$ *and* $\sum_{j=1}^{n} \omega_h = 1$.

Theorem 10.20 *Let* $\mathcal{N}_j = \left(\theta_j, \iota_j, \phi_j\right) \in SVNS(\mathcal{N})$ *where* $j \in \mathbb{N}$ *and the weights* $(\omega_1, \omega_2, ..., \omega_n)$ *of* $\mathcal{N}_n$ *are* $\omega_j \geq 0$ *and* $\sum_{j=1}^{n} \omega_h = 1$. *The SVNMTOWG aggregation operators are a mapping from* $\mathcal{N}^n \rightarrow \mathcal{N}$ *such that* $SVNMTOWG(\mathcal{N}_1, \mathcal{N}_2, ..., \mathcal{N}_n) =$

$$\begin{cases} (\theta, \iota, \phi), \text{ if } \lambda \in (0,1], (\theta_1, \theta_2, ..., \theta_n) \in [1-\lambda, 1]^n \; and \left(\iota_1, \iota_2, ..., \iota_n\right), \\ \qquad (\phi_1, \phi_2, ..., \phi_n) \in [0, \lambda]^n \\ \left(\tilde{\iota}, \tilde{i}, \tilde{f}\right), \text{ otherwise} \end{cases}$$

where $\theta = max\left(\sum_{j=1}^{n} \omega_j t_{f(j)}, 0\right)$, $\iota = min\left(1, \sum_{j=1}^{n} \omega_j i_{f(j)}\right)$, $\phi = min\left(1, \sum_{j=1}^{n} \omega_j f_{f(j)}\right)$, $\tilde{\iota} = min\left(\omega_j t_{f(j)}\right)$, $\tilde{i} = max\left(\omega_j i_{f(j)}\right)$ *and* $\tilde{f} = max\left(\omega_j f_{f(j)}\right)$.

Proof *The proof is the same as that for theorem 10.16.*

Theorem 10.21 *Let* $\mathcal{N}_j = \left(\theta_j, \iota_j, \phi_j\right) \in SVNS(\mathcal{N})$ *where* $\mathcal{N}_j = \mathcal{N}$.

Then $SVNMTOWG(\mathcal{N}_1, \mathcal{N}_2, ..., \mathcal{N}_n) = \mathcal{N}$.

Proof *The proof is the same as that for theorem 10.17.*

Theorem 10.22 *Let* $\mathcal{N}_j = \left(\theta_j, \iota_j, \phi_j\right)$, $\mathcal{N}_j^- = \left(min\left(\theta_j\right), max\left(\iota_j\right), max\left(\phi_j\right)\right)$ *and*

$\mathcal{N}_j^+ = \left(max\left(\theta_j\right), min\left(\iota_j\right), min\left(\phi_j\right)\right) \in SVNS(\mathcal{N})$.

Then $\mathcal{N}_j^- \leq SVNMTOWG(\mathcal{N}_1, \mathcal{N}_2, ..., \mathcal{N}_n) \leq \mathcal{N}_j^+$.

Proof *The proof is the same as that for theorem 10.18.*

Theorem 10.23 *Let* $\mathcal{N}_j = \left(\theta_j, \iota_j, \phi_j\right)$ *and* $\mathcal{N}_j^+ = \left(\theta_j^+, \iota_j^+, \phi_j^+\right) \in SVNS(\mathcal{N})$ *where*

$\left(\theta_j, \iota_j, \phi_j\right) \leq \left(\theta_j^*, \iota_j^+, \phi_j^+\right)$.

Then $SVNMTOWG(\mathcal{N}_1, \mathcal{N}_2, ..., \mathcal{N}_n) \leq SVNMTOWG(\mathcal{N}_1^+, \mathcal{N}_2^+, ..., \mathcal{N}_n^+)$.

Proof *The proof is the same as that for theorem 10.19.*

Definition 10.20 *Let* $\mathcal{N}_j = \left(\theta_j, \iota_j, \phi_j\right) \in SVNS(\mathcal{N})$ *where* $j \in \mathbb{N}$. *Then, the neutrosophic Mayor–Torrens hybrid weighted geometric aggregation operator for SVNS(*$\mathcal{N}$*) is defined as follows:*

$$SVNMTHWG(\mathcal{N}_1,\mathcal{N}_2,...,\mathcal{N}_n)=\tilde{\mathcal{N}}_{f(1)}^{\omega_1}\,\bar{\wedge}\,\tilde{\mathcal{N}}_{f(2)}^{\omega_2}\,\bar{\wedge}...\bar{\wedge}\,\tilde{\mathcal{N}}_{f(n)}^{\omega_n}=\prod_{j=1}^{n}\tilde{\mathcal{N}}_{f(j)}^{\omega_j},$$

where $\tilde{\mathcal{N}}_{f(j)}$ is the jth largest of the weighted neutrosophic values $\tilde{\mathcal{N}}_j(\tilde{\mathcal{N}}_j = n\omega_j\mathcal{N}_j)$ and the weights $\omega_1,\omega_2,....,\omega_n$ of $\mathcal{N}_j$ satisfy $\omega_j \geq 0$ and $\sum_{j=1}^{n}\omega_h = 1$.

Theorem 10.24 *Let* $\mathcal{N}_j = (\theta_j, \iota_j, \phi_j) \in SVNS(\mathcal{N})$ *where* $j \in \mathbb{N}$ *and the weights* $(\omega_1,\omega_2,...,\omega_n)$ *of* $\mathcal{N}_n$ *are* $\omega_j \geq 0$ *and* $\sum_{j=1}^{n}\omega_h = 1$. *The SVNMTHWG aggregation operators are a mapping from* $\mathcal{N}^n \to \mathcal{N}$ *such that* $SVNMTHWG(\mathcal{N}_1,\mathcal{N}_2,...,\mathcal{N}_n) =$

$$\begin{cases}(\theta,\iota,\phi), \text{if } \lambda\in(0,1],(\theta_1,\theta_2,...,\theta_n)\in[1-\lambda,1]^n \; and \; (\iota_1,\iota_2,...,\iota_n),\\ \qquad (\phi_1,\phi_2,...,\phi_n)\in[0,\lambda]^n\\ (\tilde{t},\tilde{i},\tilde{f}), \text{otherwise}\end{cases}$$

where $\theta = max\left(\sum_{j=1}^{n}\omega_j\tilde{t}_{f(j)},0\right)$, $\iota = min\left(1,\sum_{j=1}^{n}\omega_j\tilde{i}_{f(j)}\right)$, $\phi = min\left(1,\sum_{j=1}^{n}\omega_j\tilde{f}_{f(j)}\right)$, $\tilde{t} = min\left(\omega_j\tilde{t}_{f(j)}\right)$, $\tilde{i} = max\left(\omega_j\tilde{i}_{f(j)}\right)$ *and* $\tilde{f} = max\left(\omega_j\tilde{f}_{f(j)}\right)$.

Proof *The proof is the same as that for theorem 10.16.*

Theorem 10.25 *Let* $\mathcal{N}_j = (\theta_j,\iota_j,\phi_j) \in SVNS(\mathcal{N})$ *where* $\mathcal{N}_j = \mathcal{N}$.

Then $SVNMTHWG(\mathcal{N}_1,\mathcal{N}_2,...,\mathcal{N}_n) = \mathcal{N}$.

Proof *The proof is the same as that for theorem 10.17.*

Theorem 10.26 *Let* $\mathcal{N}_j = (\theta_j,\iota_j,\phi_j)$, $\mathcal{N}_j^- = \left(min(\theta_j), max(\iota_j), max(\phi_j)\right)$ *and* $\mathcal{N}_j^+ = \left(max(\theta_j), min(\iota_j), min(\phi_j)\right) \in SVNS(\mathcal{N})$.

Then $\mathcal{N}_j^- \leq SVNMTHWG(\mathcal{N}_1,\mathcal{N}_2,...,\mathcal{N}_n) \leq \mathcal{N}_j^+$.

Proof *The proof is the same as that for theorem 10.18.*

Theorem 10.27 *Let* $\mathcal{N}_j = (\theta_j,\iota_j,\phi_j)$ *and* $\mathcal{N}_j^+ = (\theta_j^+,\iota_j^+,\phi_j^+) \in SVNS(\mathcal{N})$ *where* $(\theta_j,\iota_j,\phi_j) \leq (\theta_j^+,\iota_j^+,\phi_j^+)$.

Then $SVNMTHWG(\mathcal{N}_1,\mathcal{N}_2,...,\mathcal{N}_n) \leq SVNMTHWG(\mathcal{N}_1^+,\mathcal{N}_2^+,...,\mathcal{N}_n^+)$

Proof *The proof is the same as that for theorem 10.19.*

10.6 ALGORITHM FOR MCDM PROBLEMS

This section defines a systematic approach to MCDM problems based on neutrosophic Mayor–Torrens aggregation operators. Let $\mathcal{A}_1, \mathcal{A}_2, \ldots, \mathcal{A}_n$ be the n alternatives, $\mathcal{Q}_1, \mathcal{Q}_2, \ldots, \mathcal{Q}_m$ be the m attributes and $\omega_1, \omega_2, \ldots, \omega_m$ be the corresponding weights of the attibutes. Let $D_1, D_2, \ldots, D_r$ be the r decision makers.

Step 1. Obtain the neutrosophic decision matrix for each decision maker D_i, i = 1,2, . . ., r (Table 10.1).

Table 10.1 Decision-Making Matrix by D_i

	$\tilde{\mathcal{Q}}_1$	$\tilde{\mathcal{Q}}_2$	...	$\tilde{\mathcal{Q}}_r$
$\tilde{\mathcal{A}}_1$	$(\tilde{t}_{11}, \tilde{i}_{11}, \tilde{f}_{11})$	$(\tilde{t}_{12}, \tilde{i}_{12}, \tilde{f}_{12})$	...	$(\tilde{t}_{1r}, \tilde{i}_{1r}, \tilde{f}_{1r})$
$\tilde{\mathcal{A}}_2$	$(\tilde{t}_{21}, \tilde{i}_{21}, \tilde{f}_{21})$	$(\tilde{t}_{22}, \tilde{i}_{22}, \tilde{f}_{22})$	...	$(\tilde{t}_{2r}, \tilde{i}_{2r}, \tilde{f}_{2r})$
.	.	.		.
.	.	.		.
.	.	.		.
$\tilde{\mathcal{A}}_n$	$(\tilde{t}_{n1}, \tilde{i}_{n1}, \tilde{f}_{n1})$	$(\tilde{t}_{n2}, \tilde{i}_{n2}, \tilde{f}_{n2})$	...	$(\tilde{t}_{nr}, \tilde{i}_{nr}, \tilde{f}_{nr})$

Step 2. Create the aggregated decision matrix in Table 10.2 by aggregating the various decision matrices according to the aggregation operators.

Table 10.2 Decision-Making Combined

	$\mathcal{Q}_1$	$\mathcal{Q}_2$	...	$\mathcal{Q}_r$
$\mathcal{A}_1$	(t_{11}, i_{11}, f_{11})	(t_{12}, i_{12}, f_{12})	...	(t_{1r}, i_{1r}, f_{1r})
$\mathcal{A}_2$	(t_{21}, i_{21}, f_{21})	(t_{22}, i_{22}, f_{22})	...	(t_{2r}, i_{2r}, f_{2r})
.	.	.		.
.	.	.		.
.	.	.		.
$\mathcal{A}_n$	(t_{n1}, i_{n1}, f_{n1})	(t_{n2}, i_{n2}, f_{n2})	...	(t_{nr}, i_{nr}, f_{nr})

Step 3. Use the developed aggregation operators to obtain the SVNN $\mathcal{A}_i$ (i= 1, 2, . . ., n) for the $\mathcal{Q}_j$ (j=1,2,..,r) alternatives, where $\omega_1, \omega_2, \ldots, \omega_r$ is the weight vector of the attributes.

Step 4. Compute the score for each neutrosophic set.

Step 5. Choose the option with the highest score after ranking the alternatives $\mathcal{A}_i$(i= 1, 2, . . ., n).

10.6.1 An Example in the Education Sector

Here we address the task of selecting an optimal university. Let D_1, D_2, . . ., D_r be experts in the education sector. Let $\mathcal{A}_1, \mathcal{A}_2, ..., \mathcal{A}_5$ be five universities, and let $\mathcal{Q}_1, \mathcal{Q}_2, ..., \mathcal{Q}_5$ represent teaching, infrastructure, research, curricular aspects and career resourses, respectively. Using the algorithm gives the results in Table 10.2. The corresponding weights for the criteria are $\omega_1 = 0.22$, $\omega_2 = 0.28$, $\omega_3 = 0.18$, $\omega_4 = 0.13$ and $\omega_5 = 0.19$.

1. The combined decision matrix of the experts using the aggregate operator is given in Table 10.3.

Table 10.3 Decision-making matrix

	$\mathcal{Q}_1$	$\mathcal{Q}_2$	$\mathcal{Q}_3$	$\mathcal{Q}_4$	$\mathcal{Q}_5$
$\mathcal{A}_1$	[0.2 0.3 0.6]	[0.5 0.2 0.3]	[0.3 0.2 0.4]	[0.7 0.2 0.2]	[0.6 0.2 0.3]
$\mathcal{A}_2$	[0.4 0.3 0.2]	[0.3 0.2 0.6]	[0.1 0.2 0.6]	[0.6 0.4 0.2]	[0.8 0.3 0.2]
$\mathcal{A}_3$	[0.3 0.4 0.5]	[0.4 0.3 0.3]	[0.5 0.5 0.5]	[0.1 0.4 0.6]	[0.3 0.6 0.7]
$\mathcal{A}_4$	[0.7 0.1 0.4]	[0.2 0.3 0.4]	[0.6 0.2 0.4]	[0.7 0.1 0.1]	[0.4 0.4 0.5]
$\mathcal{A}_5$	[0.1 0.2 0.7]	[0.3 0.4 0.7]	[0.3 0.3 0.5]	[0.6 0.2 0.1]	[0.2 0.3 0.4]

2. Evaluating the neutrosophic Mayor–Torrens weighted averaging aggregation operators and neutrosophic Mayor–Torrens weighted geometric aggregation operators gives Tables 10.4 and 10.5.

Table 10.4 Neutrosophic Mayor–Torrens Weighted Average Aggregation Operators

	SVNMTWA	SVNMTOWA	SVNMTHWA
$\mathcal{A}_1$	[0.4430 0.2220 0.3710]	[0.4890 0.2190 0.3480]	0.3959 0.2932 0.4266]
$\mathcal{A}_2$	[0.4200 0.2670 0.3840]	[0.4740 0.2960 0.3280]	[0.4188 0.2674 0.3738]
$\mathcal{A}_3$	[0.3380 0.4280 0.4950]	[0.3520 0.4440 0.5070]	[0.3519 0.4644 0.5274]
$\mathcal{A}_4$	[0.4850 0.2310 0.3800]	[0.5480 0.1950 0.3470]	[0.4662 0.2377 0.3730]
$\mathcal{A}_5$	[0.2760 0.2930 0.5290]	[0.3100 0.2720 0.4580]	[0.2730 0.3340 0.4991]

Table 10.5 Neutrosophic Mayor–Torrens Weighted Geometric Aggregation Operators

	SVNMTWG	SVNMTOWG	SVNMTHWG
$\mathcal{A}_1$	[0.4430 0.2220 0.3710]	[0.4890 0.2190 0.3480]	[0.4332 0.2172 0.3643]
$\mathcal{A}_2$	[0.4200 0.2670 0.3840]	[0.4740 0.2960 0.3280]	[0.2697 0.2627 0.3343]
$\mathcal{A}_3$	[0.3380 0.4280 0.4950]	[0.3520 0.4440 0.5070]	[0.3168 0.4307 0.4819]
$\mathcal{A}_4$	[0.4850 0.2310 0.3800]	[0.5480 0.1950 0.3470]	[0.5444 0.2247 0.3955]
$\mathcal{A}_5$	[0.2760 0.2930 0.5290]	[0.3100 0.2720 0.4580]	[0.2476 0.2795 0.5206]

3. The scores calculated for Tables 10.4 and 10.5 are given in Table 10.6.

Table 10.6 The Neutrosophic Mayor–Torrens Weighted Average Aggregation Operator Scores

	A_1	A_2	A_3	A_4	A_5
SVNMTWA	−0.1500	−0.2310	−0.5850	−0.1260	−0.5460
SVNMTOWA	−0.0780	−0.1500	−0.5990	0.0060	−0.4200
SVNMTHWA	−0.3239	−0.2224	−0.6399	−0.1445	−0.5601
SVNMTWG	−0.1500	−0.2310	−0.5850	−0.1260	−0.5460
SVNMTOWG	−0.0780	−0.1500	−0.5990	0.0060	−0.4200
SVNMTHWG	−0.1482	−0.3273	−0.5957	−0.0758	−0.5525

4. The ranking is given in Table 10.7, and A_4 is the best alternative.

Table 10.7 The Neutrosophic Mayor–Torrens Weighted Average Aggregation Operators Ranking

	Rank	Best Alternative
SVNMTWA	$A_4 > A_1 > A_2 > A_5 > A_3$	A_4
SVNMTOWA	$A_4 > A_1 > A_2 > A_5 > A_3$	A_4
SVNMTHWA	$A_4 > A_1 > A_2 > A_5 > A_3$	A_4
SVNMTWG	$A_4 > A_1 > A_2 > A_5 > A_3$	A_4
SVNMTOWG	$A_4 > A_1 > A_2 > A_5 > A_3$	A_4
SVNMTHWG	$A_4 > A_1 > A_2 > A_5 > A_3$	A_4

10.6.2 Comparison with Existing Methods

In this section, we compare the Mayor– Torrens aggregation operator with existing aggregation operators to show the efficiency of our algorithm usinge, we using the same data set provided in Table 10.3. The results are shown in Table 10.8, and these results are the same as those in Table 10.7, demonstrating the accuracy of the proposed method.

Table 10.8 Comparing the Mayor–Torrens Aggregation Operator with Existing Operators

	Ranking	Best Alternative
ST-SVNOWG [18]	$A_4 > A_1 > A_2 > A_5 > A_3$	A_4
ST-SVNOWA [18]	$A_4 > A_1 > A_2 > A_5 > A_3$	A_4
ST-SVNWG [18]	$A_4 > A_1 > A_2 > A_5 > A_3$	A_4
ST-SVNWA [18]	$A_4 > A_1 > A_2 > A_5 > A_3$	A_4
SVNHWA [14]	$A_4 > A_1 > A_2 > A_5 > A_3$	A_4

(Continued)

Table 10.8 (Continued)

	Ranking	Best Alternative
L-SVNWG [15]	$A_4 > A_1 > A_2 > A_5 > A_3$	A_4
L-SVNOWA [15]	$A_4 > A_1 > A_2 > A_5 > A_3$	A_4
L-SVNWA [15]	$A_4 > A_1 > A_2 > A_5 > A_3$	A_4
SVNOWA [12]	$A_4 > A_1 > A_2 > A_5 > A_3$	A_4
SVNWA [12]	$A_4 > A_1 > A_2 > A_5 > A_3$	A_4
NWA [13]	$A_4 > A_1 > A_2 > A_5 > A_3$	A_4
SVNYWA [19]	$A_4 > A_1 > A_2 > A_5 > A_3$	A_4
SVNYOWA [19]	$A_4 > A_1 > A_2 > A_5 > A_3$	A_4
SVNYHWA [19]	$A_4 > A_1 > A_2 > A_5 > A_3$	A_4
SVNYWG [19]	$A_4 > A_1 > A_2 > A_5 > A_3$	A_4
SVNYOWG [19]	$A_4 > A_1 > A_2 > A_5 > A_3$	A_4
SVNYHWG [19]	$A_4 > A_1 > A_2 > A_5 > A_3$	A_4

10.7 CONCLUSION

Neutrosophic set theory has become critical in decision-making since it deals with neutralities. Aggregation operators in neutrosophic sets provide a systematic way to deal with uncertainties. In this paper, we developed neutrosophic Mayor–Torrens t-norms and t-conorms and established aggregating operators to make decision making more effortless. We also discuss several theorems and compare our proposed method with existing methods. In the future, the model can be extended to decision-making problems, operations research, etc.

REFERENCES

1. L.A. Zadeh. Fuzzy sets. *Information and Control*, 8:338–353, 6, 1965.
2. Krassimir T. Atanassov. *Intuitionistic fuzzy sets*. Springer, 1999.
3. Florentin Smarandache. *A unifying field in logics: Neutrosophic logic: Neutrosophy, neutrosophic set, neutrosophic probability and statistics*, 4th edition. American Research Press, 1999.
4. Florentin Smarandache. Neutrosophic set – a generalization of the intuitionistic fuzzy set. *Journal of Defense Resources Management*, 1:107–116, 2010.
5. R.R. Yager. On ordered weighted averaging aggregation operators in multicriteria decisionmaking. *IEEE Transactions on Systems, Man, and Cybernetics*, 18:183–190, 1988.
6. Zeshui Xu. Intuitionistic fuzzy aggregation operators. *IEEE Transactions on Fuzzy Systems*, 15:1179–1187, 12, 2007.

7. Zeshui Xu. An overview of methods for determining owa weights. *International Journal of Intelligent Systems*, 20:843–865, 8, 2005.
8. Huchang Liao and Zeshui Xu. Intuitionistic fuzzy hybrid weighted aggregation operators. *International Journal of Intelligent Systems*, 29:971–993, 11, 2014.
9. Anna Bryniarska. Mathematical models of diagnostic information granules generated by scaling intuitionistic fuzzy sets. *Applied Sciences*, 12:2597, 3, 2022.
10. Guiwu Wei. Picture fuzzy aggregation operators and their application to multiple attribute decision making. *Journal of Intelligent & Fuzzy Systems*, 33:713–724, 7, 2017.
11. Muhammad Sajjad Ali Khan, Saleem Abdullah, Asad Ali, Fazli Amin, and Khaista Rahman. Hybrid aggregation operators based on pythagorean hesitant fuzzy sets and their application to group decision making. *Granular Computing*, 4:469–482, 7, 2019.
12. Juan-Juan Peng, Jian qiang Wang, Jing Wang, Hong yu Zhang, and Xiao hong Chen. Simplified neutrosophic sets and their applications in multi-criteria group decision-making problems. *International Journal of Systems Science*, 47:2342–2358, 7, 2016.
13. Jun Ye. Subtraction and division operations of simplified neutrosophic sets. *Information*, 8:51, 5 2017.
14. Peide Liu, Yanchang Chu, Yanwei Li, and Yubao Chen. Some generalized neutrosophic number hamacher aggregation operators and their application to group decision making. *International Journal of Fuzzy Systems*, 16(2), 2014.
15. Harish Garg and Nancy. New logarithmic operational laws and their applications to multiattribute decision making for single-valued neutrosophic numbers. *Cognitive Systems Research*, 52:931–946, 12, 2018.
16. Peide Liu. The aggregation operators based on archimedean t-conorm and t-norm for single-valued neutrosophic numbers and their application to decision making. *International Journal of Fuzzy Systems*, 18(5):849–863, 2016.
17. Chiranjibe Jana and Madhumangal Pal. Multi-criteria decision making process based on some single-valued neutrosophic dombi power aggregation operators. *Soft Computing*, 25:5055–5072, 1, 2021.
18. Shahzaib Ashraf, Saleem Abdullah, Shouzhen Zeng, Huanhuan Jin, and Fazal Ghani. Fuzzy decision support modeling for hydrogen power plant selection based on single valued neutrosophic sine trigonometric aggregation operators. *Symmetry*, 12:298, 2, 2020.
19. Shabeer Khan, Saleem Abdullah, Shahzaib Ashraf, Ronnason Chinram, and Samruam Baupradist. Decision support technique based on neutrosophic yager aggregation operators: Application in solar power plant locations—case study of bahawalpur, pakistan. *Mathematical Problems in Engineering*, 2020:1–21, 12, 2020.
20. D. Nagarajan, M. Lathamaheswari, Said Broumi, and J. Kavikumar. A new perspective on traffic control management using triangular interval type-2 fuzzy sets and interval neutrosophic sets. *Operations Research Perspectives*, 6:100099, 1, 2019.
21. D. Nagarajan, M. Lathamaheswari, S. Broumi, and J. Kavikumar. Dombi interval valued neutrosophic graph and its role in traffic control management. *Neutrosophic Sets and Systems*, 24:114–133, 2019.
22. Seyyed Ahmad Edalatpanah. Neutrosophic structured element. *Expert Systems*, 37, 10, 2020.
23. Ranjan Kumar, S. A. Edalatpanah, Sripati Jha, Said Broumi, Ramayan Singh, and Arindam Dey. *A multi objective programming approach to solve integer valued neutrosophic shortest path problems*. Infinite Study, 2019.

24. Ibrahim M. Hezam, Arunodaya Raj Mishra, R. Krishankumar, K. S. Ravichandran, Samarjit Kar, and Dragan Stevan Pamucar. A single-valued neutrosophic decision framework for the assessment of sustainable transport investment projects based on discrimination measure. *Management Decision*, 61:443–471, 3, 2023.
25. Pratibha Rani, Jabir Ali, Raghunathan Krishankumar, Arunodaya Raj Mishra, Fausto Cavallaro, and Kattur S. Ravichandran. An integrated single-valued neutrosophic combined compromise solution methodology for renewable energy resource selection problem. *Energies*, 14:4594, 7, 2021.
26. Erich Peter Klement, Radko Mesiar, and Endre Pap. *Triangular norms*, Vol. 8. Springer Netherlands, 2000.

Chapter 11

Similarity Measures of Fermatean Neutrosophic Sets Based on the Euclidean Distance

R. Princy, R. Radha, and Said Broumi

11.1 INTRODUCTION

The concept of the fuzzy set (FS) $A = \{< x_i, \mu_{A_{x_i}} > | \, x_i \in X\}$ in $X = \{x_1, x_2, \ldots, x_n\}$ was proposed by Zadeh [15], where the membership $\mu_{A_{x_i}}$ is a single value between zero and one. FSs have been widely applied in many fields, such as medical diagnosis, image processing, and supply decision-making [5, 11–14]. In some uncertain decision-making problems, the degree of membership is not exactly a numeric value but more an interval.

Therefore, Zadeh [15] proposed the interval-valued fuzzy set (IVFS). However, FSs and IVFSs only have membership degrees; they cannot describe the non-membership of an element belonging to a set. Then, Atanassov et al. [2] proposed the intuitionistic fuzzy set (IFS) $E = \{<x_i, \mu_E(x_i), \vartheta_E(x_i)> | x_i \in X\}$, where $\mu_E(x_i)(0 \le \mu_E(x_i) \le 1)$ and $\vartheta_E(x_i)(0 \le \vartheta_E(x_i) \le 1)$ represent the membership and the non-membership, respectively, and indeterminacy is represented by $\pi_E(x_i) = 1 - \mu_E(x_i) - \vartheta_E(x_i)$. IFSs is more effective for dealing with vague information than FSs and IVFSs.

Yager [7] introduced a general class of these sets called q-rung ortho pair fuzzy sets in which the sum of the qth power of the support for and the qth power of the support against is bonded by one. He noted that as q increases, the space-acceptable ortho pairs increase and thus give the user more freedom in expressing their belief about membership grade. When q = 2, Yager considered q-rung ortho pair fuzzy sets as Pythagorean fuzzy sets. Senapati and Yager introduced q-rung orthopair fuzzy sets as Fermatean fuzzy sets for when q = 3 [6]. Later Fermatean neutrosophic sets were proposed by Antony and Jansi [1]. In this chapter, we propose a new method of constructing similarity measures for FSs. They have important practical applications in pattern recognition, medical diagnosis, and so on.

DOI: 10.1201/9781003487104-11

11.2 PRELIMINARIES

Definition 11.2.1 [2]: Let X be a universe. An intuitionistic fuzzy set A on E can be defined as follows:

$$A=\{(x,\mu_A(x),\vartheta_A(x))/x\in X\}$$

where $\mu_A:E\to[0,1]$ and $\vartheta_A:E\to[0,1]$ such that $0\le\mu_A+\vartheta_A\le 1$ for any $x\in E$.

Here, $\mu_A(x)$ and $\vartheta_A(x)$ are the degree of membership and degree of non-membership of the element x, respectively.

Definition 11.2.2 [1]: Let X be a universe of discourse. A Fermatean neutrosophic set [FN Set] A on X is an object of the form $A=\{(x,\mu_M(x),\zeta_M(x),v_M(x))/x\in X\}$, where $\mu_M(x),\zeta_M(x),v_M(x)\in[0,1], 0\le(\mu_M(x))^3+(v_M(x))^3\le 1$ and $0\le (\zeta_M(x))^3\le 1$.

Then, $0\le(\mu_M(x))^3+(\zeta_M(x))^3+(v_M(x))^3\le 2$ for all $x\in X$.

$\mu_M(x)$ is the membership function, $\zeta_M(x)$ is the indeterminacy function and $v_M(x)$ is the non-membership function. Here $\mu_M(x)$ and $v_M(x)$ are dependent components, and $\zeta_M(x)$ is an independent component.

Definition 11.2.3 [1]: Let X be a nonempty set and I the unit interval [0,1]. Fermatean neutrosophic sets M and N of the form $M=\{(x,\mu_M(x),\zeta_M(x),v_M(x))/x\in X\}$ and $N=\{(x,\mu_N(x),\zeta_N(x),v_N(x))/x\in X\}$

1) $M^C=\{(x,v_M(x),1-\zeta_M(x),\mu_M(x))/x\in X\}$.
2) $M\cup N=\{(x,\max(\mu_M(x),\mu_N(x)),\min(\zeta_M(x),\zeta_N(x)),$ $\min(v_M(x),v_N(x))/x\in X\}$.
3) $M\cap N=\{(x,\min(\mu_M(x),\mu_N(x)),\max(\zeta_M(x),\zeta_N(x)),$ $\max(v_M(x),v_N(x))/x\in X\}$

11.3 NEW SIMILARITY MEASURES

The similarity measure is a widely used tool for evaluating the relationship between two sets. The following axiom about the similarity measure of IVSFSs should be satisfied:

Lemma: Let $X=\{x_1,x_2,\ldots,x_n\}$ be the universal set [18] if the similarity measure S(A, B) between SFSs A and B satisfies the following properties:

(1) $0\le S(A,B)\le 1$
(2) $S(A,B)=1$ if and only if $A=B$
(3) $S(A,B)=S(B,A)$.

Then, $S(A,B)$ is a genuine similarity measure.

11.3.1 The New Similarity Measures between FNSs

Definition 11.3.1.1: Let $X=\{x_1,x_2,\ldots,x_n\}$ be the universal set for any two FNSs $A=\{<x_i,\mu_{A_{x_i}},\vartheta_{A_{x_i}},\pi_{A_{x_i}}>|\,x_i\in X\}$ and $B=\{<x_i,\mu_{B_{x_i}},\vartheta_{B_{x_i}},\pi_{B_{x_i}}>|\,x_i\in X\}$; then the Euclidean distance between FNSs A and B is defined as follows:

$$D_{FNSs}(A,B)=\sqrt{\frac{\sum_{i=1}^{n}\left[\left(\mu_A^{\ 3}(x_i)-\mu_B^{\ 3}(x_i)\right)^2+\left(\vartheta_A^{\ 3}(x_i)-\vartheta_B^{\ 3}(x_i)\right)^2+\left(\pi_A^{\ 3}(x_i)-\pi_B^{\ 3}(x_i)\right)^2\right]}{3n}} \quad (11.1)$$

Now, we construct new similarity measures of FNSs based on the Euclidean distance measures.

Definition 11.3.1.2: Let $X=\{x_1,x_2,\ldots,x_n\}$ be the universal set for any two FNSs $A=\{<x_i,\mu_{A_{x_i}},\vartheta_{A_{x_i}},\pi_{A_{x_i}}>|\,x_i\in X\}$ and $B=\{<x_i,\mu_{B_{x_i}},\vartheta_{B_{x_i}},\pi_{B_{x_i}}>|\,x_i\in X\}$; the similarity measure of FNSs between A and B is defined as follows:

$$S_{1SFSs}(A,B)=\frac{\sum_{i=1}^{n}\left(min\left(\mu_A^{\ 3}(x_i),\mu_B^{\ 3}(x_i)\right)+min\left(\vartheta_A^{\ 3}(x_i),\vartheta_B^{\ 3}(x_i)\right)+min\left(\pi_A^{\ 3}(x_i),\pi_B^{\ 3}(x_i)\right)\right)}{\sum_{i=1}^{n}\left(max\left(\mu_A^{\ 3}(x_i),\mu_B^{\ 3}(x_i)\right)+max\left(\vartheta_A^{\ 3}(x_i),\vartheta_B^{\ 3}(x_i)\right)+max\left(\pi_A^{\ 3}(x_i),\pi_B^{\ 3}(x_i)\right)\right)} \quad (11.2)$$

The similarity measure S_{1FNSs} satisfies the properties in lemma.

Next, we propose a new method of constructing a new similarity measure for FNSs and the Euclidean distance that can be defined as follows:

Definition 11.3.1.3: Let $X=\{x_1,x_2,\ldots,x_n\}$ be the universal set for any two FNSs $A=\{<x_i,\mu_{A_{x_i}},\vartheta_{A_{x_i}},\pi_{A_{x_i}}>|\,x_i\in X\}$ and $B=\{<x_i,\mu_{B_{x_i}},\vartheta_{B_{x_i}},\pi_{B_{x_i}}>|\,x_i\in X\}$; a new similarity measure $S^*_{1FNSs}(A,B)$ is defined as follows:

$$S^*_{1FNSs}(A,B)=\frac{1}{2}\left(S_{1FNSs}(A,B)+1-D_{FNSs}(A,B)\right). \quad (11.3)$$

The proposed similarity measure of FNSs satisfies the theorem.

Theorem 11.3.1.4: The similarity measure $S^*_{1FNSs}(A,B)$ between $A=\{<x_i,\mu_{A_{x_i}},\vartheta_{A_{x_i}},\pi_{A_{x_i}}>|\,x_i\in X\}$ and $B=\{<x_i,\mu_{B_{x_i}},\vartheta_{B_{x_i}},\pi_{B_{x_i}}>|\,x_i\in X\}$ satisfies the following properties:

(11.3.1.4.1) $0\le S^*_{1FNSs}(A,B)\le 1$
(11.3.1.4.2) $S^*_{1FNSs}(A,B)=1$ if and only if $A=B$
(11.3.1.4.3) $S^*_{1FNSs}(A,B)=S^*_{1FNSs}(B,A)$.

Proof:

1) Because $D_{FNSs}(A,B)$ is a Euclidean distance measure, obviously, $0\le D_{FNSs}(A,B)\le 1$. Furthermore, according to lemma, we know that

$0 \le S_{1FNSs}(A,B) \le 1$. Then, $0 \le \frac{1}{2}\left(S_{1FNSs}(A,B)+1-D_{FNSs}(A,B)\right) \le 1$; that is, $0 \le S^{*}_{1FNSs}(A,B) \le 1$.

2) If $S^{*}_{1FNSs}(A,B)=1$, we have $S_{1FNSs}(A,B)+1-D_{FNSs}(A,B)=2$; that is, $S_{1FNSs}(A,B)=1+D_{FNSs}(A,B)$. Because $D_{FNSs}(A,B)$ is the Euclidean distance measure $0 \le D_{FNSs}(A,B) \le 1$. Furthermore, $0 \le S_{1FNSs}(A,B) \le 1$, then $S_{1FNSs}(A,B)=1$ and $D_{FNSs}(A,B)=0$ should be established at the same time if the Euclidean distance measure $D_{FNSs}(A,B)=0, A=B$ is obvious. According to lemma 1, when $S_{1FNSs}(A,B)=1, A=B$; so if $S^{*}_{1FNSs}(A,B)=1, A=B$ is obtained.
   On the other hand, when $A=B$, formulae (11.3.1.4.1) and (11.3.1.4.2) give $D_{FNSs}(A,B)=0$ and $S_{1FNSs}(A,B)=1$, respectively. Furthermore, we obtain $S^{*}_{1FNSs}(A,B)=1$.
3) $S^{*}_{1FNSs}(A,B)=S^{*}_{1FNSs}(B,A)$ is straightforward.
   From Theorem 1, we know that the proposed new similarity measure $S^{*}_{1FNSs}(A,B)$ is a genuine similarity measure. On the other hand, cosine similarity is also important. The cosine similarity measure between FNSs is as follows:

Definition 11.3.1.5: Let $X=\{x_1,x_2,\ldots,x_n\}$ be the universal set for any two FNSs $A=\{<x_i,\mu_{A_{x_i}},\vartheta_{A_{x_i}},\pi_{A_{x_i}}>|\, x_i \in X\}$ and $B=\{<x_i,\mu_{B_{x_i}},\vartheta_{B_{x_i}},\pi_{B_{x_i}}>|\, x_i \in X\}$; the cosine similarity measure of FNSs between A and B is defined as follows:

$$S_{2FNSs}(A,B) = \frac{1}{n}\sum_{i=1}^{n}\frac{\left(\left(\mu_A^3(x_i)\mu_B^3(x_i)\right)+\left(\vartheta_A^3(x_i)\vartheta_B^3(x_i)\right)+\left(\pi_A^3(x_i)\pi_B^3(x_i)\right)\right)}{\sqrt{\left(\mu_A^3(x_i)\right)^2+\left(\vartheta_A^3(x_i)\right)^2+\left(\pi_A^3(x_i)\right)^2}\sqrt{\left(\mu_B^3(x_i)\right)^2+\left(\vartheta_B^3(x_i)\right)^2+\left(\pi_B^3(x_i)\right)^2}} \tag{11.4}$$

Now, we are going to propose another similarity measure of FNSs based on the cosine similarity measure and the Euclidean distance D_{FNSs}. It considers the similarity measure not only from the point of view of algebra but also from the point of view of geometry and can be defined as:

Definition 11.3.1.6: Let $X=\{x_1,x_2,\ldots,x_n\}$ be the universal set for any two FNSs $A=\{<x_i,\mu_{A_{x_i}},\vartheta_{A_{x_i}},\pi_{A_{x_i}}>|\, x_i \in X\}$ and $B=\{<x_i,\mu_{B_{x_i}},\vartheta_{B_{x_i}},\pi_{B_{x_i}}>|\, x_i \in X\}$; a new similarity measure $S^{*}_{2FNSs}(A,B)$ is defined as follows:

$$S^{*}_{2FNSs}(A,B)=\frac{1}{2}\left(S_{2FNSs}(A,B)+1-D_{FNSs}(A,B)\right). \tag{11.5}$$

Theorem 11.3.1.7: The similarity measure $S^{*}_{2FNSs}(A,B)$ between $A=\{<x_i,\mu_{A_{x_i}},\vartheta_{A_{x_i}},\pi_{A_{x_i}}>|\, x_i \in X\}$ and $B=\{<x_i,\mu_{B_{x_i}},\vartheta_{B_{x_i}},\pi_{B_{x_i}}>|\, x_i \in X\}$ satisfies the following properties:

(11.3.1.7.1) $0 \le S^{*}_{2FNSs}(A,B) \le 1$
(11.3.1.7.2) $S^{*}_{2FNSs}(A,B) = 1$ if and only if $A = B$
(11.3.1.7.3) $S^{*}_{2FNSs}(A,B) = S^{*}_{2FNSs}(B,A)$.

Proof:

1) Because $D_{FNSs}(A,B)$ is an Euclidean distance measure, obviously, $0 \le D_{FNSs}(A,B) \le 1$. Furthermore, according to lemma, we know that $0 \le S_{FNSs}(A,B) \le 1$. Then, $0 \le \frac{1}{2}\left(S_{2FNSs}(A,B) + 1 - D_{FNSs}(A,B)\right) \le 1$, i.e., $0 \le S^{*}_{1FNSs}(A,B) \le 1$

2) If $S^{*}_{2FNSs}(A,B) = 1$, we have $S_{2FNSs}(A,B) + 1 - D_{FNSs}(A,B) = 2$; that is, $S_{2FNSs}(A,B) = 1 + D_{FNSs}(A,B)$. Because $D_{FNSs}(A,B)$ is the Euclidean distance measure $0 \le D_{FNSs}(A,B) \le 1$. Furthermore, $0 \le S_{2FNSs}(A,B) \le 1$, then $S_{2FNSs}(A,B) = 1$ and $D_{FNs}(A,B) = 0$ should be established at the same time. When $S_{2FNSs}(A,B) = 1$, we have $\mu_A(x_i) = k\mu_B(x_i), \vartheta_A(x_i) = k\vartheta_B(x_i), and\, \pi_A(x_i) = k\pi_B(x_i)$ (k is a constant). When the Euclidean distance measure $D_{FNSs}(A,B) = 0, A = B$, then $A = B$ is obtained.
On the other hand, when $A = B$, according to formulae (11.3.1.7.1) and (11.3.1.7.2) if $A = B$, $D_{FNSs}(A,B) = 0$ and $S_{2FNSs}(A,B) = 1$ are obtained respectively. Furthermore, we can get $S^{*}_{2FNSs}(A,B) = 1$.
3) $S^{*}_{2FNSs}(A,B) = S^{*}_{2FNSs}(B,A)$ is straightforward.
Thus $S^{*}_{2FNSs}(A,B)$ satisfies all the properties of Theorem 11.3.1.7.

In the next section, we apply the proposed new similarity measures to a medical diagnosis decision problem; numerical examples are also given to illustrate the application and effectiveness of the proposed new similarity measures.

11.4 APPLICATIONS OF THE PROPOSED SIMILARITY MEASURES

11.4.1 The Proposed Similarity Measures between FNSs for Medical Diagnosis

We first give a numeric example medical diagnosis to illustrate the feasibility of the proposed new similarity measure $S^{*}_{1FNSs}(A,B)$ and $S^{*}_{2FNSs}(A,B)$ between FNSs.

Example 11.4.1: Suppose a set of diagnosis $Q = \{Q_1$ (viral fever), Q_2 (malaria), Q_3 (typhoid), Q_4 (gastritis), Q_5 (stenocardia)$\}$ and a set of symptoms $S = \{S_1$ (fever), S_2 (headache), S_3 (stomach), S_4 (cough), S_5 (chestpain)$\}$. Assume a patient P_1

Table 11.1 XXX

	S_1	S_2	S_3	S_4	S_5
Q_1	[0.4, 0.6, 0.0]	[0.3, 0.2, 0.5]	[0.1, 0.3, 0.7]	[0.4, 0.3, 0.3]	[0.1, 0.2, 0.7]
Q_2	[0.7, 0.3, 0.0]	[0.2, 0.2, 0.6]	[0.0, 0.1, 0.9]	[0.7, 0.3, 0.0]	[0.1, 0.1, 0.8]
Q_3	[0.3, 0.4, 0.3]	[0.6, 0.3, 0.1]	[0.2, 0.1, 0.7]	[0.2, 0.2, 0.6]	[0.1, 0.0, 0.9]
Q_4	[0.1, 0.2, 0.7]	[0.2, 0.2, 0.4]	[0.8, 0.2, 0.0]	[0.2, 0.1, 0.7]	[0.2, 0.1, 0.7]
Q_5	[0.1, 0.1, 0.8]	[0.0, 0.2, 0.8]	[0.2, 0.0, 0.8]	[0.3, 0.1, 0.8]	[0.8, 0.1, 0.1]

Table 11.2 XXX

	Q_1	Q_2	Q_3	Q_4	Q_5
$S^*_{1FNSs}(P_1, Q_i)$	0.5980	**0.6801**	0.5729	0.3919	0.3820
$S^*_{2FNSs}(P_1, Q_i)$	0.4277	**0.4581**	0.4024	0.3514	0.3155

has all the symptoms in the process of diagnosis; the FNS-evaluated information about P_1 is

$$P_1(\text{Patient}) = \{< S_1, 0.8, 0.2, 0.1 >, < S_2, 0.6, 0.3, 0.1 >, < S_3, 0.2, 0.1, 0.8 >, \\ < S_4, 0.6, 0.5, 0.1 >, < S_5, 0.1, 0.4, 0.6 >\}$$

The diagnosis information $Q_i\,(i = 1, 2, \ldots, 5)$ with respect to symptoms $S_i\,(i = 1, 2, \ldots, 5)$ also can be represented by the FNSs, which is shown in Table 11.1.

By applying formulae (11.3) and (11.5) we can obtain the similarity measure values $S^*_{1FNSs}(P_1, Q_i)$ and $S^*_{2FNSs}(P_1, Q_i)$; the results are shown in Table 11.2.

From the two similarity measures S^*_{1FNSs} and S^*_{2FNSs}, we can conclude that the patient P_1 has malaria (Q_2). The proposed two similarity measures are feasible and effective.

11.4.2 Comparative Analysis of Existing Similarity Measures

To illustrative the effectiveness of the proposed similarity measures for medical diagnosis, we changed the existing similarity measures for FNSs and applied the existing similarity measures for comparative analyses.

First, we introduce the existing similarity measures between FNSs as follows:

Let $A = \{< x_i, \mu_{A_{x_i}}, \vartheta_{A_{x_i}}, \pi_{A_{x_i}} > | x_i \in X\}$ and $B = \{< x_i, \mu_{B_{x_i}}, \vartheta_{B_{x_i}}, \pi_{B_{x_i}} > | x_i \in X\}$ be two FNSs in $X = \{x_1, x_2, \ldots, x_n\}$; the existing measures between A and B are defined as follows:

1) Broumi et al. [4] proposed the similarity measure SM_{FNS}:

$$SM_{FNS}(A,B) = 1 - D_{FNS}(A,B) \tag{11.6}$$

2) Sahin and Ahmet [5] proposed the similarity measure SD_{FNS}:

$$SD_{FNS} = \frac{1}{1 + D_{FNS}(A,B)} \tag{11.7}$$

3) Ye [13] proposed the improved cosine similarity measure SC_{1FNS} and SC_{2FNS}:

$$SC_{1FNS}(A,B) = \frac{1}{n}\sum_{i=1}^{n} cos\left[\frac{\pi.max\left(\left|\mu_A^2(x_i) - \mu_B^2(x_i)\right|, \left|\vartheta_A^2(x_i) - \vartheta_B^2(x_i)\right|, \left|\pi_A^2(x_i) - \pi_B^2(x_i)\right|\right)}{2}\right] \tag{11.8}$$

$$SC_{2FNS}(A,B) = \frac{1}{n}\sum_{i=1}^{n} cos\left[\frac{\pi.\left(\left|\mu_A^2(x_i) - \mu_B^2(x_i)\right| + \left|\vartheta_A^2(x_i) - \vartheta_B^2(x_i)\right| + \left|\pi_A^2(x_i) - \pi_B^2(x_i)\right|\right)}{6}\right] \tag{11.9}$$

4) Yang et al. [10] proposed the similarity measure $SY_{FNS}(A,B)$:

$$SY_{FNS}(A,B) = \frac{SC_{FNS}(A,B)}{SC_{FNS}(A,B) + D_{FNS}(A,B)} \tag{11.10}$$

Example 11.4.2.1: We apply formulae (11.2), (11.4) and (11.6)—(11.10) to calculate Example 1 again; the similarity measure values between P_1 and Q_i $(i = 1,2,..,5)$ are shown in Table 11.3.

Table 11.3 XXX

	Q_1	Q_2	Q_3	Q_4	Q_5
SM_{FNS}	0.8003	**0.8314**	0.7449	0.6388	0.6007
SD_{FNS}	0.8335	**0.8557**	0.7967	0.7346	0.7146
SC_{1FNS}	0.8555	**0.9325**	0.6469	0.7324	0.6391
SC_{2FNS}	0.9648	**0.9759**	0.7531	0.885	0.8585
SY_{FNS}	0.8107	**0.8468**	0.7171	0.6697	0.6154
S_{1FNS}	0.3958	**0.5289**	0.4010	0.1451	0.1633
S_{2FNS}	0.0551	**0.0849**	0.0600	0.0191	0.0304

As we can see from Table 11.3, the patient P_1 is still diagnosed with malaria (Q_2); the result is the same as that from the proposed similarity measures in this paper, which means the proposed measures are feasible and effective.

11.5 CONCLUSION

The proposed similarity measures in this chapter have some advantages in solving multiple criteria decision-making problems. They are constructed based on the existing similarity measures and Euclidean distance, and they not only satisfy the axiom of similarity measures but also consider them from the points of view of algebra and geometry. Furthermore, they can be applied more widely in the field of decision-making problems.

REFERENCES

[1] C. Antony Crispin Sweety and R. Jansi, "Fermatean neutrosophic sets", International Journal of Advanced Research in Computer and Communication Engineering, Vol. 10, No. 6, June 2021.
[2] K. T. Atanassov, "Intuitionistic fuzzy sets", Fuzzy Sets and Systems, Vol. 20, No. 1, 1986, pp. 87–96.
[3] K. T. Atanassov and G. Gargov, "Interval valued intuitionisitc fuzzy sets and systems", Fuzzy Sets and Systems, Vol. 31, pp. 343–349, 1989.
[4] S. Broumi and F. Smarandache, "Several similarity measures of neutrosophic sets", Neutrosophic Sets and Systems, Vol. 1, No. 1, pp. 54–62, 2013.
[5] R. Sahin and K. Ahmet, "On similarity and entropy of neutrosophic soft sets", Journal of Intelligent & Fuzzy Systems, Vol. 29, 2014.
[6] T. Senapati and R. R. Yager, "Fermatean fuzzy sets", Journal of Ambient Intelligence and Humanized Computing, 2019a.
[7] R. R. Yager, "Pythagorean membership grades in multicriteria decision making", IEEE Transactions on Fuzzy Systems, Vol. 22, pp. 958–965, 2014.
[8] R. R. Yager, "Generalized orthopair fuzzy sets", IEEE Transactions on Fuzzy Systems, Vol. 25, No. 5, pp. 1222–1230, 2017.
[9] Y. Yang and F. Chiclana, "Intuitionistic fuzzy sets: Spherical representation and distances", International Journal of Intelligent Systems, Vol. 24, No. 4, pp. 399–420, 2009.
[10] Y. Yang, R. Zhang and J. Guo, "A multi-attribute decision-making approach based on hesitant neutrosophic sets", Fuzzy Systems and Mathematics, Vol. 31, No. 2, pp. 114–122, 2017.
[11] J. Ye, "Cosine similarity measures for intuitionistic fuzzy sets and their applications", Mathematical and Computer Modelling, Vol. 53, No. 1, pp. 91–97, 2011.
[12] J. Ye, "Vector similarty meaures of simplified neutrosophic sets and their application in multicriteria decision making", International Journal of Fuzzy Systems, Vol. 16, No. 2, pp. 204–211, June 2014.

[13] J. Ye, "Improved cosine similarity measures for simplified neutrosophic sets for medical diagnosis", Artificial Intelligence in Medicine, Vol. 63, No. 3, pp. 171–179, 2015.
[14] J. Ye, "Multicriteria decision-making method based on cosine similarity measure between interval-valued fuzzy sets with risk preferences", Economic Computation and Economic Cybernetics Studies and Research, Vol. 50, No. 4, pp. 205–215, 2016.
[15] L. A. Zadeh, "Fuzzy sets", Information and Control, Vol. 8, pp. 338–353, 1965.

Index